Theoretical and Computational Approaches to Interface Phenomena

Theoretical and Computational Approaches to Interface Phenomena

Edited by

Harrell Lee Sellers
South Dakota State University
Brookings, South Dakota

and

Joseph Thomas Golab
Amoco Research Center
Naperville, Illinois

Springer Science+Business Media, LLC

Library of Congress Cataloging-in-Publication Data

Theoretical and computational approaches to interface phenomena / edited by Harrell Lee Sellers and Joseph Thomas Golab.
p. cm.
"Proceedings of an International Conference on Theoretical and Computational Approaches to Interface Phenomena, held August 2-4, 1993, in Brookings, South Dakota'--T.p. verso.
Includes bibliographical references and index.

1. Interfaces (Physical sciences)--Congresses. 2. Interfaces (Physical sciences)--Computer simulation--Congresses. I. Sellers, Harrell Lee. II. Golab, Joseph Thomas. III. International Conference on Theoretical and Computational Approaches to Interface Phenomena (1993 : Brookings, S.D.)
QC173.4.I57T48 1995
541.3'3--dc20 95-1410
CIP

Proceedings of an International Conference on Theoretical and Computational Approaches to Interface Phenomena, held August 2–4, 1993, in Brookings, South Dakota

ISBN 978-1-4899-1321-0 ISBN 978-1-4899-1319-7 (eBook)
DOI 10.1007/978-1-4899-1319-7

Originally published by Plenum Press, New York in 1994
Softcover reprint of the hardcover 1st edition 1994

PREFACE

Many chemical processes that are important to society take place at boundaries between phases. Understanding these processes is critical in order for them to be subject to human control. The building of theoretical or computational models of them puts them into a theoretical framework in terms of which the behavior of the system can be understood on a detailed level. Theoretical and computational models are often capable of giving descriptions of interfacial phenomena that are more detailed, on a molecular level, than can be obtained through experimental observation. Advances in computer hardware have also made possible the treatment of larger and chemically more interesting systems.

The study of interfacial phenomena is a multi-disciplinary endeavor which requires collaboration and communication among researchers in different fields and across different types of institutions. Because there are many important problems in this field much effort is being expended to understand these processes by industrial laboratories as well as by groups at universities. Our conference titled "Theoretical and Computational Approaches to Interface Phenomena" held at South Dakota State University, August 2-4, 1993 brought together over thirty scientists from industry and academia and three countries in the western hemisphere to discuss the modeling of interfacial phenomena. Important topics discussed at this conference and within the pages of this book are: heterogeneous catalysis, surface diffusion, electrochemical phenomena, solvation, self-assembling organic films, chemisorption processes and manipulation of surface properties. It is our hope that this book will contribute to the activity level and the development of the field.

We are indebted to the sponsors of the conference without whose help the conference would not have been possible. We thank Cray Research, Inc. for their high level of participation. The conference was sponsored by the IBM Corporation, Biosym Technologies, the South Dakota EPSCoR program of the National Science Foundation, Amoco Chemical Company and South Dakota State University.

Harrell Sellers
Brookings, South Dakota

Joseph Golab
Naperville, Illinois

May 1994

CONTENTS

Acronyms ix

Reaction Path Approach to Dynamics at a Gas-Solid Interface: Quantum Tunneling Effects for an Adatom on a non-rigid Metallic Surface 1
S.E. Wonchoba, W.-P. Hu and D. G. Truhlar

Catalysis Modeling Employing Ab Initio and Bond Order Conservation - Morse Potential Methods 35
H. Sellers

Computer Simulations of Excitable Reaction-Diffusion Systems 57
M.R. Hoffmann and S.P. Müller

Molecular Dynamics Computer Simulations of Charged Metal Electrode - Aqueous Electrolyte Interfaces 75
M.R. Philpott and J.N. Glosli

Molecular Dynamics Computer Simulations of Aqueous Solution/Platinum Interface 101
M.L. Berkowitz and L. Perera

Diffusion Mechanisms of Flexible Molecules on Metallic Surfaces 119
M. Silverberg

Computer Simulation of Solvation in Supercritical Fluids 131
G.S. Anderson, K.M. Hegvik and M.R. Hoffmann

Structure-Function Modeling in Blood Coagulation: Interfaces, Biology and Chemistry 139
M.N. Liebman

Domains and Superlattices in Self-Assembled Monolayers of Long-Chain Molecules 149
J. Hautman and M.L. Klein

Manipulating Wetting and Ordering at Interfaces by Adsorption of Impurities 161
D.J. Olbris and Y. Shnidman

Density Functional Description of Metal-Metal and Metal-Ligand Bonds . 187
D.R. Salahub, M. Castro, R. Fournier, P. Calaminici, N. Godbout, A. Goursot, C. Jamorski, H. Kobayashi, A. Martínez, I. Pápai, E. Proynov, N. Russo, S. Sirois, J. Ushio and A. Vela

Density Functional Studies of Boron Substituted Zeolite ZSM-5 219
M.S. Stave and J.B. Nicholas

Index 245

ACRONYMS

ACE	Allis-Chalmers Enthusiast
AE	All Electron
AFM	Atomic Force Microscopy
ATC	Amoco Technology Company
BDA	Bond Dihedral Angle
BDE	Bond Dissociation Energy
BOC	Bond Order Conservation; Bond Order Constraint
BOC-MP	Bond Order Conservation - Morse Potential
BP	Becke Perdew (Exchange Correlation Functional)
BSSE	Basis Set Superposition Error
CCDB	Cambridge Crystalographic Data Bank
CD	Charge Density; Crystalographic Diffraction
CD-SCSAG	Centrifugal Dominant Small Curvature Semi-classical Adiabatic Ground State
CI	Configuration Interaction
CN	Crank-Nicholson
CS	(Hard Sphere Expansion) Conformal Solution
CSOV	Constrained Space Orbital Variance
DB	Dynamical Bottleneck
DFT	Density Functional Theory
DN	Double Numerical
DNP	Double Numerical with Polarization
DMP	Dynamic Morse Potential
DZVP	Double Zeta plus Valence Polarization
EAM	Embedded Atom Method
ECP	Effective Core Potential
EMBL	European Molecular Biology Laboratory
ESCA	Electron Spectroscopy for Chemical Analysis
ESR	Electron Spin Resonance
EV	Excluded Volume
FCC	Face Centered Cubic
FIX	Factor IX
FTCS	Forward Time Center Space
FT-IR	Fourier Transform - Infrared Spectroscopy
GB	GenBank Database
GIXS	Grazing Incidence X-ray Scattering
G-L	Greer-Levitt (Algorithm)
GS	Ground State
GTO	Gaussian Type Orbital
GVB	Generalized Valence Bond
HF	Hartree-Fock
HFS-LCAO	Hartree-Fock-Slater Linear Combination of Atomic Orbitals
HMB	Hemoglobin Mutation Bank
IHP	Inner Helmholtz Plane
IP	Ionization Potential

IR	Infrared (Spectroscopy)
IRAS	Infrared Reflection Absorption Spectroscopy
KS	Kohn-Sham (Functions,Theory)
K-S	Kabsch - Sanders (Algorithm)
LC	Local Composition
LCAO	Linear Combination of Atomic Orbitals
LCGTO-DFT	Linear Combination of Gaussian Type Orbitals - Density Functional Theory
LCGTO - LSD	Linear Combination of Gaussian Type Orbitals - Local Spin Density (Approximation)
LDA	Local Density Approximation
LDP	Linear Distance Plot
LEED	Low Energy Electron Diffraction
LSDA	Local Spin Density Approximation
MC	Monte Carlo (Method)
MCP	Model Core Potential
MCSCF	Multiconfigurational Self-Consistent Field (Approximation)
MD	Molecular Dynamics (Classical)
MEP	Minimum Energy Path
MES	Minimum Energy Site
MNDO	Modified Neglect of Differential Overlap
MO	Molecular Orbital
MP2	Møller-Plesset (Second Order) Perturbation Theory
MRSDCI	Multi-Reference Singles and Doubles Configuration Interaction
NCBI	National Center for Biotechnology Informatics
NL	Non-Local
NMR	Nuclear Magnetic Resonance
NN	Nearest Neighbor
NNB	Nearest Neighbor Bonding
OHP	Outer Helmholtz Plane
PA	Proton Affinity
PC	Point Charge
PDB	Protein Database (Data Bank)
PDE	Partial Differential Equation
PEP	Pauli Exclusion Principle
PERI	Protein Engineering Research Institute
PES	Potential Energy Surface
PIR	Protein Identification Resource
PI-TST	Path Integral Transition State Theory
PMF	Potential Mean Force
PN	Patri Net
RECP	Relativistic Effective Core Potential
RHF	Restricted Hartree-Fock (Theory)
SAM	Self-Assembled(ing) Monolayers
SCF	Self-Consistent Field (Theory)
SCT	Small Curvature Tunneling
SEXAFS	Surface Extended X-ray Absorption
STM	Scanning Tunneling Microscopy
TM	Transition Metal
TMA	Trimethylamine
TST	Transition State Theory
TST-QEP	Transition State THeory with Quantum Effective Potentials
VWN	Vosko, Wilk and Nusair (Exchange Correlation Functional)
XC	Exchange Correlation
XPS	X-ray Photoelectron Spectroscopy
XSW	X-ray Standing Wave

REACTION PATH APPROACH TO DYNAMICS AT A GAS-SOLID INTERFACE: QUANTUM TUNNELING EFFECTS FOR AN ADATOM ON A NON-RIGID METALLIC SURFACE

Steven E. Wonchoba, Wei-Ping Hu, and Donald G. Truhlar

Department of Chemistry and Supercomputer Institute
University of Minnesota, Minneapolis, MN 55455-0431

1. INTRODUCTION

Chemical reactions occurring on metal surfaces are of great technological importance, especially for catalysis.[1–6] Diffusion of reagents on the surface is a critical step in many such reactions.[1,2,7–9] Surface diffusion is also important in molecular beam epitaxy, chemical vapor deposition, and controlled growth of thin films.[10] Diffusion of hydrogen atoms is particularly interesting from a theoretical point of view because of the large quantum mechanical tunneling contributions to this process.[11–38] Laser-induced thermal desorption, field emission fluctuation, and linear optical diffraction techniques have been used to study hydrogen diffusion on several metals, including Ni, W, Ru, Pt, Rh, and Cu.[39–62] Theoretical studies of these processes can complement the data available from these experiments and can eventually be used to study subsurface and bulk diffusion processes more accurately than may be allowed by current experiments. These subsurface and bulk processes are fundamental for energy storage and fuel cell development, hydrogen embrittlement, and the possibility of subsurface hydrogen in catalysis.

Under a broad range of conditions one can model surface diffusion of adsorbed atoms as a unimolecular chemical reaction in which the chemisorption or physisorption bonds of the adatom at an initial site are broken and new bonds are formed at another site. The system composed of the adatom bound at the initial site is called the reactant, and the system composed of the adatom bound at the final site is called the product.

In Section 2 we provide an overview of the reaction-path approach for calculating rate constants of chemical reactions that involve large tunneling effects. We assume that the nuclear

Theoretical and Computational Approaches to Interface Phenomena
Edited by H.L. Sellers and J.T. Golab, Plenum Press, New York, 1994

motion is governed by an effective potential. Section 2.1 discusses the Born-Oppenheimer-Huang electronic adiabatic approximation for an adatom on a metal, which provides the justification for this assumption. The reaction-path approach involves two steps: variational transition state theory with quantized vibrations for the overbarrier reactive flux, reviewed in section 2.2, and multidimensional semiclassical approximations for tunneling, reviewed in Section 2.3. The semiclassical tunneling method presented here assumes small curvature of the reaction path in isoinertial coordinates, which is a reasonable assumption for processes such as diffusion of hydrogen on metals, although more general tunneling approaches are available for other cases. In section 2.4 we discuss the embedded cluster method which is used to model the hydrogen-metal systems studied with these theoretical dynamical methods.

In section 3, we present recent results of applying these methods to hydrogen and deuterium diffusion on Cu(100) and to hydrogen diffusion on Ni(100).

2. METHODS

2.1. Born-Oppenheimer-Huang Approximation for a Metal

Often the existence of a potential energy function governing the atomic motions in a system to be simulated is taken as a given. But this question merits further thought for systems involving metals.

For molecular and insulating solids, the existence of potential energy functions governing the internuclear (interatomic) motion is usually justified by the perturbational method used by Born and Oppenheimer[63] or the variational method used by Born and Huang.[64] Either method provides a justification for the separation of electronic and nuclear motion in which the electronic motion adjusts adiabatically to the nuclear motions. At this point a comment is in order on the validity of the electronic adiabatic approximation for metals. Perturbation theory shows that the leading nonadiabatic corrections are of the order of the vibrational excitation energies divided by the electronic excitation energies.[65] Let the maximum vibrational frequency in wavenumbers be $\overline{\upsilon}_{\max}$, and let the smallest electronic excitation energy be $\Delta E_{\min}$. Then, denoting Planck's constant by h and the speed of light by c, $hc\overline{\upsilon}_{\max}/\Delta E_{\min}$ might be thought to measure the importance of nonadiabatic effects, and—although this ratio is small for closed-shell molecules, insulators, and semiconductors—it is not small for metals, which are conductors. But it has been argued that it would be more appropriate for many properties of a metal to use $hc\overline{\upsilon}_{\max}/\langle E\rangle$, where $\langle E\rangle$ is the average allowed electronic excitation energy. For most of the conduction band, $\langle E\rangle$ is on the order of $0.5(E_F - E_0)$, where E_F is the Fermi energy, and E_0 is the energy at the bottom of the conduction band, since excitations with smaller excitation energy are typically blocked by the Pauli Exclusion Principle.[65] Using this argument leads to the criterion that the electronic adiabatic approximation is expected to be useful if $2hc\overline{\upsilon}_{\max}/(E_F - E_0)$ is small, and this is typically reasonably well satisfied. Thus properties of a metal, e.g., cohesive energy and normal modes of vibration, that depend on all the valence electrons, and not just those in orbitals near the Fermi level, should be reasonably well described by an electronically adiabatic treatment,[65] and we will adopt such a procedure here. With this justification, we assume that an effective potential is available (the actual forms used for the potential energy functions are provided in Sections 3.1 and 3.2), and we proceed to consider how the dynamics may be modeled.

2.2. Variational Transition State Theory

The fundamental assumption of classical transition state theory (TST) is that any system that crosses a dynamical bottleneck (DB) will do so only once over the course of its trajectory. This is called the no-recrossing assumption.[66–68] Using this assumption, TST calculates the forward rate as the equilibrium one-way flux through the DB from reactants to products.

In order to calculate the reactive flux in a convenient way, we first transform to isoinertial coordinates. Isoinertial coordinates are ones for which the kinetic energy has the same reduced mass for every square term and there are no cross terms. The simplest example is mass-scaled cartesians,[67] which are similar to the mass-weighted cartesians[69] of infrared spectroscopy (the difference being that mass-scaled cartesians have units of length whereas mass-weighted cartesians have units of mass$^{1/2}$ length). We use mass-scaled cartesians in which all the reduced masses are μ. We then calculate the minimum energy path (MEP) from reactants to products, and we define a reaction coordinate, s, as the distance along this MEP. The MEP begins at the saddle point, at which $s = 0$, and follows the path of steepest descent in isoinertial coordinates towards the reactants (s = negative) and the products (s = positive). The reactants and products are called minimum energy sites (MESs), and we define the value of s at the reactant MES as s^{R}.

We then define a dividing surface orthogonal to the MEP at each s, such that it separates phase space into a reactant region and a product region. This dividing surface is a generalized transition state, and it will be used as a trial DB. Conventional TST corresponds to calculating the equilibrium one-way flux through the trial DB at $s = 0$, which is the highest energy point on the MEP. In generalized TST, we calculate the equilibrium one-way fluxes for other trial surfaces. First consider the result if the calculation is carried out by classical mechanics, which we denote by subscript C. Then, for temperature T and trial DB at s, the calculated rate constant is[66–68]

$$k_{\mathrm{C}}^{\mathrm{GT}}(T,s) = \sigma \frac{k_B T}{h} \frac{Q_{\mathrm{C}}^{\mathrm{GT}}(T,s)}{Q_{\mathrm{C}}^{\mathrm{R}}(T)} \exp\left(\frac{-V_{\mathrm{MEP}}(s)}{k_B T} \right) \tag{1}$$

where the superscript GT indicates a value corresponding to a generalized transition state, σ is the symmetry factor accounting for the number of equivalent paths from a particular reactant site to products [4 for a hydrogen atom diffusing on a (100) face centered cubic (FCC) surface], k_B

is Boltzmann's constant, $V_{MEP}(s)$ is the potential energy of the system along the MEP at s with zero of energy such that $V_{MEP}(s^R) = 0$, and $Q_C^{GT}(T,s)$ and $Q_C^R(T)$ are the partition functions of the generalized transition state and reactant species, respectively. Notice that since the generalized transition state corresponds to a definite value of s, it does not include any reaction-coordinate motion. The remaining modes, i.e., those included in $Q_C^{GT}(T,s)$, are called transverse modes. Thus, if N is the number of atoms, the number, F, of degrees of freedom of the reactant is $3N$, but the generalized transition state has $3N - 1$ degrees of freedom.

As mentioned above, conventional TST places the DB at the saddle point on the PES, i.e., $s = 0$ in Eq. (1), but this is not usually the best DB because this treatment neglects recrossing, and the no-recrossing assumption is not completely correct. The interpretation of Eq. (1) for a unimolecular reaction is that it represents, for a canonical ensemble at temperature T, the local flux in the reactants to products direction through the dividing surface at s divided by the concentration of reactants. Any trajectory that crosses the DB from the reactant side toward products, even if it started on the product side, whether it ends as reactant or product, and whether or not it returns to the DB again and again, and hence gets counted again, will be counted as part of the forward flux, and the resulting rate constant may thus be overestimated. The goal of canonical variational transition state theory (CVT) is to find the best generalized transition state, i.e., the one that minimizes recrossing, which is the root of all the overcounting. Minimizing Eq. (1) with respect to s would lead to the best possible upper limit to the classical rate constant. However, in a canonical ensemble the rate is dominated by systems with energy close to the minimum required to pass through the DB region. For reactions in the threshold regime, the adiabatic separation of transverse coordinates from s is a good approximation,[71–73] so quantum effects on these modes can be included by quantizing transverse modes with fixed s.[74–76] Doing this and also quantizing the reactant modes (which requires no special justification) yields

$$k^{GT}(T,s) = \sigma \frac{k_B T}{h} \frac{Q^{GT}(T,s)}{Q^R(T,s)} \exp\left(\frac{-V_{MEP}(s)}{k_B T} \right) \tag{2}$$

where $Q^{GT}(T,s)$ and $Q^R(T,s)$ are now defined as sums over states, in contrast to $Q_C^{GT}(T,s)$ and $Q_C^R(T)$, which are phase space integrals. The resulting expression in Eq. (2) is a hybrid rate constant, since all modes except the reaction coordinate have been quantized, but the reaction coordinate motion is still classical In quantizing the partition functions, we assume that the reaction coordinate does not couple with any of the other modes and that the total partition function can be separated into electronic and vibrational factors (for processes at a gas-solid interface, there is no rotation or translation, because all modes that would be rotations or translations for a gas-phase species are actually vibrations due to the presence of the surrounding metallic lattice; see section 2.4 discussing the embedded cluster method).

The CVT estimate of the rate constant is obtained by minimizing the calculated rate constant of Eq. (2) with respect to s. The position, s, along the MEP at which the variationally optimized dividing surface (the variational transition state) at temperature T is ultimately placed is called $s_*^{\mathrm{CVT}}(T)$, and the CVT rate constant is defined as

$$k^{\mathrm{CVT}}(T) = \sigma \frac{k_B T}{h} \frac{Q^{\mathrm{CVT}}(T)}{Q^{\mathrm{R}}(T)} \exp\left(\frac{-V_{\mathrm{MEP}}^{\mathrm{CVT}}(T)}{k_B T} \right) \tag{3}$$

where superscript CVT denotes values determined at $s_*^{\mathrm{CVT}}(T)$.

In the harmonic approximation,

$$Q^{\mathrm{CVT}}(T) = \prod_{m=1}^{F-1} \left[\sum_{v=0}^{\infty} \exp\left(\frac{-\varepsilon_v^{\mathrm{CVT},m}(T)}{k_{\mathrm{B}} T} \right) \right] \tag{4a}$$

$$= \prod_{m=1}^{F-1} \frac{\exp\left(\frac{-hc\overline{\upsilon}_m^{\mathrm{CVT}}(T)}{2k_{\mathrm{B}}T} \right)}{1 - \exp\left(\frac{-hc\overline{\upsilon}_m^{\mathrm{CVT}}(T)}{k_{\mathrm{B}}T} \right)} \tag{4b}$$

where $\varepsilon_v^{\mathrm{CVT},m}(T)$ is the vibrational energy of level v of CVT mode m at temperature T, and $\overline{\upsilon}_m^{\mathrm{CVT}}(T)$ is the frequency of CVT mode m in wavenumbers. Similarly,

$$Q^{\mathrm{R}}(T) = \prod_{m=1}^{F} \left[\sum_{v=0}^{\infty} \exp\left(\frac{-\varepsilon_v^{\mathrm{R},m}}{k_{\mathrm{B}} T} \right) \right] \tag{5a}$$

$$= \prod_{m=1}^{F} \frac{\exp\left(\frac{-hc\overline{\upsilon}_m^{\mathrm{R}}}{2k_{\mathrm{B}}T} \right)}{1 - \exp\left(\frac{-hc\overline{\upsilon}_m^{\mathrm{R}}}{k_{\mathrm{B}}T} \right)} \tag{5b}$$

where $\varepsilon_v^{\mathrm{R},m}$ is the vibrational energy of level v of reactant mode m with frequency $\overline{\upsilon}_m^{\mathrm{R}}$ in wavenumbers.

2.3 Small-Curvature Tunneling Approximations

The neglect of tunneling in the motion along the reaction path often underestimates the true rate constant, especially at low temperatures for processes in which the reaction coordinate is dominated by hydrogenic motion.[76,77] To account for tunneling along the reaction coordinate, we assume that when tunneling occurs, the system is at a low enough temperature that it passes through the DB in the ground state or in a state where the effective potential has almost the same shape as the ground state.[66,78] Under such conditions the effective potential for tunneling is the vibrationally adiabatic ground-state potential energy curve, defined as

$$V_a^G(s) = V_{MEP}(s) + \varepsilon_{transv}^G(s) \tag{6}$$

where $V_{MEP}(s)$ is the potential energy of the system at s on the MEP, and $\varepsilon_{transv}^G(s)$ is the sum over the zero point energies of all transverse modes at s. The maximum of $V_a^G(s)$ is called V_a^{AG} and is the threshold energy (that energy above which the transmission probability is unity and below which it is zero) when the transverse modes are in the quantal ground state, and the reaction coordinate is classical. For surface processes modeled by the embedded cluster approach (see section 2.4), there are a total of (N_p + 1) moving atoms, where N_p is the number of moving metal atoms and the additional atom is the hydrogen atom. This yields ($3N_p$ + 2) transverse modes, which, as mentioned below Eq. (2), are all vibrational.

To include quantal effects on the reaction coordinate, k^{CVT} in Eq. (3) is multiplied by a ground-state transmission coefficient, $\kappa^{CVT/G}$, which accounts for tunneling along the reaction path. The rate constant including tunneling is given by

$$k^{CVT/G}(T) = \kappa^{CVT/G}(T)k^{CVT}(T). \tag{7}$$

The transmission coefficient is the product of two factors.[78] The first factor is the ratio of the Boltzmann average of the quantum probability for transmission through $V_a^G(s)$ to the Boltzmann average of the classical transmission probability for $V_a^G(s)$. The second factor is the ratio of the Boltzmann average of the classical transmission probability with V_a^{AG} as the threshold energy to that with $V_a^G\left[s = s_*^{CVT}(T)\right]$ as the threshold energy. The product of these factors gives the following expression for the ground-state transmission coefficient at the CVT level:

$$\kappa^{CVT/G}(T) = \frac{\int_{V_a^G\left(s=s^R\right)}^{\infty} dE\, P^G(E)\exp(-E/k_BT)}{\int_{V_a^G\left[s=s_*^{CVT}(T)\right]}^{\infty} dE\, \exp(-E/k_BT)} \tag{8}$$

where the lower limit of the integral in the numerator is approximate for a model in which $V_a^G(s)$ increases monotonically from $s = s^R$ to the barrier top, and where $P^G(E)$ is the quantum transmission probability at energy E for the effective potential. This transmission probability is approximated semiclassically by[79]

$$P^G(E) = 1/\{1 + \exp[2\theta(E)]\} \tag{9}$$

where $\theta(E)$ is the magnitude of the imaginary action integral for barrier transmission and depends upon the tunneling approximation used.

Several methods by which $P^{\mathrm{G}}(E)$ may be approximated are available.[67,68,78,80–89] For most processes, tunneling is not localized to the DB region, and the tunneling path is significantly shorter than the MEP. The best tunneling path should be chosen in principle as that path which minimizes the imaginary action integral.[84] When the curvature along the reaction path is small, we use a small-curvature tunneling approximation,[76,81,82,86–88] in particular the centrifugal-dominant small-curvature semiclassical adiabatic ground-state (CD-SCSAG) method,[86,88] to estimate this effect. This approximation is abbreviated SCT (small-curvature tunneling) for brevity.

In Eq. (8), the tunneling energies are selected from a continuum. However, in a unimolecular reaction, such as the case when an adatom diffuses across a solid surface between minimum energy sites, the reaction coordinate motion is initially restricted to discrete energy levels in the potential energy well, and each site-to-site hop is more accurately represented as a transition initiating in a discrete energy eigenstate rather than from a continuum energy state. This discretization becomes very important at low temperatures when excited states have small occupancies compared to the ground state. To account for this, when such quantization may be important we replace[23] the integral in Eq. (8) by a sum over discrete energy levels E_v^{R} below $V_{\mathrm{a}}^{\mathrm{AG}}$ where the energy levels correspond to exciting the reaction-coordinate mode (F) of the reactant. Then, in Eq. (8), only energies above $V_{\mathrm{a}}^{\mathrm{AG}}$ are treated as a continuum, and the ground-state transmission coefficient becomes

$$\kappa^{\mathrm{CVT/QG}}(T)=\frac{\sum_{v=0}^{M}\frac{\mathrm{d}E_v^{\mathrm{R}}}{\mathrm{d}v}P^{\mathrm{G}}\left(E_v^{\mathrm{R}}\right)\exp\left(-E_v^{\mathrm{R}}/k_BT\right)+\int_{V_{\mathrm{a}}^{\mathrm{AG}}}^{\infty}\mathrm{d}E\,P^{\mathrm{G}}(E)\exp(-E/k_BT)}{\int_{V_{\mathrm{a}}^{\mathrm{G}}\left[s=s_*^{\mathrm{CVT}}(T)\right]}^{\infty}\mathrm{d}E\,\exp(-E/k_BT)} \tag{10}$$

where the Q has been introduced in the superscript to indicate that the reaction coordinate energy levels have been quantized. The upper limit, M, of the summation is the number of excited energy eigenstates, E_v^{R}, of the reaction coordinate motion below $\mathrm{V}_{\mathrm{a}}^{\mathrm{AG}}$.

A complication occurs in practice because the nature of the reaction coordinate changes along the reaction path. Near the barrier top and over most of the reaction path, the reaction coordinate is a mostly hydrogenic mode. For H diffusion on Ni(100), for example, $s^{\mathrm{R}}\approx-1.5$ Å, and the reaction coordinate is mostly hydrogenic from $s\approx-1.2$ Å to $s\approx1.2$ Å, and there are two transverse hydrogenic modes. For $s\lessapprox-1.2$ Å, though, the amount of metal-motion character in the reaction coordinate increases until at $s=s^{\mathrm{R}}$, there are three hydrogenic transverse modes, with frequencies $\overline{\upsilon}_1^{\mathrm{R}}>\overline{\upsilon}_2^{\mathrm{R}}=\overline{\upsilon}_3^{\mathrm{R}}$. In the harmonic approximation, a very good approximation to the effective potential along the reaction coordinate in the vicinity of the reactant is

$$V_{\mathrm{eff}}(s)=V_{\mathrm{a}}^{\mathrm{G}}\left(s=s^{\mathrm{R}}\right)+\frac{1}{2}hc\left(\overline{\upsilon}_F^{\mathrm{R}}-\overline{\upsilon}_3^{\mathrm{R}}\right)+\frac{1}{2}f_3\left(s-s^{\mathrm{R}}\right)^2 \tag{11}$$

In Eq. (11), f_3 is the force constant corresponding to $\overline{\upsilon}_3^{\mathrm{R}}$ and is equal to $\mu\left(2\pi c\overline{\upsilon}_3^{\mathrm{R}}\right)^2$, and $\overline{\upsilon}_F^{\mathrm{R}}$ is the frequency in wavenumbers of the reactant normal mode that corresponds to the reaction coordinate, which is the lowest-frequency mode at the reactant MES. The energy levels of the reactant, computed from this potential, are

$$E_v^{\mathrm{R}} = V_{\mathrm{a}}^{\mathrm{G}}\left(s = s^{\mathrm{R}}\right) + \frac{1}{2}hc\overline{\upsilon}_F^{\mathrm{R}} + vhc\overline{\upsilon}_3^{\mathrm{R}}, \quad v = 0, 1, ... \tag{12}$$

Then, from Eq. (12),

$$\frac{\mathrm{d}E_v^{\mathrm{R}}}{\mathrm{d}v} = hc\overline{\upsilon}_3^{\mathrm{R}}. \tag{13}$$

When Eq. (10) is used to calculate the transmission coefficient, the small-curvature tunneling approximation is abbreviated SCTQ. Equation (7) is now replaced by

$$k^{\mathrm{CVT/QG}}(T) = \kappa^{\mathrm{CVT/QG}}(T)k^{\mathrm{CVT}}(T). \tag{14}$$

In the SCTQ and SCT approximations, the imaginary action integral in Eq. (9) is given by[81,82,86,88]

$$\theta(E) = (2\pi/h)\int_{s_0}^{s_1} \mathrm{d}s \sqrt{2\mu_{\mathrm{eff}}(s)\left[V_{\mathrm{a}}^{\mathrm{G}}(s) - E\right]} \tag{15}$$

where s_0 and s_1 are the limits of the tunneling path, and μ_{eff} is an effective reduced mass which accounts for reaction-path curvature. The physical interpretation is that the effective reduced mass, μ_{eff}, in Eq. (15) is smaller than the inertial reduced mass, μ, because the tunneling path is shorter than the MEP. The effective reduced mass is calculated as[86,88]

$$\mu_{\mathrm{eff}}(s) = \mu \min\begin{cases} \exp\left\{-2a(s) - [a(s)]^2 + (\mathrm{d}t/\mathrm{d}s)^2\right\} \\ 1 \end{cases} \tag{16}$$

where

$$a = C^{3/2}\left\{\sum_{m=1}^{F-1}[C_m(s)]^2[t_m(s)]^{-4}\right\}^{-1/4}, \tag{17}$$

$$t = C^{1/2}\left\{\sum_{m=1}^{F-1}[C_m(s)]^2[t_m(s)]^{-4}\right\}^{-1/4}, \tag{18}$$

$$C = \left\{ \sum_{m=1}^{F-1} [C_m(s)]^2 \right\}^{1/2}, \tag{19}$$

C_m is the reaction-path curvature component along generalized normal mode m, and $t_m(s)$ is the ground-state concave-side vibrational turning point of mode m.

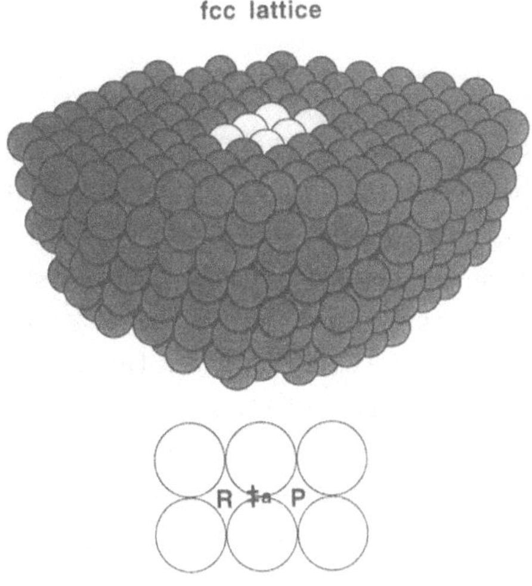

Figure 1. Model of the (100) surface of a face centered cubic (FCC) lattice. The upper figure shows a 6-atom embedded cluster; the lower figure is a close-up looking down on these 6 atoms. R and P are fourfold minimum energy sites for the adatom, and ‡a is a twofold transition state.

For interpretative purposes we note that including only the first term in the numerator of Eq. (10) yields the tunneling rate constant, k_{tun}(T), when substituted in Eq. (14), and including only the second term yields the overbarrier rate constant, k_{over}(T). The fraction of reaction that occurs by tunneling is

$$F_{tun}(T) = \frac{k_{tun}(T)}{k_{tun}(T) + k_{over}(T)}. \tag{20}$$

The diffusion process on an FCC crystal face consists of the hydrogenic atom hopping from an equilibrium four-fold minimum energy binding site, through a two-fold transition state, to another MES. The (100) surface and the stationary points of interest are shown in Figure 1. Assuming that the hops between the MESs are uncorrelated, meaning that the H atom remains at each fourfold site long enough to become thermalized, and therefore the previous history has no effect on each subsequent hop, the surface diffusion coefficient is given by[90]

$$D(T) = \left(\lambda^2 / 4\right) k_{\text{uni}}(T) \tag{21}$$

where λ is the lateral distance between two MESs (which, for the (100) surface, is equal to $R_0/\sqrt{2}$, where R_0 is the lattice constant), and $k_{\text{uni}}(T)$ is the hopping rate constant.

2.4. Embedded Cluster Method

The hydrogen-metal systems in this study are modeled by the embedded cluster method.[24,29] The systems consist of a single hydrogenic atom and a finite lattice of metal atoms, stacked as (100) FCC planes. A solid-state cluster, consisting of an increasing number of moving atoms (N_p), is surrounded by a set of immovable lattice atoms, fixed at geometries defined by the bulk lattice constant, turning all isolated-molecule rotations and translations into vibrations. We start with $N_p = 0$, i.e., a rigid metal lattice, and we run the dynamics calculations for this simple (but often unrealistic) system. We then increase N_p, allowing lattice atoms near the representative site of diffusion to move. When these metal atoms are allowed to move, their motion couples to the reaction-coordinate, and this has an effect on the dynamics. As we further increase N_p, and atoms further away from the diffusion site are allowed to move, the coupling to the reaction-coordinate subsides, and the rate constant eventually converges. The movable atoms are chosen as those which fall within the boundaries of either of two hemispheres which have equal radii and are centered at representative sites which depend upon the system being studied. The radii are expanded until the desired number of atoms, N_p, are enclosed within the hemispheres. The centers of the hemispheres are somewhat arbitrary, but need to be well defined for consistency within the study, and they naturally must be near the reactant and product MESs for the specific process; otherwise the coupling of lattice motion to the reaction-coordinate will not converge in a physical manner.

The full lattice is created large enough that all movable atoms in the largest cluster considered (N_{max} = 56 for Cu and N_{max} = 36 for Ni) are surrounded by all interacting neighbors in all directions as far as the distance at which the potential is cut off. The result is that each movable atom is in the environment necessary to be treated as part of an infinite metal lattice.

For the Cu(100) system, the expansion spheres are centered precisely at the reactant and product MESs (as determined by the rigid system with $N_p = 0$), and the full lattice consists of 324 atoms: 78, 70, 60, 52, 38, and 26 lattice atoms in the first through sixth planes, descending down perpendicular to the (100) top surface. For the Ni(100) system, which is part of a larger project studying subsurface diffusion, the spheres are centered at octahedral subsurface sites immediately below two adjacent surface atoms. The full lattice consists of 666 atoms: 100, 98, 100, 98, 78, 78, 58, 38, and 18 lattice atoms in the first through ninth planes. These values are somewhat larger than for Cu because the Ni potential energy function has a larger cutoff distance.

Since convergence with respect to N_p was found to be relatively rapid for both systems, we limit the presentation of results and discussion here to the rigid ($N_p = 0$) and fully converged ($N_p = N_{max}$) systems unless otherwise specified. We refer readers to journal articles for some intermediate results.[29,93] The $N_p = 56$ metal cluster for Cu consists of 20, 18, 16, and 2 moving atoms in the first through fourth planes, and the $N_p = 36$ metal cluster for Ni consists of 12, 16, and 8 atoms in the first, second, and third planes, respectively. Since there is one adatom, the total number of moving atoms, N, equals $N_p + 1$.

3. APPLICATIONS

We used the POLYRATE[86,94] code to calculate hopping rate constants for H and D on both Cu(100) and Ni(100). We made calculations without tunneling, as well as with SCT and SCTQ tunneling corrections. All vibrational energies are calculated in the harmonic approximation in the present work.

3.1. H/Cu(100) and D/Cu(100)

The potential energy for the H/Cu system is approximated as a sum of pair potentials. As a consequence of the Born-Oppenheimer approximation, these pair potentials are independent of isotopic mass, i.e., the same for D/Cu as for H/Cu. For both the H-Cu and Cu-Cu interaction potential, we used a Morse-spline function of the following form:

$$V(R) = \begin{cases} D_e\left\{\left[1-\exp(-\alpha(R-R_e))\right]^2-1\right\}, & R \le R_c - D_c \\ \sum_{i=3}^{5} C_i(R-R_c-D_c)^i, & R_c - D_c \le R \le R_c + D_c \\ 0, & R > R_c + D_c \end{cases} \quad (22)$$

where R is the Cu-Cu or H-Cu interatomic distance, D_e, α, and R_e are Morse parameters, and R_c, D_c, and C_i control the spline for smoothing the potential cutoff, $R_c + D_c$. The parameters of the H-Cu potential were based on earlier work by Valone *et al.*[19] and Truong *et al.*[26] For the Cu-Cu interaction,[24–26] D_e and α were chosen[24] to match features of the Lennard-Jones pair potential of Halicoğlu and Pound,[91] and R_e was chosen[24] to yield an interatomic spacing consistent with the bulk lattice.

For both the Cu-Cu and H-Cu interactions, R_C and the spline width, $2D_C$, were chosen in such a way that the cutoff was smooth and did not introduce any spurious behavior in the frequencies. The parameters C_i were chosen to make the potential function and its first and second derivatives continuous at $R = R_C - D_C$. The values of all pair potential parameters used in Eq. (22) for both the Cu-Cu and H-Cu interactions are listed in Table 1, and the functions are plotted in Figure 2.

Table 1. Potential parameters used for the Cu-Cu and H-Cu interaction in Eq. (22).

Parameter	Cu-Cu	H-Cu
α (Å^{-1})	2.287	1.43
D_e (kcal mol^{-1})	9.4378	7.2875
R_e (Å)	2.578942	2.34
R_C (Å)	5.157883	7.02
C_3 (kcal mol^{-1} Å^{-3})	447.2	154.619
C_4 (kcal mol^{-1} Å^{-4})	6231.8	2168.84
C_5 (kcal mol^{-1} Å^{-5})	23372.7	8156.16
D_C (Å)	0.0529	0.0529

Although this potential energy function is not quantitatively accurate,[29] it has been widely used, and it is qualitatively realistic. Thus it has become a prototype potential to use with new theoretical methods.

The determination of the energetically optimized bulk lattice constant for the assumed potential energy function was a very important process. When lattice atoms move from their original bulk lattice positions defined by the lattice constant, they do so as a result of their proximity to the exposed surface (surface atoms behave differently than bulk atoms because they have different numbers of neighbor atoms) and/or the presence of the adatom. If the lattice constant is not energetically optimized, then when an atom is allowed to move, it will do so not only at the surface but also in the bulk. Therefore, the effect of unfreezing of an atom may not be a result of physical coupling to the process under study but rather a result of inconsistent lattice spacing. As a result, the effect of allowing lattice atoms to move will be overestimated. A lattice constant properly optimized for the potential being used prevents these artifacts.

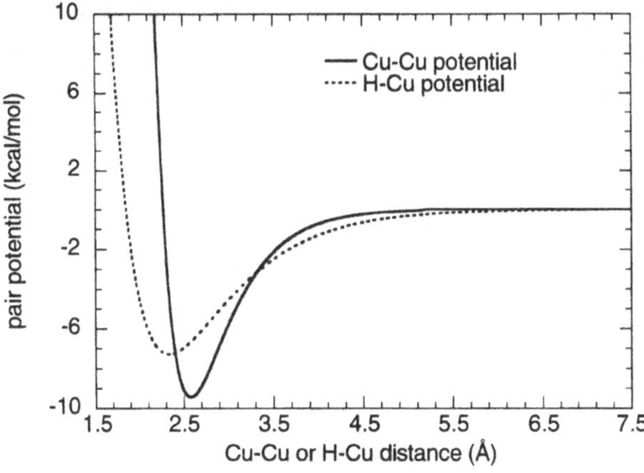

Figure 2. H-Cu and Cu-Cu pair interaction defined by Eq. (22) vs. interatomic distance.

The potential energy of a bulk atom is calculated as one half of the sum of all pair potentials between a selected bulk atom and all other atoms in the system, using the assumed potential energy function for Cu-Cu interaction. The optimum, energetically minimized lattice constant is that which yields the minimum energy for a bulk atom. This value was determined[29] to be 3.5818 Å, which is close to the experimental bulk lattice constant of 3.61 Å.[92] To avoid the artifacts discussed above, the energetically minimized lattice constant was used for all Cu(100) results presented here unless otherwise specified.

Table 2. Binding Energies and barrier heights (kcal/mol) for H on Cu(100).

	Fixed lattice ($N_p = 0$)	Relaxed lattice ($N_p = 56$)
Binding Energy		
4-fold site	50.3	50.3
2-fold site	38.4	38.9
Barrier Height		
classical	11.9	11.4
vibrationally adiabatic ground-state	10.6	9.9

Calculated binding energies and barrier heights for H on Cu(100) are given in Table 2. The binding energies are referenced to a hydrogen atom fully separated from the surface, and they do not include zero point contributions. The classical barrier height is calculated as the energy difference between the twofold saddle point and the fourfold reactant MES, and the vibrationally adiabatic ground-state barrier is calculated as the energy difference between V_a^{AG} and $V_a^G(s = s^R)$. Diffusion coefficients for the H/Cu system with the optimized lattice constant are given for a range of temperatures at the CVT (no tunneling), CVT/SCT, and CVT/SCTQ levels for both rigid ($N_p = 0$) and moving ($N_p = 56$) lattices in Table 3. The results for the D/Cu system are given in Table 4.

Table 3. CVT, CVT/SCT, and CVT/SCTQ diffusion coefficients (cm^2/s) for H on Cu(100) surfaces with $N_p = 0$ and $N_p = 56$ at a variety of temperatures. Numbers in parentheses are powers of 10.

	$N_p = 0$			$N_p = 56$		
T (K)	CVT	CVT/SCT	CVT/SCTQ	CVT	CVT/SCT	CVT/SCTQ
40	2.55(-61)	2.00(-23)	4.15(-22)	5.02(-58)	4.49(-22)	1.02(-20)
50	9.27(-50)	3.03(-23)	4.15(-22)	4.03(-47)	6.83(-22)	1.02(-20)
60	4.89(-42)	4.55(-23)	4.15(-22)	7.78(-40)	1.04(-21)	1.03(-20)
80	2.33(-32)	1.16(-22)	4.23(-22)	1.06(-30)	2.84(-21)	1.06(-20)
100	1.57(-26)	4.96(-22)	6.99(-22)	3.42(-25)	1.35(-20)	1.80(-20)
120	1.25(-22)	9.83(-21)	1.01(-20)	1.65(-21)	2.37(-19)	2.25(-19)
200	9.17(-15)	2.18(-14)	9.75(-15)	4.57(-14)	1.53(-13)	1.04(-13)
250	2.25(-12)	3.75(-12)	1.94(-12)	8.37(-12)	1.80(-11)	1.11(-11)
300	9.06(-11)	1.27(-10)	7.63(-11)	2.79(-10)	4.75(-10)	2.92(-10)
400	9.64(-9)	1.16(-8)	8.24(-9)	2.34(-8)	3.16(-8)	2.12(-8)
500	1.64(-7)	1.83(-7)	1.43(-7)	3.45(-7)	4.19(-7)	3.03(-7)
600	1.09(-6)	1.18(-6)	9.72(-7)	2.10(-6)	2.41(-6)	1.86(-6)
800	1.20(-5)	1.25(-5)	1.09(-5)	2.07(-5)	2.23(-5)	1.84(-5)
1000	5.08(-5)	5.23(-5)	4.71(-5)	8.26(-5)	8.68(-5)	7.49(-5)

To evaluate the effect of tunneling on this process, we calculate the ratio of the diffusion coefficients calculated including tunneling to those calculated without tunneling (i.e., CVT/SCT to CVT and CVT/SCTQ to CVT). This quantity is given in Table 5. At high temperatures, tunneling does not contribute significantly to the diffusion coefficient. But as the temperature decreases, tunneling becomes increasingly more important, and at temperatures $\leq$ 100 K, virtually the entire process proceeds by tunneling.

Table 4. CVT, CVT/SCT, and CVT/SCTQ diffusion coefficients (cm^2/s) for D on Cu(100) surfaces with $N_p = 0$ and $N_p = 56$ at a variety of temperatures. Numbers in parentheses are powers of 10.

	$N_p = 0$			$N_p = 56$		
T (K)	CVT	CVT/SCT	CVT/SCTQ	CVT	CVT/SCT	CVT/SCTQ
40	1.92(-63)	4.19(-33)	5.18(-32)	2.34(-60)	2.49(-31)	3.05(-30)
50	1.86(-51)	7.25(-33)	5.20(-32)	5.50(-49)	4.39(-31)	3.06(-30)
60	1.88(-43)	1.55(-32)	5.52(-32)	2.18(-41)	9.78(-31)	3.34(-30)
80	2.02(-33)	9.72(-31)	1.04(-30)	7.28(-32)	6.90(-29)	8.40(-29)
100	2.23(-27)	1.73(-26)	1.78(-26)	3.99(-26)	5.89(-25)	7.91(-25)
120	2.45(-23)	7.76(-23)	6.25(-23)	2.77(-22)	1.50(-21)	2.53(-21)
200	3.43(-15)	4.73(-15)	3.24(-15)	1.56(-14)	2.81(-14)	5.61(-14)
250	1.01(-12)	1.23(-12)	9.11(-13)	3.50(-12)	5.12(-12)	9.65(-12)
300	4.60(-11)	5.24(-11)	4.11(-11)	1.33(-10)	1.73(-10)	3.06(-10)
400	5.58(-9)	6.00(-9)	5.05(-9)	1.30(-8)	1.51(-8)	2.37(-8)
500	1.01(-7)	1.06(-7)	9.27(-8)	2.07(-7)	2.28(-7)	3.32(-7)
600	6.53(-7)	7.25(-7)	6.53(-7)	1.33(-6)	1.42(-6)	1.96(-6)
800	8.02(-6)	8.15(-6)	7.55(-6)	1.37(-5)	1.42(-5)	1.81(-5)
1000	3.48(-5)	3.51(-5)	3.30(-5)	5.60(-5)	5.74(-5)	6.96(-5)

Table 5. Ratio of tunneling to non-tunneling diffusion coefficients for H on a moving ($N_p = 56$) Cu(100) surface. Numbers in parentheses are powers of 10.

T(K)	$\frac{\text{CVT/SCT}}{\text{CVT}}$	$\frac{\text{CVT/SCTQ}}{\text{CVT}}$
40	8.94(35)	2.03(37)
50	1.69(25)	2.53(26)
60	1.34(18)	1.32(19)
80	2.68(9)	1.00(10)
100	3.95(4)	5.26(4)
120	1.44(2)	1.36(2)
200	3.35	2.28
250	2.15	1.33
300	1.70	1.05
400	1.35	0.91
500	1.21	0.88
600	1.15	0.89
800	1.07	0.89
1000	1.05	0.91

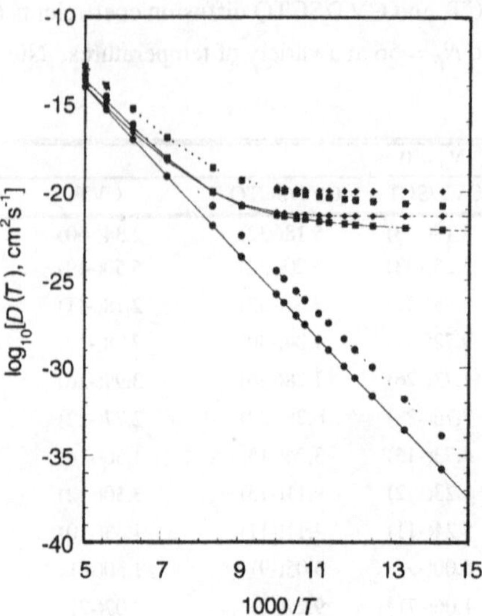

Figure 3. Arrhenius plot for H diffusion on Cu(100) Solid lines are for $N_p = 0$, dashed lines are for $N_p = 56$. CVT, CVT/SCT, and CVT/SCTQ diffusion coefficients represented by circles, squares, and diamonds, respectively.

Tables 3 and 4 show that the CVT/SCTQ diffusion coefficient becomes independent of T at very low temperatures. This effect is also shown in the Arrhenius plot of the data in Figure 3. For both rigid and moving surfaces, the plot of the diffusion coefficient with quantized reactant states levels off at approximately 90 K. To quantify the analysis, we next define two temperatures of interest on the Arrhenius plot.

First, we note that the Arrhenius plot of the CVT/SCTQ diffusion coefficients in Figure 3 consists largely of two approximately linear regions (T above approximately 120 K and T below approximately 90 K). An intermediate transition region (T between 90 K and 120 K) of high curvature joins the two linear regions. We define the transition temperature, T_{tr}, as the point of maximum curvature (analytically, the point of maximum second derivative) of this intermediate transition region on the Arrhenius plot.

We will next derive an approximate analytic formula for T_{tr}. We begin by noting that at low temperatures, the denominators of Eqs. (4b) and (5b) closely approach unity, and by using Eqs. (6) and (12) and the convention that $V_{\text{MEP}}\left(s^{\text{R}}\right) = 0$, the ratio of partition functions in Eq.

(3) can be approximated as

$$\frac{Q^{\mathrm{CVT}}(T_{\mathrm{low}})}{Q^{\mathrm{R}}(T_{\mathrm{low}})} \approx \exp\left(-\frac{1}{k_{\mathrm{B}}T_{\mathrm{low}}}\left\{V_{\mathrm{a}}^{\mathrm{G}}\left[s_*^{\mathrm{CVT}}(T_{\mathrm{low}})\right] - V_{\mathrm{MEP}}^{\mathrm{CVT}}(T_{\mathrm{low}}) - E_0^{\mathrm{R}}\right\}\right) \tag{23}$$

where T_{low} denotes a low temperature.

Next, we approximate the transmission coefficient at low temperature by assuming that only states below $V_{\mathrm{a}}^{\mathrm{AG}}$ make a significant contribution to the diffusion coefficient (i.e., that the contribution due to the states in the energy continuum is negligible). This approximation, along with Eqs. (12) and (13), allows us to write the ground-state transmission coefficient, Eq. (10), as

$$\kappa^{\mathrm{CVT/QG}}(T_{\mathrm{low}}) = \frac{1}{k_{\mathrm{B}}T_{\mathrm{low}}}\exp\left\{\frac{1}{k_{\mathrm{B}}T_{\mathrm{low}}}V_{\mathrm{a}}^{\mathrm{G}}\left[s_*^{\mathrm{CVT}}(T_{\mathrm{low}})\right]\right\}hc\bar{\upsilon}_3^{\mathrm{R}}\sum_{v=0}^{M}P^{\mathrm{G}}\left(E_v^{\mathrm{R}}\right)\exp\left(-\frac{E_v^{\mathrm{R}}}{k_{\mathrm{B}}T_{\mathrm{low}}}\right) \tag{24}$$

where the denominator of Eq. (10) has been integrated directly. Substituting Eqs. (12) and (23) into Eq. (3) and then substituting Eqs. (3) and (24) into Eq. (7) gives the following expression for the low-temperature rate constant:

$$k(T_{\mathrm{low}}) = \sigma c\bar{\upsilon}_3^{\mathrm{R}}\sum_{v=0}^{M}P^{\mathrm{G}}\left(E_v^{\mathrm{R}}\right)\exp\left(\frac{-vhc\bar{\upsilon}_3^{\mathrm{R}}}{k_{\mathrm{B}}T_{\mathrm{low}}}\right) \tag{25}$$

and, from Eq. (23),

$$D(T_{\mathrm{low}}) = \frac{\lambda^2}{4}\sigma c\bar{\upsilon}_3^{\mathrm{R}}\sum_{v=0}^{M}P^{\mathrm{G}}\left(E_v^{\mathrm{R}}\right)\exp\left(\frac{-vhc\bar{\upsilon}_3^{\mathrm{R}}}{k_{\mathrm{B}}T_{\mathrm{low}}}\right). \tag{26}$$

For the H/Cu system, in the harmonic approximation of Eq. (12), there are a total of 4 states below $V_{\mathrm{a}}^{\mathrm{AG}}$. Therefore, $M = 3$ in Eqs. (25) and (26). For the D/Cu system, $M = 5$. Equation (25) is also the expression for the tunneling rate constant, $k_{\mathrm{tun}}(T_{\mathrm{low}})$.

Using Eq. (26) as the expression for the low temperature diffusion coefficient, the point of maximum curvature of the Arrhenius plot (Figure 3) occurs when

$$\left.\frac{\mathrm{d}^3\left[D(T_{\mathrm{low}})\right]}{\mathrm{d}(1/T_{\mathrm{low}})^3}\right|_{T_{\mathrm{low}}=T_{\mathrm{tr}}} = 0. \tag{27}$$

Making a parabolic approximation to the vibrationally adiabatic potential energy curve,[95] assuming that E_{a}, the activation energy for the diffusion of the adatom on the surface, is much larger than $hc\bar{\upsilon}_3^{\mathrm{R}}$, and solving for T in Eq. (27) yields the following expression for the transition temperature:[93]

$$T_{\text{tr}} = \frac{hc\left|\overline{\upsilon}^{\ddagger}\right|}{2\pi k_{\text{B}}} \tag{28}$$

where $\left|\overline{\upsilon}^{\ddagger}\right|$ is the magnitude of the imaginary frequency at the top of the potential barrier. Since Eqs. (25) and (26) only differ by a constant, the same result would be obtained if Eq. (25) were used in the third derivative in Eq. (27).

A second temperature of interest on the Arrhenius plot is the point at which the diffusion coefficient levels off and becomes visually temperature independent. At very low temperatures, the $v = 0$ (ground-state) term in Eqs. (25) and (26) dominates the other (excited-state) terms, and the expression for the overall rate constant approaches the low-temperature limit,

$$k(T) \xrightarrow[T\to 0]{} \sigma c \overline{\upsilon}_3^{\text{R}} P^{\text{G}}\left(E_0^{\text{R}}\right) \tag{29}$$

or

$$D(T) \xrightarrow[T\to 0]{} \frac{\lambda^2}{4} \sigma c \overline{\upsilon}_3^{\text{R}} P^{\text{G}}\left(E_0^{\text{R}}\right). \tag{30}$$

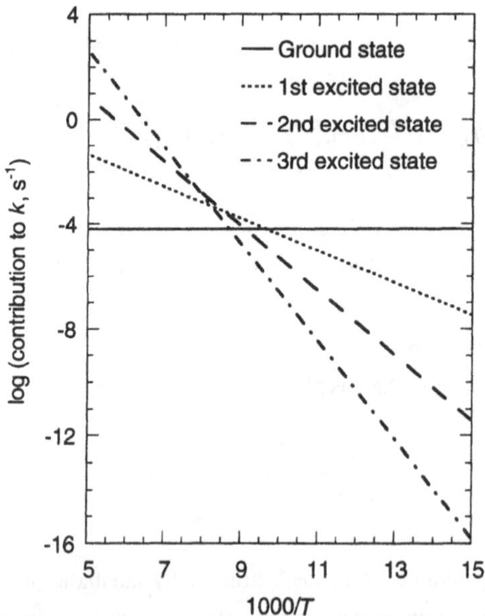

Figure 4. Logarithm to the base 10 of the contribution to the rate constant of each quantized state with energy below V_{a}^{AG} versus inverse temperature for H diffusion on Cu(100) with $N_p = 56$.

In Figure 4, the logarithm to the base 10 of the contribution to the overall rate constant of each of the quantized states for the moving ($N_p = 56$) surface is plotted using the same abscissa as was used for the Arrhenius plot in Figure 3. Note that the contribution of the ground state is independent of temperature, and is, in fact, the low-temperature limit to the rate constant given in Eq. (29). At roughly 90 K (the temperature at which the Arrhenius plot levels off), the contribution of the ground state is approximately one order of magnitude greater than the contribution of the first excited state, and it is several orders of magnitude greater than the contributions of the higher excited states (see Figure 4). As a result, we empirically define a level-off temperature, T_0, as the temperature at which the ground-state contribution to the rate constant is greater than the first excited state contribution by one order of magnitude. Making the same approximations that led to Eq. (28) then gives the following expression for the level-off temperature:

$$T_0 = \frac{hc\overline{\upsilon}_3^{\mathrm{R}}}{k_{\mathrm{B}}\left[2\pi\left(\frac{\overline{\upsilon}_3^{\mathrm{R}}}{\left|\overline{\upsilon}^{\ddagger}\right|}\right) + \ln(10)\right]}. \tag{31}$$

The precise values of the transition temperature and the level-off temperature are dependent upon their definitions. However, the definitions given above are reasonable. It can be shown[93] that when the $V_{\mathrm{a}}^{\mathrm{G}}$ is parabolic [which is a reasonably valid approximation[93,95] and was used to derive Eqs. (28) and (31)], then the transition temperature, Eq. (28), corresponds to the temperature at which all states contribute to the rate constant equally. For H on Cu(100) with $N_p = 56$, $\overline{\upsilon}_3^{\mathrm{R}} = 983$ cm^{-1}. Using coordinates scaled to a mass μ of 1 amu, we fit the $V_{\mathrm{a}}^{\mathrm{G}}$ curve to a parabola from $s = -1.3$ Å to $s = 1.3$ Å, where $s^{\mathrm{R}} \approx -1.7$ Å, and we determined the magnitude of the imaginary frequency, $\left|\overline{\upsilon}^{\ddagger}\right|$, from the second derivative of this curve, i.e.,

$$\left|\overline{\upsilon}^{\ddagger}\right| = \frac{1}{2\pi c}\left[\frac{\mathrm{d}^2 V_{\mathrm{a}}^{\mathrm{G}}(s)}{\mathrm{d}s^2} \Big/ \mu\right]^{1/2}. \tag{32}$$

This yields $\left|\overline{\upsilon}^{\ddagger}\right| = 458$ cm^{-1}. Then, Eqs. (28) and (31) yield $T_{\mathrm{tr}} = 105$ K and $T_0 = 90$ K, both of which are visually consistent with the Arrhenius plot in Figure 3, indicating that the assumptions made in the derivations are valid.

We note that computing $\left|\overline{\upsilon}^{\ddagger}\right|$ from the second derivative of $V_{\mathrm{MEP}}(s)$ at its maximum instead of from a global fit to $V_{\mathrm{a}}^{\mathrm{G}}(s)$ would have yielded 461 cm^{-1}, resulting in about the same level of agreement.

We can also make comparisons to previous results obtained with other theoretical methods, in particular to the calculations of Sun and Voth[37], who used path integral transition state theory (PI-TST) and to those of Valone, Voter, and Doll[19], who used transition state theory with quantum effective potentials (TST-QEP). We note that these comparisons suffer from some serious limitations. First, previous work[29] has suggested that the embedded clusters of lattice atoms used in the PI-TST and TST-QEP studies were too small to yield converged rate constants. Second, both studies used the experimental lattice constant, rather than the one energetically optimized one for the assumed potential energy function, to create the Cu lattices. The effects of this have been discussed earlier. Third, the previous studies were limited to a smaller range of temperatures. Despite these limitations, the comparisons are interesting.

In Table 6, the present calculated diffusion coefficients are compared to PI-TST and TST-QEP results. All values listed in this table were calculated using a specially constructed lattice with the experimental lattice constant, $R_0 = 3.61$ Å, to duplicate the system used in the PI-TST and TST-QEP calculations. In general, the agreement is very good for temperatures greater than or equal to 120 K, but as the temperature is lowered below this, the CVT/SCQT results predict a faster onset of the temperature independence of the diffusion coefficient than do the PI-TST and TST-QEP results. Table 6 also gives the ratios of the diffusion coefficients calculated for a moving surface to those calculated for a rigid surface. The number of moving lattice atoms, N_{moving}, is 56 for the CVT/SCTQ results, 30 for the PI-TST results, and 36 (all surface layer atoms) for the TST-QEP results. These values are also in good agreement for temperatures greater than or equal to 120 K, and the disagreement of the $T = 100$ K results can be traced to the faster onset of temperature independence seen in the CVT/SCTQ results.

Table 6. Comparisons of CVT/SCT and CVT/SCQT results to PI-TST results.[37] and TST-QEP results.[19] The CVT results in this table were calculated using the experimental lattice constant of 3.61 Å to get a better comparison between the two methods. The number of atoms in the moving lattice, N_{moving}, is 56 for the CVT results, 30 for the PI-TST results, and 36 (all surface atoms) for the TST-QEP results. Numbers in parentheses are powers of 10.

	T(K)	CVT/SCT	CVT/SCQT	PI-TST	TST-QEP
$D(N_p = 0)$	100	2.2(-23)	3.2(-23)	1.2(-24)	3.5(-25)
	120	4.8(-22)	5.2(-22)	3.1(-22)	...
	200	3.3(-15)	4.3(-15)	6.9(-15)	1.0(-14)
	300	3.3(-11)	3.7(-11)	6.9(-11)	7.9(-11)
$D(N_{moving})/D(N_p = 0)$	100	17	8.0	42	7.4
	120	21	18	18	...
	200	7.7	3.0	5.5	2.4
	300	4.1	2.3	3.3	1.9

We now consider the results obtained using the energetically minimized lattice constant. The onset of temperature independence can be seen in the CVT/SCTQ results shown in Figure 3. Figure 5 shows the logarithm to the base 10 of the diffusion coefficient ratio described above. The ratios are plotted for the CVT, CVT/SCT, and CVT/SCTQ levels of theory and compared to the PI-TST and TST-QEP results which are given in Table 6. The CVT ratios logarithmically increase with decreasing temperature, and the CVT/SCT and CVT/SCQT ratios level off at approximately 110 K. The PI-TST and TST-QEP ratios, however, do not level off, and this may be a consequence of the use of the non-energetically optimized lattice constant in the latter studies (see previous discussion). Finally, we again stress that all of these comparisons must be interpreted with the disclaimers discussed above. For more extensive comparisons of CVT/SCT, PI-TST, and TST-QEP results, we refer readers to Ref. 29.

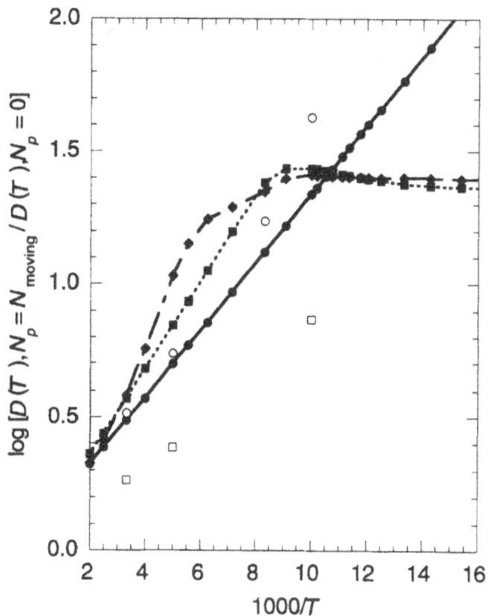

Figure 5. Logarithm to the base 10 of the ratio of diffusion coefficients for H on moving Cu(100) to those for H on rigid Cu(100). CVT, CVT/SCT, and CVT/SCTQ are represented by solid circles, squares, and diamonds, respectively. PI-TST and TST-QEP results are represented by hollow circles and squares, respectively. N_{moving} = 56 for CVT results, 30 for PI-TST results, and 36 (all surface atoms) for TST-QEP results.

Finally, we evaluate the importance of variational optimization of the dividing surface. Conventional TST places a dividing surface orthogonal to the MEP passing through the saddle point, which in a classical world would yield an upper bound to the rate constant. We can evaluate the importance of variational placement of the dividing surface to minimize the classical rate constant by comparing the conventional TST rate constant [i.e., the rate constant determined with $s = 0$ in Eq. (1)] to the CVT rate constant [i.e., the rate constant determined with $s = s_*^{\text{CVT}}$ in Eq. (1)]. At all temperatures for both H and D on moving Cu(100) with N_p = 56, the conventional TST rate constant is exactly equal to the CVT rate constant, indicating that variational optimization of the dividing surface is not important for these systems. For the rigid (N_p = 0) system, variational optimization makes a slight difference in the calculated rate constants. The effects are greater at lower temperatures and are greater for H than for D. Specifically, variational optimization decreases the H/Cu diffusion coefficient by 34% and 5% at 40 K and 1000 K, respectively. The D/Cu diffusion coefficient is decreased by 8% and 4% at these temperatures upon variational optimization.

3.2. H/Ni(100)

For a system with only one non-metal atom, the total potential energy is estimated by the embedded atom method (EAM)[96,97] as

$$V=\sum_i\left[F_i(\overline{\rho}_i)+\tfrac{1}{2}\sum_{j\neq i}\phi_{ij}(R_{ij})\right]. \tag{33}$$

In this expression, the summations over i and j are over all atoms in the system, $F_i(\overline{\rho}_i)$ is the energy required to embed atom i into a vacancy at a point where the electron density due to all other atoms in the system is $\overline{\rho}_i$, ϕ_{ij} is a pair potential, and R_{ij} is the internuclear distance between atoms i and j. The expressions for the electron densities are given in other references.[93,96]

The energy to embed a hydrogen atom at a point where the density is ρ is given by[98]

$$F_{\mathrm{H}}(\rho)=\alpha_{\mathrm{H}}\rho\exp(-\beta_{\mathrm{H}}\rho) \tag{34}$$

and that to embed a nickel atom is given by[98]

$$F_{\mathrm{Ni}}(\rho)=\begin{cases}A\rho\exp(-\alpha\rho)+B\rho^3\exp(-\beta\rho)+C\rho\exp(-\gamma\rho), & 0\le\rho\le\rho_c-\Delta\\ A_s(\rho-\rho_c)^5+B_s(\rho-\rho_c)^4+C_s(\rho-\rho_c)^3+D_s, & \rho_c-\Delta<\rho\le\rho_c\\ D_s, & \rho_c<\rho\end{cases} \tag{35}$$

where the coefficients have been adjusted for the present work to agree as closely as possible with experimental results as discussed below. The repulsive pair potential, ϕ_{ij}, is defined as

$$\phi_{ij}(R_{ij})=\frac{CZ_i(R_{ij})Z_j(R_{ij})}{R_{ij}} \tag{36}$$

where C is a constant, and $Z(R_{ij})$ is the effective charge for the atom defined by Foiles *et al.*, as[99]

$$Z(R_{ij})=Z_0(1+\beta R^{\gamma})\exp(-\alpha R) \tag{37}$$

where the parameters Z_0, α, β, and γ have again been determined to match experiment. The cutoffs for the electron density contributions and the pair potential are established by a smoothing function which has been defined previously.[100]

The work presented here used EAM parameter set 5 (EAM5). For a complete description of and discussion of the evolution of this parameter set, see reference 93. For the Ni(100) surface, EAM5 accurately reproduces experimental hydrogen vibrational frequencies and surface binding energies. Table 7 lists the surface binding energies, hydrogen-nickel equilibrium interatomic distances, hydrogen distances above the surface plane, and hydrogen vibrational frequencies calculated by EAM5 and compares them to literature values quantities.[38,101–111] The EAM5 potential also accurately reproduces the bulk values[112–116] of the monovacancy formation energy (E_{1V}^{F}) and the sublimation energy (E_s). The energetically minimized bulk lattice constant, R_0, for EAM5 is 3.5211 Å, fit to the experimental value of 3.52 Å.[92] These three bulk quantities are compared to experiment in Table 8.

Ni, like Cu, crystallizes as an FCC structure, and the picture in Figure 1 applies here as well as to Cu. Diffusion coefficients for H on Ni(100) at the CVT, CVT/SCT, and CVT/SCTQ levels are given in Table 9 for the rigid ($N_p = 0$) and moving ($N_p = 36$) Ni(100) surfaces.

As for Cu(100), the CVT/SCTQ diffusion coefficients for Ni(100) level off at low T and become independent of temperature. Figure 6 is an Arrhenius plot of the CVT/SCTQ diffusion coefficients. The plot also shows experimental results performed at a variety of surface coverages varying from $\theta = 0.12$ to 1.0 and the results of one other theoretical study. George *et al.*[50] and Mullins *et al.*[57] measured the diffusion coefficient with laser-induced thermal desorption at several temperatures between 211 and 283 K. Lin and Gomer[49] used the field emission fluctuation technique,[41] and Zhu and co-workers[60,61] used linear optical diffraction techniques to examine this process at lower temperatures (between 75 and 200 K). Mattsson *et al.*[38] studied the process with path integral techniques with an EAM potential function and used numerical Monte Carlo techniques to evaluate the path integrals. Their study covered a temperature range down to 25 K, but to preserve the resolution of the Arrhenius plot and enable adequate visual comparison to other results, only results down to 40 K are shown in Figure 6.

Diffusion coefficients are plotted for the George *et al.*,[50] Mullins *et al.*,[57] and Zhu *et al.*[60] experiments and for the calculations of Mattsson *et al.*.[38] For the Lee *et al.*[61] experiments, the data were fit to an Arrhenius form,

$$D(T) = D_0 \exp\left(\frac{-E_a}{RT}\right) \tag{38}$$

where R is the gas constant, and E_a is the activation energy, which can be determined from Eq. (38) as

$$E_a(T) = -R\frac{d\{\ln[D(T)]\}}{d(1/T)}. \tag{39}$$

Table 7. Binding energies (kcal/mol), hydrogen-nickel interatomic distance (Å), hydrogen height above the surface (Å), and hydrogen vibrational frequencies (cm^{-1}) calculated in this study [for a rigid ($N_p = 0$) and moving ($N_p = 36$) Ni(100) surface] compared to experimental and calculated values from the literature. In the literature column, experimental values are listed first, followed by calculated values in parentheses.

	Binding energy			$R_{H\text{-}Ni}$			$R_{H\text{-}surface}$			Frequencies		
Site	$N_p = 0$	$N_p = 36$	Lit.	$N_p = 0$	$N_p = 36$	Lit.	$N_p = 0$	$N_p = 36$	Lit.	$N_p = 0$	$N_p = 36$	Lit.
H(a)	64.73	64.76	64.4 ± 0.6[a]	1.84	1.83	1.82–1.84[b]	0.53	0.50	0.5 ± 0.1[b]	767	753	589[c]
			64.6 ± 0.9[d]			1.9–2.0[e]			0.9–1.0[e]			597[a]
			(62)[f]			(1.78)[g]			(0.3)[g]			621[h]
			(62)[i]			(1.8)[j]			(0.32)[j]			(532)[k]
			(69)[g]			(1.91)[f]			(0.8)[i]			(588)[g]
			(79)[j]			(1.92)[i]						(613)[f]
												(637)[i]
												(686)[l]
												(726)[j]
										517	524	387[c]
												(645)[g]
‡a	60.73	60.76	(63)[g]	1.56	1.56		0.94	0.93		1277	1270	(1428)[g]
			(77)[j]							438	449	
										290i	292i	

[a]Christmann, Schober, Ertl, and Neumann, Ref. 102

[b]Stensgaard and Jakobsen (D/Ni), Ref. 108

[c]Mårtensson, Nyberg, and Andersson, Ref. 111

[d]Lapujoulade and Neil, Ref. 101

[e]Rieder and Wilsch, Ref. 106

[f]Nordlander, Holloway, and Nørskov, Ref. 107

[g]Upton and Goddard, Ref. 104

[h]Anderson, Ref. 103

[i]Nørskov, Ref. 105

[j]Umrigar and Wilkins, Ref. 109

[k]Karlsson, Mårtensson, Andersson, and Nordlander, Ref. 110

[l]Mattsson, Engberg, and Wahnström, Ref. 38

Table 8. EAM5 calculated values for bulk lattice quantities compared to experiment.

Quantity	EAM5	experiment
monovacancy formation energy	1.66 eV	1.39–1.70 eV[a]
sublimation energy	4.43 eV	4.45 eV[b]
bulk lattice constant	3.5211 Å	3.52 Å[c]

[a] Refs. 112–115

[b] Ref. 116

[c] Ref. 92

Table 9. CVT, CVT/SCT, and CVT/SCQT diffusion coefficients (cm^2/s) for H on rigid ($N_p = 0$) and moving ($N_p = 36$) Ni(100) surfaces at a variety of temperatures. Numbers in parentheses are powers of 10.

	$N_p = 0$			$N_p = 36$		
T (K)	CVT	CVT/SCT	CVT/SCQT	CVT	CVT/SCT	CVT/SCQT
40	3.43(-25)	2.67(-15)	1.88(-14)	3.01(-25)	2.59(-15)	1.83(-14)
50	7.38(-21)	5.29(-15)	1.92(-14)	6.68(-21)	5.24(-15)	1.88(-14)
60	5.89(-18)	1.31(-14)	2.39(-14)	5.47(-18)	1.31(-14)	2.36(-14)
80	2.67(-14)	3.41(-13)	3.41(-13)	2.54(-14)	3.44(-13)	3.49(-13)
100	4.37(-12)	1.55(-11)	1.21(-11)	4.26(-12)	1.55(-11)	1.26(-11)
120	1.35(-10)	3.01(-10)	2.03(-10)	1.34(-10)	3.02(-10)	2.12(-10)
200	1.49(-7)	1.92(-7)	1.35(-7)	1.51(-7)	1.97(-7)	1.40(-7)
250	1.28(-6)	1.50(-6)	1.14(-6)	1.31(-6)	1.54(-6)	1.17(-6)
300	5.43(-6)	6.08(-6)	4.85(-6)	5.58(-6)	6.25(-6)	5.01(-6)
400	3.35(-5)	3.57(-5)	3.04(-5)	3.46(-5)	3.67(-5)	3.13(-5)
500	9.97(-5)	1.04(-4)	9.16(-5)	1.03(-4)	1.07(-4)	9.47(-5)
600	2.06(-4)	2.11(-4)	1.91(-4)	2.12(-4)	2.19(-4)	1.98(-4)
800	5.04(-4)	5.12(-4)	4.76(-4)	5.22(-4)	5.30(-4)	4.93(-4)
1000	8.59(-4)	8.68(-4)	8.20(-4)	8.90(-4)	8.99(-4)	8.49(-4)

Then Eq. (38) yields the pre-exponential factor, $D_0(T)$. Notice that both $E_a(T)$ and $D_0(T)$ are treated as functions of temperature. The two-part linear plot for this data in Figure 6 is derived from the pre-exponential factors and activation energies in two temperature ranges (120–160 K and 160–200 K for hydrogen, and 120–170 K and 170–200 K for deuterium) reported in Ref. 61. For the Lin and Gomer[49] experiments, the data in the high-temperature region of the plot in Figure 6 is derived from the pre-exponential factor and activation energy of the data in that region, and the data in the low-temperature region is estimated from the fit for this data in Ref. 49. The current results are calculated in the low-coverage single-adatom limit.

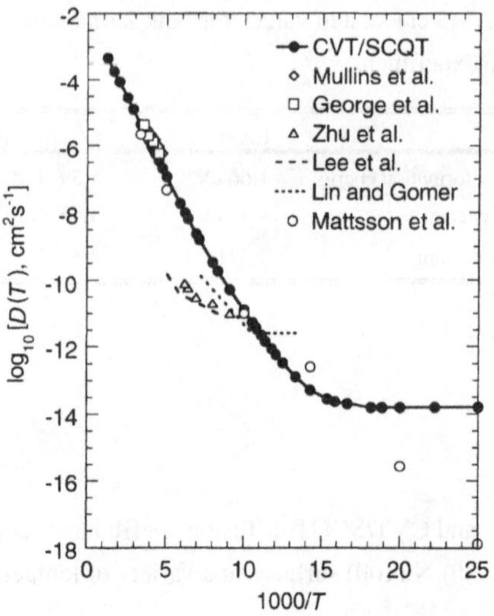

Figure 6. Arrhenius plot for H diffusion on Ni(100) compared to experimental measurements and theoretical calculations.

Our results are in excellent agreement with those of George *et al.*[50] and Mullins *et al.*[57] and with the high-temperature results of Mattsson *et al.*,[38] and they are in reasonable agreement with those of Lin and Gomer.[49] All low-temperature measurements and calculations (i.e., all except the Mullins *et al.*,[57] and George *et al.*[50] experiments) showed a leveling off of the Arrhenius plot similar to that which was discussed in the previous section for H on Cu(100).

Lin and Gomer[49] find the transition temperature to occur at about 100 K, and Zhu and co-workers[60,61] find it to occur at approximately 160 K. The results of the path integral studies by Mattsson *et al.*[38] showed the transition temperature to occur at about 40 K, somewhat low in comparison to experiment, and this transition temperature is not shown in Figure 6. Using Eq. (31) and a parabolic fit to the V_a^{AG} curve as a function of s in mass-scaled coordinates with $\mu = 1$ amu from $s = -1.2$ Å to $s = 1.2$ Å, where $s^R \approx -1.5$ Å, we calculate $|\overline{\upsilon}^{\ddagger}|$ for this process to be 288 cm^{-1}. Using this value with Eq. (28) yields a transition temperature, $T_{tr} = 66$ K, which is, to our knowledge, the only theoretical approximation to predict a transition temperature so close to the experimentally reported values.

We note that computing $|\overline{\upsilon}^{\ddagger}|$ from the second derivative of $V_{MEP}(s)$ at its maximum instead of from a global fit to $V_a^G(s)$ would have yielded 292 cm^{-1} (see Table 7), resulting in the same level of agreement.

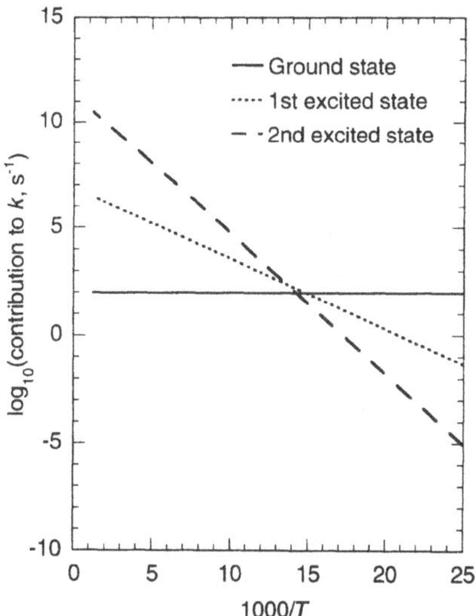

Figure 7. Logarithm to the base 10 of contributions to the rate constant of each quantized state with energy below V_a^{AG} versus inverse temperature for H diffusion on Ni(100) with $N_p = 36$.

We analyze the level-off regime by examining the individual contributions to the rate constant of each of the quantized states below V_a^{AG} as we did with the Cu(100) studies. In Figure 7, the logarithm of the contributions to the rate constant of each of the quantized states is plotted against the same abscissa used in the Arrhenius plot in Figure 6. Again we find that the onset of visual temperature independence on the Arrhenius plot coincides with the temperature at which the ground-state contribution to the rate constant is greater than the first excited state contribution to the rate constant by about one order of magnitude. Lin and Gomer[49] find the level-off temperature to coincide with the transition temperature at about 100 K. Zhu and co-workers[60,61] do not observe a level-off temperature in their experiments. They present results for the diffusion coefficient measured at temperatures as low as 120 K, at which point the diffusion coefficient is still not visually temperature independent, therefore we can only state that the level-off temperature in these experiments is below 120 K. Mattsson *et al.*[38] found the level-off temperature to coincide with the transition temperature at 40 K. For this process, we calculate $\overline{\upsilon}_3^R = 524$ cm^{-1} with the EAM5 potential surface. Using Eq. (31) with this quantity and $\left|\overline{\upsilon}^{\ddagger}\right|$ given above yields a level-off temperature of $T_0 = 55$ K. This temperature is consistent with the Arrhenius plot, justifying the model of Eq. (31) for this process.

We also make one very important remark about these results in terms of Eq. (20). Lin and Gomer[49] and Zhu and co-workers[60,61] suggest that the transition temperature marks a shift from activated over-barrier diffusion to non-activated tunneling diffusion. This hypothesis is based on the observation that at the transition temperature, the activation energy shifts dramatically from a high value, which is close to the assumed classical barrier height of the reaction, to a much lower value, indicating that at this temperature, tunneling begins to dominate the reaction. To determine whether this is an accurate interpretation of the transition temperature, we calculated $k_{\text{tun}}(T)$ from Eq. (25) [M = 2 for H diffusion on Ni(100)] where the energy eigenstates, E_0^{R}, E_1^{R}, and E_2^{R}, were determined by Eq. (12) ($\overline{\upsilon}_3^{\text{R}}$ = 524 cm^{-1} for this process with N_p = 36 as mentioned above), and the transmission probabilities, $P^{\text{G}}\left(E_v^{\text{R}}\right)$, were calculated in the SCTQ approximation as 1.87 x 10^{-12}, 1.36 x 10^{-7}, and 3.06 x 10^{-3}, for v = 0, 1, and 2, respectively. At the transition temperature, T_{tr} = 66 K, k_{tun} is equal to 2.44 x 10^2, and $k_{\text{tun}}(T_{\text{tr}})+k_{\text{over}}(T_{\text{tr}})$ is equal to 2.51 x 10^2, which yields $F_{\text{tun}}(T_{\text{tr}})$ = 0.97 from Eq. (20). This indicates that at the transition temperature, 97% of the process still occurs by a tunneling mechanism. Similar calculations at temperatures above T_{tr} indicate that tunneling still dominates well above the transition temperature. For example, the tunneling contributes about 55% of the hops when T = 125 K. Therefore, we conclude that the transition temperature does not mark the shift from over-barrier activated diffusion to tunneling diffusion, but rather it indicates the point at which excited-state contribution to the rate constant becomes negligible compared to the ground-state contribution.

The data in Figure 6 are fit to an Arrhenius form, Eqs. (38) and (39), and the pre-exponential factors and activation energies for various temperature ranges are extracted from the plot and compared to experiment in Table 10. As expected, our activation energies and pre-exponential factors agree very well with those of Mullins *et al.*[57] and of George *et al.*[50] in the temperature ranges of those experiments. Only below the transition temperatures of the other low-temperature experiments do our calculations disagree significantly with any of the experimental values. This is, of course, simply an artifact of various locations of the transition temperature. That is, since our calculated transition temperature is lower than the experimental ones, our low-temperature diffusion coefficients are also lower than experiment, because our diffusion coefficients continue to decrease after the experimental ones have leveled off.

Finally, we note that variational optimization of the dividing surface is not important for the H/Ni(100) surface diffusion process. For all temperatures studied and for both N_p = 0 and N_p = 36, we find s_*^{CVT} = 0, so the CVT rate constant is precisely equal to the conventional TST rate constant.

Table 10. Activation energies (E_a, kcal/mol) and pre-exponential factors (D_0, cm^2/s) for surface diffusion of H on Ni(100) with 36 moving Ni atoms for several temperature ranges compared to experimental values. Powers of 10 in parentheses.

	E_a		D_0	
T(K)	CVT/SCTQ	Experiment	CVT/SCTQ	Experiment
40–50	0.011		2.09 (-14)	
55–65	0.41		8.36 (-13)	
70–80	1.9		6.79 (-8)	
85–95	2.9		2.64 (-5)	
100–140	3.5	3.2[a]	5.33 (-4)	8. (-6)[a]
120–170	3.8	1.2[b]	1.66 (-3)	1.5 (-9)[b]
156–161	3.9	3.5[c]	2.69 (-3)	8. (-6)[c]
170–200	4.1	3.5[b]	4.28 (-3)	1.1 (-6)[b]
211–263	4.3	3.5 ± 0.3[d]	6.17 (-3)	2.5 (-3)[d]
223–283	4.3	4 ± 0.5[e]	6.55 (-3)	4.5 (-3)[e]
300–400	4.4		7.64 (-3)	

[a]Lin and Gomer, Ref. 49

[b]Lee, Zhu, Deng, and Linke, Ref. 61

[c]Zhu, Lee, Wong, and Linke, Ref. 60

[d]Mullins, Roop, Costello, and White, Ref. 57

[e]George, DeSantolo, and Hall, Ref. 50

4. CONCLUDING REMARKS

In this chapter, we reviewed canonical variational transition state theory with the small-curvature tunneling approximation with quantized reactant states. We presented results obtained by such calculations for the diffusion of H and D on Cu(100) and for diffusion of H on Ni(100). Where applicable, we compared the calculated diffusion coefficients to experimental values and to other theoretical calculations, and in most cases we found very close agreement. We showed that at very low temperatures, the diffusion coefficient loses its temperature dependence for both interfaces, and this is in agreement with three sets of experimental results on the Ni(100) surface. We showed that this onset of temperature independence occurs when the ground-state contribution to the diffusion coefficient is greater than that of the first excited state by about one order of magnitude.

We also showed that the EAM5 embedded atom parameter set for the H on Ni system gives very good agreement with experimentally determined energetics, H-Ni interatomic distances, and vibrational frequencies for the (100) surface as well as with quantities defining the lattice itself (monovacancy formation energy, sublimation energy, and bulk lattice constant). Current and future work is being aimed at fitting a new embedded atom parameter set so that these experimental quantities are also accurately reproduced for the Ni(111) crystal face without damaging current agreement with the (100) face and bulk quantities. An accurate potential energy surface for the H/Ni system (gas-solid interface and bulk) will allow us to study subsurface processes[117] and eventually to carry out calculations for other reactions such as the Ni(111) surface catalyzed reaction of H with methyl radical.

ACKNOWLEDGMENTS

This work was supported in part by the National Science Foundation through grant no. CHE89-22048.

REFERENCES

1. C.N. Satterfield, "Heterogeneous Catalysis in Practice," McGraw-Hill, New York (1980).
2. G. Somorjai, "Chemistry in Two Dimensions," Cornell University, New York (1981).
3. G.C. Bond, "Heterogeneous Catalysis," 2nd ed., Oxford University, Oxford (1987).
4. I.M. Campbell, "Catalysis at Surfaces," Chapman and Hall, Ltd., New York (1988).
5. R.A. van Santen, "Theoretical Heterogeneous Catalysis," World Scientific, Singapore (1991).
6. F. Ruette, ed., "Quantum Chemistry Approaches to Chemisorption and Heterogeneous Catalysis," Kluwer, Dordrecht (1992).
7. V.T. Binh, ed., "Surface Mobilities on Solid Materials," Plenum, New York (1983).
8. M. Boudart and G. Djéga-Mariadassou, "Kinetics of Heterogeneous Catalytic Reactions," Princeton University, Princeton, NJ (1984).
9. E. Shustorovich, ed., "Metal-Surface Reaction Energetics," VCH, New York (1991).
10. M.G. Lagally, Atom motion on surfaces, *Physics Today*, November 1993: 24.
11. K. Kitahara, H. Metiu, J. Ross, and R. Sibley, *J. Chem. Phys.* 65: 2871 (1976).
12. S. Efrima and H. Metiu, *J. Chem. Phys.* 69: 2286 (1978).
13. S. Efrima and H. Metiu, *J. Chem. Phys.* 69: 5113 (1978).
14. K. Haug, G. Wahnström, and H. Metiu, *J. Chem. Phys.* 92: 1083 (1990).
15. K. Haug and H. Metiu, *J. Chem. Phys.* 94: 3251 (1991).
16. J.P. Sethna, *Phys. Rev. B* 24: 698 (1981).
17. K.A. Muttalib and J. Sethna, *Phys. Rev. B* 32: 3462 (1985).
18. S.M. Valone, A.F. Voter, and J.D. Doll, *Surf. Sci.* 155: 687 (1985).
19. S.M. Valone, A.F. Voter, and J.D. Doll, *J. Chem. Phys.* 85: 7480 (1986).
20. J.D. Doll and A.F. Voter, *Annu. Rev. Phys. Chem.* 38: 413 (1987).
21. S.W. Rick, D.L. Lynch, and J.D. Doll, *J. Chem. Phys.* 99: 8183 (1993).
22. J.G. Lauderdale and D.G. Truhlar, *J. Amer. Chem. Soc.* 107: 4590 (1985).

23. J.G. Lauderdale and D.G. Truhlar, *Surf. Sci.* 164: 558 (1985).

24. J.G. Lauderdale and D.G Truhlar, *J. Chem. Phys.* 84: 1843 (1986).

25. T.N. Truong and D.G. Truhlar, *J. Phys. Chem.* 91: 6229 (1987).

26. T.N. Truong and D.G. Truhlar, *J. Chem. Phys.* 88: 6611 (1988).

27. T.N. Truong, D.G. Truhlar, J.R. Chelikowsky, and M.Y. Chou, *J. Phys. Chem.* 94: 1973 (1990).

28. T.N. Truong and D.G. Truhlar, *J. Chem. Phys.* 93: 2125 (1990).

29. S.E. Wonchoba and D.G. Truhlar, *J. Chem. Phys.* 99: 9637 (1993).

30. R. Jaquet and W.H. Miller, *J. Phys. Chem.* 89: 2139 (1985).

31. K.F. Freed, *J. Chem. Phys.* 82: 5264 (1985).

32. (a) A. Auerbach, K. F. Freed, and R. Gomer, *J. Chem. Phys.* 86: 2356 (1987); (b) M.D. Miller, *Surf. Sci.* 127: 383 (1983); (c) Q. Niu, *J. Stat. Phys.* 65: 317 (1991); (d) R. Nieminen, *Nature* 356: 289 (1992).

33. R.M. Stratt, *Phys. Rev. Lett.* 55: 1443 (1985).

34. K.B. Whaley, A.N. Nitzan, and R.B. Gerber, *J. Chem. Phys.* 84: 5181 (1986).

35. P.D. Reilly, R.A. Harris, and K.B. Whaley, *J. Chem. Phys.* 97: 6875 (1992).

36. B.M. Rice, B.C. Garrett, M.L. Koszykowski, S.M. Foiles, and M.S. Daw, *J.Chem. Phys.* 92: 775 (1990).

37. Y.-C. Sun and G.A. Voth, *J. Chem. Phys.* 98: 7451 (1993).

38. T. R. Mattsson, U. Engberg, and G. Wahnström, *Phys. Rev. Lett.* 71: 2615 (1993).

39. R. Wortman, R. Gomer, and R. Lundy, *J. Chem. Phys.* 27: 1099 (1957).

40. R. DiFoggio and R. Gomer, *Phys. Rev. Lett.* 44: 1258 (1980).

41. G. Mazenko, J.R. Banavar, and R. Gomer, *Surf. Sci.* 107: 459 (1981).

42. R. DiFoggio and R. Gomer, *Phys. Rev. B* 25: 3490 (1982).

43. R. Gomer, *Comments on Solid State Phys.* 10: 253 (1983).

44. R. Gomer, *Vacuum* 33: 537 (1983).

45. R. Gomer, *in:* "Dynamics on Surfaces," B. Pullman, J. Jortner, A. Nitzan, and B. Gerber, eds., Reidel, Dordrecht (1984), p. 203.

46. C. Dharmadhikari and R. Gomer, *Surf. Sci.* 143: 223 (1984).

47. S.C. Wang and R. Gomer, *J. Chem. Phys.* 83: 4193 (1985).

48. E.A. Daniels, J.C. Lin, and R. Gomer, *Surf. Sci.* 204: 129 (1988).

49. T.-S. Lin and R. Gomer, *Surf. Sci.* 255: 41 (1991).

50. S.M. George, A.M. DeSantolo, and R.B. Hall, *Surf. Sci.* 159: L425 (1985).

51. (a) C. H. Mak, J.L. Brand, A.A. Deckert, and S.M. George, *J. Chem. Phys.* 85: 1676 (1986); (b) C. H. Mak, J.L. Brand, B.G. Koehler, and S.M. George, *Surf. Sci.* 188: 312 (1987); (c) C. H. Mak, J.L. Brand, B.G. Koehler, and S.M. George, *Surf. Sci.* 191: 108 (1987); (d) C. H. Mak, J.L. Brand, B.G. Koehler, and S.M. George, *J. Chem. Phys.* 87: 2340 (1987); (e) J.L. Brand, A.A. Deckert, and S.M. George, *Surf. Sci.* 194: 457 (1988); (f) C. H. Mak, A.A. Deckert, and S.M. George, *J. Chem. Phys.* 89: 5242 (1988); (g) C. H. Mak, and S.M. George, *Chem. Phys. Lett.* 135: 381 (1987); (h) A.A. Deckert, J.L. Brand, M.V. Arena, and S.M. George, *J. Vac. Sci. Technol.* A6: 794 (1988); (i) A.A. Deckert, J.L. Brand, M.V. Arena, and S.M. George, *Surf. Sci.* 208: 441 (1989); (j) C. H. Mak, B.G. Koehler, and S M. George, *J. Vac. Sci. Technol.* A6: 856 (1988); (k) J.L. Brand, A.A. Deckert, M.V.

Arena, and S.M. George, *J. Phys. Chem.* 92: 5136 (1990); (l) E.D. Werste, M.V. Arena, A.A. Deckert, and S.M. George, *Surf. Sci.* 233: 293 (1990); (m) M.V. Arena, A.A. Deckert, J.L. Brand, and S.M. George, *J. Phys. Chem.* 94: 6792 (1990).

52. R. Viswanathan, D.R. Burgess, J.P.C. Stair, and E. Weitz, *J. Vac. Sci. Technol.* 20: 605 (1982).

53. (a) D.R. Mullins, B. Roop, and J.M. White, *Chem. Phys. Lett.* 129: 511 (1986); (b) B. Roop, S.A. Costello, D.R. Mullins, and J.M. White, *J. Chem. Phys.* 86: 3003 (1987).

54. S.M. George, *in:* "Physical Methods of Chemistry," 2nd Ed., Vol. 9A, B.W. Rossiter and R.C. Baetzold, eds., John Wiley and Sons, New York (1993), p. 474.

55. E.G. Seebauer and L.D. Schmidt, *Chem. Phys. Lett.* 123: 129 (1986).

56. E.G. Seebauer, A.C.F. Konig, and L.D. Schmidt, *J. Chem. Phys.* 88: 6597 (1988).

57. D.R. Mullins, B. Roop, S.A. Costello, and J.M. White, *Surf. Sci.* 186: 67 (1987).

58. C.-H. Hsu, B.E. Larson, M. El-Batanouny, C.R. Willis, and K.M. Martini, *Phys. Rev. Lett.* 66: 3164 (1991).

59. C. Astaldi, A. Bianco, S. Modesti, and E. Touatti, *Phys. Rev. Lett.* 68: 90 (1992).

60. X.D. Zhu, A. Lee, A. Wong, and U. Linke, *Phys. Rev. Lett.* 68: 1862 (1992).

61. A. Lee, X.D. Zhu, L. Deng, and U. Linke, *Phys. Rev. B* 46: 15472 (1992).

62. A. Lee, X.D. Zhu, A. Wong, L. Deng, and U. Linke, *Phys. Rev. B* 48: 11256 (1993).

63. M. Born and J.R. Oppenheimer, *Ann. Phys.* 84: 457 (1927).

64. M. Born and K. Huang, "Dynamical Theory of Crystal Lattices," Oxford University, New York (1956).

65. G.V. Chester, *Adv. Phys.* 10: 357 (1961).

66. D.G. Truhlar and B.C Garrett, *Accts. Chem. Research* 13: 440 (1980).

67. D.G. Truhlar, A.D. Isaacson, and B.C. Garrett, *in:* "Theory of Chemical Reaction Dynamics," M. Baer, ed., CRC Press, Boca Raton, FL (1985), Vol. 4, pp 65-137.

68. S.C. Tucker and D.G. Truhlar, *in:* "New Theoretical Methods for Understanding Organic Reactions," J. Bertrán and I.G. Czizmadia, eds., Kluwer, Dordrecht (1989), pp 219-346.

69. E.B. Wilson, Jr., J.C. Decius, and P.C. Cross, "Molecular Vibrations," McGraw-Hill, New York (1958), p. 14.

70. S. Glasstone, K.J. Laidler, and H. Eyring, "Theory of Rate Processes," McGraw-Hill, New York (1944).

71. D.G. Truhlar and A. Kuppermaan, *J. Chem. Phys.* 52: 3841 (1970).

72. D.G. Truhlar and A. Kuppermaan, *J. Chem. Phys.* 56: 2232 (1972).

73. J.M. Bowman, A. Kuppermaan, J.T. Adams, and D.G. Truhlar, *Chem. Phys. Lett.* 20: 229 (1973).

74. B.C. Garrett and D.G. Truhlar, *J. Chem. Phys.* 70: 1593 (1979).

75. B.C. Garrett and D.G. Truhlar, *J. Phys. Chem.* 83: 1079 (1979).

76. D.G. Truhlar, A.D. Isaacson, R.T. Skodje, and B.C. Garrett, *J. Phys. Chem.* 86: 2252 (1982).

77. D.G. Truhlar, W.L. Hase, and J.T. Hynes, *J. Phys. Chem.* 87: 2664, 5523(E) (1983).

78. B.C. Garrett, D.G. Truhlar, R.S. Grev, and A.W. Magnuson, *J. Phys. Chem.* 84: 1730 (1980).

79. J. Heading, "An Introduction to Phase Integral Methods," Methuen, London (1961).

80. B.C. Garrett and D.G. Truhlar, *J. Phys. Chem.* 83: 2921 (1979).

81. R.T. Skodje, D.G. Truhlar, and B.C. Garrett, *J. Phys. Chem.* 85: 3019 (1981).

82. R.T. Skodje, D.G. Truhlar, and B.C. Garrett, *J. Chem. Phys.* 77: 5955 (1982).

83. B.C. Garrett, D.G. Truhlar, A.F. Wagner, and T.H. Dunning, Jr. *J. Chem. Phys.* 78: 4400 (1983).

84. B.C. Garrett and D.G. Truhlar, *J. Chem. Phys.* 79: 4931 (1983).

85. B.C. Garrett, N. Abusalbi, D.J. Kouri, and D.G. Truhlar, *J. Chem. Phys.* 83: 2252 (1985).

86. D.-h. Lu, T.N. Truong, V.S. Melissas, G.L. Lynch, Y.-P. Liu, B.C. Garrett, R. Steckler, A.D. Isaacson, S.N. Rai, G.C. Hancock, J.G. Lauderdale, T. Joseph, and D.G. Truhlar, *Computer Phys. Commun.* 71: 235 (1992).

87. D.G. Truhlar, D.-h. Lu, S.C. Tucker, X. G. Zhao, A. González-Lafont, T.N. Truong, D. Maurice, Y.-P. Liu, and G.C. Lynch, *ACS Symp. Ser.* 502: 16 (1992).

88. Y.-P. Liu, G. Lynch, T.N. Truong, D.-h. Lu, D.G. Truhlar, and B.C Garrett, *J. Amer. Chem. Soc.* 115: 2408 (1993).

89. Y.-P. Liu, D.-h. Lu, A. González-Lafont, D.G. Truhlar, and B.C. Garrett, *J. Amer. Chem. Soc.* 115: 7806 (1993).

90. A.F. Voter and J.D. Doll, *J. Chem. Phys.* 80: 5832 (1984).

91. T. Halicoğlu and G.M. Pound, *Phys. Stat Solidi A* 30: 619 (1975).

92. C. Kittel, "Introduction to Solid State Physics," 6th Ed., John Wiley & Sons, New York (1986).

93. S.E. Wonchoba, W.-P. Hu, and D.G. Truhlar, to be submitted.

94. Y.-P. Liu, G.C. Lynch, W.-P. Hu, V.S. Melissas, R. Steckler, B.C. Garrett, D.-h. Lu, T.N. Truong, A.D. Isaacson, S.N. Rai, G.C. Hancock, J.G. Lauderdale, T. Joseph, and D.G. Truhlar, *QCPE Bull.* 13: 28 (1993).

95. R.T. Skodje and D.G.Truhlar, *J. Phys. Chem.* 85: 624 (1981).

96. M. Daw and M. Baskes, *Phys. Rev. B* 29: 6443 (1984).

97. M.S. Daw, S.M. Foiles, and M.I. Baskes, *Mat. Sci. Rep.* 9: 251 (1993).

98. T. N. Truong, D.G. Truhlar, and B.C. Garrett, *J. Phys. Chem* 93: 8227 (1989).

99. S.M. Foiles, M.I. Baskes, and M.S. Daw, *Phys. Rev. B* 33: 7983 (1986).

100. T.N. Truong and D.G. Truhlar, *J. Phys. Chem.* 94: 8262 (1990).

101. J. Lapujoulade and K.S. Nail, *Surf. Sci.* 35: 288 (1973).

102. K. Christmann, O. Schober, G. Ertl, and H. Neumann, *J. Chem. Phys.* 60: 4528 (1974).

103. S. Anderson, Chem. *Phys. Lett.* 55: 185 (1978).

104. T.H. Upton and W.A. Goddard, *Phys. Rev. Lett.* 42: 472 (1979).

105. J.K. Nørskov, *Phys. Rev. Lett.* 48: 1620 (1982).

106. K.H. Rieder and H. Wilsch, *Surf. Sci.* 131: 245 (1983).

107. P. Nordlander, S. Holloway, J.K. Nørskov, *Surf. Sci.* 136: 59 (1984).

108. I. Stensgaard and F. Jakobsen, *Phys. Rev. Lett.* 54: 711 (1985).

109. C. Umrigar and J.W. Wilkins, *Phys. Rev. Lett.* 54: 1551 (1985).

110. P.-A. Karlsson, A.-S. Mårtensson, S. Andersson, and P. Nordlander, *Surf. Sci.* 175: L759 (1986).

111. A.-S. Märtensson, C. Nyberg, and S. Andersson, *Surf. Sci.* 205: 12 (1988).

112. (a) A. Seeger, D. Schumacher, W. Schilling, and J. Diehl, eds., "Vacancies and Interstitials in Metals," North Holland Pub. Co., Amsterdam (1970), p 36; (b) H. Bakker, *Phys. Stat. Solidi* 28: 569 (1968).

113. R.A. Johnson, *Phys. Rev.* 145: 423 (1965).

114. A.A. Mamalui, T.D. Ostinskaya, V.A. Pervakov, and V.I. Khomkevich, *Sov. Phys. Solid State* 10: 2290 (1969).

115. W. Wycisk, M. Feller-Kniepmeier, *J. Nucl. Mater.* 69–70: 616 (1978).

116. C.J. Smith, ed., "Metal Reference Book," 5th ed., Betterworths, London (1976), p 186.

117. K.J. Maynard, A.D. Johnson, S.P. Daley, and S.T. Ceyer, *Faraday Discuss. Chem. Soc.* 91: 437 (1991).

CATALYSIS MODELING EMPLOYING AB INITIO AND BOND ORDER CONSERVATION–MORSE POTENTIAL METHODS

Harrell Sellers

Department of Chemistry
South Dakota State University
Brookings, SD 57007
and
National Center for Supercomputing Applications
Beckman Institute of Science and Technology
University of Illinois
Urbana, IL 61801

INTRODUCTION

One probably cannot know such a thing accurately, but, it has been estimated that the worldwide value of products, of which the production relies on the technology of catalysis, is in excess of one trillion dollars per year [1]. It has also been estimated that one sixth of the value of all the goods manufactured in the United States involves catalytic processes [2]. In this author's mind these are probably conservative estimates, particularly when one consideres the indirect effects of catalyzed processes (for example the present health standards in developed countries due to drugs synthesized via catalytic steps and such things as present vehicular transportation requiring fuels which depend on catalysis for production). Certainly our society and the societies of other developed nations depend on catalyzed processes. It is not really possible to quantify the impact on the human race from processes that depend on catalysis.

In this chapter we will discuss the modeling of several processes that are considered to be catalyzed by metal surfaces. First, we will discuss the pertinent details of the bond order conservation - Morse potential model, since it is probable that these details are not completely familiar to the reader. Since most chemists these days are familiar with ab initio quantum chemical calculations on some level we will not present a discussion of them, but, rather we will describe the computational details of our own calculations. Then we will give the results of calculations on the chemisorption of alkane thiols which form self-assembled coatings. Lastly, we will present extentions of the bond order conservation model that yield reactive potential energy surfaces that can be used in molecular dynamics or quantum simulations of reactions on metal surfaces. We then give example calculations for the dissociation of H_2 on the Ni(111) surface.

THE BOND ORDER CONSERVATION MODEL

The bond order conservation (BOC) model is applied in several ways. For atomic chemisorption on metal surfaces each two center interaction between a metal atom and the adatom is described by a Morse potential with an associated Pauling-type bond order [3-5]:

$$E_i(Z) = Q_0 (Z_i^2 - 2 Z_i) \qquad 1)$$

$$Z_i = \exp\{-(R_i - R_{0,i})/a\} \qquad 2)$$

Equation 1 is a Morse potential as a function of the Pauling-type bond order between the adatom and metal atom i (Z as defined by equation 2). The pair interaction energy, E_i, has the zero of energy coinciding with zero bond order and the energy minimum occuring when $R = R_0$. For atomic chemisorption on a pure metal the reference distance in equation 2 can be considered to be essentially the minimum energy distance for the adatom in an ontop site. In the case that the particular pair interaction under consideration is a bond in an admolecule, one usually takes R_0 to be the equilibrium bond distance. In the applications of the BOC model we do not consider equilibrium bond orders greater than 1 even for systems such as CO or N_2 which are considered to be multiply bonded systems in the usual chemical sense. One can consider this to be a bond order 'normalization to unity' [3,4] or simply a choice of $R_0 = R_{eq}$ that requires Z = 1 at the minimum energy geometry. Johnston [6] has discussed this 'degree of freedom' in the meaning of the bond order (or the meaning of 'single bond').

For atomic chemisorption processes Shustorovich [3,4] has minimized the total interaction energy, $\Sigma_i E_i$, subject to the constraint that the total bond order remains at unity; $\Sigma_i Z_i = 1$ to obtain expressions for the enthalpy of atomic chemisorption. In Shustorovich's development [3,4] some simplifying assumptions were made, such as considering only nearest neighbors for the chemisorption site, in order to obtain very simple algebraic expressions for the chemisorption energy. While computationally simple, the formulas of Shustorovich yield quite accurate results when properly applied. The expression for the atomic chemisorption energy obtained from the constrained minimization of the total energy is:

$$Q_A = Q_{0A} (2 - 1/n) \qquad 3)$$

where n is the number of metal atoms in the chemisorption site (1 for ontop, 2 for bridge, 3 for three atom hollow, etc.). In actual practice Q_A is usually taken to be the experimental heat of chemisorption for atom A on the present metal of interest and Q_{0A} is obtained by solving equation 3 for Q_{0A}. If equation 3 is to be believed we should expect that atoms will chemisorb in the highest coordination site possible (n is a maximum in equation 3). Shustorovich has discussed conditions under which this might not be the case [3]. For open surfaces and small adsorbates, for example H atoms, a low coordination might be more favorable or perhaps the surface may reconstruct. In these cases we consider that the best picture is not one having the adatom bound to the surface but rather that a molecule has formed involving the adatom and some few metal atoms and that this new species is chemisorbed. Coverage effects might also cause a lower coordination preference. The distances between the adatom and the surface metal atoms making up the chemisorption site are [3]:

$$r_n = r_0 + a \ln \{n\} \qquad 4)$$

These distances are not the perpendicular adatom distance to the surface except for the case of n=1. Shustorovich [3] has also given formula for the vibrational frequencies for atoms in various chemisorption sites in terms of these parameters.

For the chemisorption of diatomics and polyatomics that are treated as pseudodiatomics one considers each binding situation independently. For a diatomic, AB, a potential energy function can be written down as:

$$\Sigma_i E_{A,i} + \Sigma_i E_{B,i} + E_{AB} \qquad 5)$$

and the energy terms appearing in the sums are defined similar to that of equation 1. The E_{AB} term is the Morse potential associated with the AB bond:

$$E_{AB} = D_{AB} (Z^2_{AB} - 2 Z_{AB}) \qquad 6)$$

Since there are only attractive (and no repulsive) long range forces in the BOC model, there are no repulsions to account for the orientation of the molecule. So here we simply assume an orientation for the chemisorbed molecule and compute the chemisorption energy. For weak chemisorption in an upright position in an n-fold site the minimization of equation 5 under the constraint that the total bond order is conserved to unity yields:

$$Q_{AB,n} = Q^2_{oA} / ((Q_{oA}/n) + D_{AB}) \qquad 7)$$

for $D_{AB} > Q_{oA} (n - 1) / n$. No terms appear that depend on Q_{oB} in equation 7 because they are neglected, since atom B is oriented away from the surface and presumably the interaction between atom B and the surface atoms is small at that distance. For the parallel chemisorption of AB in which each atom of the diatomic, AB, coordinates with one metal atom of the surface, the chemisorption energy obtained from the constrained variational procedure is [4]:

$$Q_{AB} = a\, b\, (a + b) + D_{AB} (a - b) / (a\, b + D_{AB} (a + b)) \qquad 8)$$

with

$$a = Q^2_{oA} (Q_{oA} + 2\, Q_{oB}) / (Q_{oA} + Q_{oB})^2 \qquad 9)$$

$$b = Q^2_{oB} (Q_{oB} + 2\, Q_{oA}) / (Q_{oA} + Q_{oB})^2 \qquad 10)$$

Shustorovich gives additional expressions for the chemisorption of a chelate structure [4].

Table I contains a collection of atomic heats of chemisorption in kcal/mol, Q_A, that are used to obtain the Q_{0A} for the evaluation of the molecular heats of chemisorption from equations 7 and 8-10. These atomic heats of chemisorption will also be used below to calculate the heats of chemisorption for more strongly bound specie. We point out that the values for atomic carbon are usually an estimate of some kind [4]. Table 2 contains a list of computed and experimental chemisorption energies for small molecules on various metals.

Table 1. Atomic heats of chemisorption in kcal/mol for various metals.

system	Q_A	ref	system	Q_A	ref	system	Q_A	ref
H/W(110)	68	48	H/Fe(110)	64	48	H/Fe(111)	62	4
H/Ru(001)	67	49	H/Rh(111)	61	50	H/Ir(111)	58	51
H/Ni(111)	63	48	H/Ni(110)	63	4	H/Pd(111)	62	48
H/Pt(111)	61	52	H/Cu(111)	56	48	H/Cu(100)	56	4
H/Ag(111)	<52	53	H/Au(111)	46	7	H/Au(110)	<58	54
O/W(110)	125	55	O/Ru(001)	100	56	O/Rh(111)	102	57
O/Ir(111)	93	48	O/Ni(111)	115	61	O/Ni(100)	130	62
O/Pd(111)	87	48	O/Pt(111)	85	48	O/Cu(111)	103	63
O/Ag(111)	80	48	O/Au(110)	<75	54			
N/W(110)	155	48	N/Ir(111)	127	48	N/Ni(111)	135	48
N/Pd(111)	130	48	N/Pt(111)	116	58	N/Fe(110)	138	4
N/Fe(100)	140	4	Fe(111)	139	4			
C/Ni(111)	171	43	C/Pd(111)	160	4	C/Pt(111)	150	4
C/Au(111)	140	8						
S/Au(111)	80	7	S/Ni(111)	112	60	S/Ni(100)	112	60
S/Pt(111)	92	60	S/Pt(100)	99	60	S/Cu(100)	92	60
S/Ag(111)	78	60	S/Ag(100)	81	60			

Some of the data are taken from Shustorovich [3,4] and some are our own data [7,8]. We have found that accuracy at the 1-2 kcal/mol level is usual. We have also found in quantum mechanical calculations that closed shell molecules such as water and methane thiol prefer to bind to the ontop site [7,8] as is predicted by the BOC model [3,4]. Although equation 7 yields the result that higher coordination is always preferred, this is due to terms that are neglected. Shustorovich discusses [3,4] the preference of molecules with high D_{AB} for low coordination sites. A prediction of equation 7 is regarding which atom is the contact atom in the chemisorption. Whichever atom, A or B, has the greatest value of Q_{0X} will yield the largest chemisorption energy and will be the favored surface contact atom.

Table 2. Molecular Heats of Chemisorption in kcal/mol, Q_{AB}, for Weakly Bound Molecules.

System	site	mode	$Q_{AB,calc}$	$Q_{AB,exp}$	ref.
CO/Ni(111)	ontop	upright	29	27	48
CO/Fe(110)	ontop	upright	38	36	74
NO/Pt(111)	bridge	upright	26	27	48
NO/Pd(111)	bridge	upright	32	31	48
O_2/Pt(111)	bridge	parallel	11	9	64
O_2/Ag(110)	bridge	parallel	10	10	65
H_2O/Pt(111)	ontop	upright	11	12	66
H_2O/Ag(111)	ontop	upright	9	10	67
NH_3/Ni(111)	ontop	upright	18	20	68
CH_3OH/Ni(111)	ontop	upright	18	14	69
CH_3OH/Pt(111)	ontop	upright	11	11	70
CH_3OH/Pd(111)	ontop	upright	12	11	71
CH_3OH/Rh(111)	ontop	upright	14	12	72
CH_3OH/Ag(111)	ontop	upright	10	11	70
CH_4/Rh(111)	ontop	upright	6	6	73
$HSCH_3$/Au(111)	ontop	upright	11	12	47
H_2S/Au(111)	ontop	upright	10	-	-

In the case of strong diatomic (or pseudo-diatomic) chemisorption (usually not a closed shell adsorbate) the energy expression can be reformulated in terms of Morse a potential that describes the interaction between the asorbate contact atom and the surface rather than a sum of pairwise interactions between the contact atom and the metal surface atoms. One then considers an overall bond order between the surface and the adsorbate atoms in the bond order conservation condition:

$$Z_A + Z_B + Z_{AB} = 1 \quad 11)$$

where Z_A is the overall bond order between adsorbate atom A and the surface. Once again neglecting the B - surface interaction for upright chemisorption yields the energy expression:

$$E = Q_A (Z^2_A - 2 Z_A) + D_{AB} (Z^2_{AB} - 2 Z_{AB}) \quad 12)$$

where Q_A is the experimental heat of chemisorption for atom A and D_{AB} is the AB bond enthalpy. The constrained variational procedure gives for the heat of chemisorption of AB [3,4]:

$$Q_{AB} = Q^2_A / (Q_A + D_{AB}) \tag{13}$$

with the equilibrium AB bond order given by [5]:

$$Z^{eq}_{AB} = D_{AB} / (D_{AB} + Q_A) \tag{14}$$

This procedure of reformulating the energy expression and using the experimental atomic heat of chemisorption for the Q_A Morse parameter works because the strongly bound molecule (radical) will bind to the surface in the same way as do as atoms. We employ equation 7 for all closed shell molecules and equation 13 for molecular radicals that retain at least one localized unpaired electron such as OH, OOH, CH_3O, CH and CH_2. For the situation of intermediate-strength bonding encountered for monovalent carbon radicals like CH_3, CH_2OH and HCO Shustorovich has suggested averaging the values obtained from equation 7 (with n=3) and equation 13.

The chemisorption energies of small polyatomic molecules are obtained by treating them as pseudo-diatomics. In this process one chooses the contact atom based on the greatest atomic chemisorption energy and steric effects. For example, in the case of $HSCH_3$ we choose the contact atom to be the sulfur rather than the carbon even though atomic carbon will have a higher heat of chemisorption than atomic sulfur on all metals. In the methane thiol molecule the carbon atom is sterically hindered to the point that it cannot be the contact atom. Then the D_{AB} that is used in the above expressions for the chemisorption energy is the energy required to pull everything off the contact atom. For example in the chemisorption calculations for H_2O, the D_{AB} is taken to be 220 kcal/mol. We will demonstrate the accuracy of equation 13 by employing these molecular heats of chemisorption to calculate activation barriers to compare with experimental values.

The expressions for the activation energies of the dissociation, recombination and disproportionation reactions were developed by Shustorovich [4] as an interpolation between well defined limits. (It was wrongly stated by Wang and Pollard recently [9] that the BOC model assumes a zero AB bond order at the transition state.) For the reactions AB -----> A + B and A + BC -----> AB + C, the expressions for the activation barriers *relative to surface adsorbed reactants* are, respectively:

$$E^f_a = 1/2 (\Delta H + Q_A Q_B / (Q_A + Q_B) \tag{15}$$

with ΔH for the dissociation reaction being:

$$\Delta H = D_{AB} + Q_{AB} - Q_A - Q_B \tag{16}$$

and

$$E^f_a = 1/2 (\Delta H + Q_{AB} Q_C / (Q_{AB} + Q_C) \tag{17}$$

with ΔH for the disproportionation reaction being:

$$\Delta H = Q_A + Q_{BC} + D_{BC} - D_{AB} - Q_{AB} - Q_C \qquad 18)$$

The activation barriers for the reverse of the dissociation reaction is obtained from equation 15 but using the ΔH for the recombination. Table 3 provides the data for comparisons between the prediction of the BOC model and experimental determinations. The data of table 3 is intended to be convincing to the reader that the BOC model is capable of giving accurate predictions of the forward and reverse activation barriers for a wide variety of reactions on metal surfaces. We stress that we have not chosen to reproduce here only the favorable cases. We have always found that the BOC model gives accurate predictions when it is properly applied. More examples can be found in the work of Shustorovich and co-workers [3,4,10-15].

Table 3. Computed and experimental activation barriers, in kcal/mol, for reactions on metal surfaces.

reaction (metal)	$E^f_{a,calc}$	$E^r_{a,calc}$	$E^f_{a,exp}$	$E^r_{a,exp}$	ref
$H_2O + O \longrightarrow 2\,OH$ (Pt)	10.5	18.9	10.2	18 ± 3	51,75
$H + O \longrightarrow OH$ (Pt)	9.7	25.7	≤ 13	≤29	51
$H + OH \longrightarrow H_2O$ (Pt)	11.3	18.9	16 ± 5	25 ± 5	51
$CH_3O + H \longrightarrow CH_3OH$ (Rh)	13.3		13.8		72
$CH_4 \longrightarrow CH_3 + H$ (Pt)	17.7		18.4		76
$C_2H_6 \longrightarrow C_2H_5 + H$ (Pt)	13.0		16.4		76
$HSCH_3 \longrightarrow SCH_3 + H$ (Au)	16.9		18.0		47

The observed activation barriers for reactions on metal surfaces are effective barriers and contain contributions from diffusion. Shustorovich has given expressions, which we have reproduced in the literature [16], for the effective reaction rate constant and activation barrier in terms of the intrinsic and diffusion activation barriers (Freeman and Doll have given similar expressions [17,18]):

$$k_{app} = D\, k_{int} / (D^* + k_{int}) \qquad 19)$$

$$E_{app} = (E_{int}\, D^* + E_{dif}\, k_{int}) / (D^* + k_{int}) \qquad 20)$$

where D is the diffusion constant for the adsorbed reactants, D^* is the diffusion constant for the precursor state, $A^* + B^*$ and k_{int} is the intrinsic reaction rate constant for the recombination reaction. In general the intrinsic activation barriers are larger than the diffusion barriers and the apparent activation barrier is approximated to a high degree by the intrinsic one. Only in the case of very low intrinsic activation barriers will we have to consider the diffusional contributions.

The previous discussion has been relevant to the zero coverage limit. Coverage effects cause modifications in several respects. Firstly, the binding energy of an adsorbate atom to a surface decreases with increasing coverage. This is due to the fact that at higher coverages the metal atoms that make up the surface interact with more than one adsorbate. Secondly, adsorbate - adsorbate interactions can cause changes in the apparent adsorbate binding energy, Q_{AB}. When these interactions are stabilizing influences responsible for such phenomena as island formation, they bring about an increase in the apparent Q_{AB}. For example we have shown that the hydrogen bonding between the hydroxyl radicals together with the decreased O atom binding energy upon increased coverage accounts for the observed behavior of the chemisorption energy of hydroxyl radicals [19]. Thirdly, spectator adsorbates can promote certain reactions by blocking competing reactions by simply taking up needed binding sites on the surface [16]. Shustorovich has described in detail the way in which the BOC model can account for the first type of coverage effect [3,4]. The BOC energy expression must be modified, as we have done for the modeling of hydroxyl chemisorption [19], to account for the second type of coverage effect, and, some sort of reaction simulation must be done to quantify the effect of the third type of coverage effect.

We have attempted to give examples convincing to the reader that the BOC model, despite its apparent simplicity, yields quite accurate predictions. In addition to the comparisons with experimental data we have performed high quality ab initio calculations to determine the extent to which the results of the Schrödinger equation would agree with the BOC model for the dissociation of H_2, O_2 and F_2 on the close-packed mercury surface [20,21]. We have found that our ab initio reaction energy profiles can be fit very well by the BOC energy expression [20,21].

Encouraged by these results we have modified the BOC model of Shustorovich [22,23] in such a way that will allow us to obtain reactive energy profiles and subsequently perform reaction simulations with these energy surfaces. Our first modification involves the form of the Morse potential. We describe the interaction between the surface and an adsorbate atom with a single Morse potential, but, we allow the Morse parameters to be functions of the position of the adsorbate atom over the surface. We have called these potentials dynamic Morse potentials (DMP) [20,21] because of their dependence on the location of the adsorbate. The form of the DMPs is:

$$E = -D\,(2Z - Z^2) \tag{21}$$

$$Z = \exp\{-(R - R_0)/\alpha\} \tag{22}$$

$$D = D_0\,\Sigma_i \exp\{-\mu\,\rho^{k}{}_i\} \tag{23}$$

$$R_0 = R_{00}\,\Sigma_i\,\exp\{-\lambda\,\rho^{k}{}_i\}\,] \tag{24}$$

$$\alpha = \alpha_0\,\Sigma_i\,(1 - \exp\{-\nu\,\rho^{k}{}_i\}\,)$$

or 25)

$$\alpha = \alpha_0 \Sigma_i \exp\{-\nu \rho^k_i\}$$

where R is the perpendicular distance between the atom and the surface; α, α_0, μ, λ, ν, D_0 and R_{00} are fitting parameters that help describe the topology of the interaction potential surface. The quantity, ρ_i, is the projection onto the plane of the metal surface of the vector between the atom and metal atom i. In practice we often put k = 2 or 4 and $\alpha = \alpha_0$. Equations 23 and 24 describe the variation of the binding energy and minimum energy atom-surface distance as a function of position in the plane of the metal surface and can be obtained by fitting to ab initio calculations or the BOC expressions of Shustorovich [3,4]. Certainly other functional forms could be substituted if they are found to be more suitable. The idea is to incorporate the topology of the energy surface as a function of position above the plane of the surface into the Morse parameters.

The BOC-MP model of Shustorovich [3,4] is not applicable to the entire reaction coordinate for a surface dissociation reaction due to the fact that the total bond order is different for the reactant state than it is for the product state. For example, the total bond order is taken to be unity when the AB molecule is chemisorbed, but when the AB bond is broken, the AB bond order is zero, and, the atom-surface bond orders each, Z_A and Z_B, should go to unity for the completely relaxed system. The fact that the BOC-MP model of Shustorovich [3,4 is so successful with a bond order normalization of unity for describing chemisorption energies and activation barriers is very strong evidence supporting the notion that the total bond order should be equal to one at the equilibrium geometry of the chemisorbed system (presuming that there is a barrier to dissociation). Clearly, the normalization of unity is not valid beyond the transition state because beyond the transition state in a dissociation reaction the system is more product-like than reactant-like and the bond order normalization should reflect this.

What we do to avoid the discontinuity that appears in the Shustorovich model for the dissociation of the AB bond is to constrain the bond order between the two physically well defined limits (N = 1 for reactant state and N = 2 for the product state) throughout the dissociation reaction, rather than require the bond order to be strictly conserved. We construct a new normalization condition that has the correct limiting behavior:

$$Z_A + Z_B + Z_{AB} = N = 1 + \exp\{-\gamma \Delta^i\} \qquad 26)$$

and

$$\Delta = Z_{AB}/(1 - Z_{AB}) \qquad 27)$$

γ is a constant and i is the exponent of Δ which is a well defined function of Z_{AB}. With this definition the BOC normalization varies smoothly between one and two as Z_{AB} varies between 1 and zero. The γ and i parameters are in essence fitting parameters and have no real physical meaning in the sense that they are not observable quantities. They do, however, effect the position and characteristics of the barrier. We have discussed these parameters in detail [22 and we find that a value of i=2 is usually advantageous.

We now develop the expressions for the energy surface on which we will perform our MD simulations. It is critical that we avoid the ambiguities of associating geometry parameters with bond order that are encountered off the 'reaction coordinate' [5]. Our present philosophy is that the mapping of bond order onto the geometry parameters is not well defined (although mappings can be contrived) unless the bond orders to the surface are such that they minimize the energy expression under the bond order conservation condition below. So, in our MD method we begin the development of the reactive potential surface by considering the energy surface of minimum energy with respect to the adsorbate-surface bond orders. We emphasize here that this is *not* a restriction of the adsorbates to move in a plane parallel to the surface. The adsorbate positions on this minimum energy surface are such that the energy is a minimum with respect to these adsorbate - surface bond orders. The adsorbates rise and fall above the surface of the metal according to how the potential varies with surface morphology. We will consider displacements off this minimum energy surface momentarily.

We consider the following situation: molecule A-B is the reactant and the reaction is: AB(ad) -----> A(ad) + B(ad) where the 'ad' indicates a surface adsorbed species. The energy expression is:

$$V = Q_A (Z_A^2 - 2Z_A) + Q_B (Z_B^2 - 2Z_B) + D_{AB} (Z^2_{AB} - 2 Z_{AB}) \qquad 28)$$

Equation 28 is the sum of the DMP energies above for the interactions of A and B with the surface and the Morse potential for the AB bond. In calculations involving molecular specie, the dynamic Morse potential parameters could be further fit to best reproduce the experimental chemisorption energy.

Our bond order conservation condition is [22]:

$$Z_1 + Z_2 + Z = N \qquad 29)$$

where N is defined by equations 26 and 27. In the present formulation the bond orders to the surface are obtained for any given Z by requiring that the energy be a minimum with respect to Z_1 and Z_2 under the condition of equation 29. We emphasize again that, since we can relate the adsorbate-surface bond orders (Z_A and Z_B) to the perpendicular distance between the adsorbate and the surface (equation 22), the adsorbate heights above the metal surface are determined when Z_A and Z_B are determined. (Again, this height changes with respect to surface morphology as prescribed by equation 24.)

To accomplish the minimization we apply the method of Lagrangian multipliers. The Lagrangian function is given by:

$$L = V - \alpha (Z_A + Z_B + Z_{AB} - N) \qquad 30)$$

Requiring that $\partial L / \partial Z_i = 0$ for i = 1,2 and $\partial L / \partial \alpha = 0$ yields the following matrix equation:

$$\mathbf{A}\, \zeta = Y \qquad 31)$$

where

$$\mathbf{A} = \begin{matrix} 2Q_A & 0 & -1 \\ 0 & 2Q_B & -1 \end{matrix} \qquad 32)$$

The ζ vector is a column vector having elements: Z_A, Z_B, and α; The Y column vector has elements: $2Q_A$, $2Q_B$, and, N - Z. The solution to equation 31 is obtained by inverting the **A** matrix analytically giving for the elements of the ζ vector:

$$Z_A = (Q_A - Q_B)/(Q_A + Q_B) + Q_B (N - Z_{AB})/(Q_A + Q_B) \qquad 33)$$

$$Z_B = (Q_B - Q_A)/(Q_A + Q_B) + Q_A (N - Z_{AB})/(Q_A + Q_B) \qquad 34)$$

$$\alpha = (2 - N + Z_{AB}) Q_A Q_B/(Q_A + Q_B) \qquad 35)$$

These equations define the 'minimum energy' reactive energy surface. This energy surface is essentially four dimensional, since the perpendicular distances of the two reacting fragments has been 'folded into' the other degrees of freedom. The adsorbate-surface bond orders, Z_A and Z_B, are given in equations 33 and 34 as functions of thermodynamic quantities and the AB bond order, Z_{AB}, which in turn is a function of the AB distance, R_{AB}.

If the metal surface were isoenergetic so that the Q parameters above were constants, there would be no need to assign cartesian coordinates to the A and B fragments. However, due to the use of our dynamic Morse potentials which account for the morphology of the surface, the Q parameters are functions of the adsorbate positions above the surface, and, the cartesian coordinates of the adsorbates are needed. Therefore, the dynamic Morse potentials require that we orient the AB bond relative to the plane of the surface. (Also the inclusion of motions off the minimum potential energy surface will require the specification of the orientation of the AB bond.) There are two ways that one might reasonably proceed. One might adopt a fixed orientation for the AB bond, for example parallel to the surface, or one can consider the perpendicular distance to be a function of the in-plane adsorbate coordinates (by in-plane we mean the cartesian adsorbate coordinates that are in the plane of the surface). In the latter of the two procedures the relationships between Z_A, Z_B and the perpendicular adsorbate distances (equation 22) are invoked which, together with the adsorbate position dependence of the Q parameters, orients the tilt of the AB bond relative to the surface. This also requires that all bond orders be greater than zero. The cartesian coordinates of the adsorbates, A and B, in the plane of the surface are then determined by choosing a direction for the projection of the AB bond onto the surface. We have performed calculations (not given herein) employing both fixed and variable bond orientation in our four dimensional calculations and have found that many reaction paths are virtually invariant to this, however, one can imagine paths that cross ontop sites, for example, in which a non-trivial difference would be observed. Also researchers have shown [24-29] that fixed orientation calculations can lead to loss of important information. We will return to this point below. In our dynamic Morse potentials the reference distance of the adsorbate over the surface is also a function of the surface morphology (equation 24). We obtain these important functions from either ab initio calculations or from a numerical fitting to the functional form given by Shustorovich for the distance of the adsorbate above the surface as a function of the surface site [3].

In order to account for coupling to the perpendicular degrees of freedom we modify the 'minimum energy' surface given above by adding a truncated Taylor series representation of the potential for perpendicular displacements. (For brevity we represent this minimum energy function described above as V^{BOC}.) The full (six dimensional) energy expression is given by:

$$V = V^{BOC} + Q_A Z_A Y^2_A / \alpha^2_A + Q_B Z_B Y^2_B / \alpha^2_B \qquad 36)$$

where $Q_A \; Z_A \; / \; \alpha^2_A$ is one-half the Pauling-Badger force constant for the perpendicular displacement of fragment A. This choice of the perpendicular force constant has the advantage that it gets smaller as the atom-surface bond order decreases. Our convention is that the plane of the metal surface is the xz plane and the surface normal is parallel to the y axis. Y_A is the displacment of fragment A along the surface normal measured with respect to the reference height above the surface, $y_A - y_{A,\,ref}$, of which the reference distance, $y_{A,ref}$, is a function of the x and z coordinates of fragments A and B. It is important to take into account that the perpendicular reference distances move with the x and z cartesian coordinates of the fragments. This is essentially the mechanism by which energy is transferred among these degrees of freedom. For example when equation 36 is differentiated with respect to the x cartesian of fragment A, the Pauling-Badger force constants, $Q_A \; Z_A / \alpha^2_A$, and the derivatives of the reference height contribute substantially to the force. These are the terms: (Y^2_A / α^2_A) (d/dx) $(Q_A \; Z_A)$ and $(Q_A \; Z_A \; / \; \alpha^2_A)$ (d/dx) (Y^2_A). This second term can be expressed as $Y_A (\partial^2 V / \partial y_A^2) (dY_A/dx)$. This quantity accounts for the non-zero force constant between the x and y degrees of freedom, since it is equal to $Y_A \; F_{xy}$ where F_{xy} is the quadratic coupling force constant. Equation 36 then specifies the potential function that is a function of all six cartesian dimensions of the fragments A and B.

We give as an example of molecular dynamics employing the above six dimensional BOC energy surface the dissociation of H_2 on Ni(111). The main property that we compute is a quantity that is essentially a reaction rate constant. In the discussion below we consider that a 'reaction channel' for the surface dissociation is specified when the initial position on the surface and the initial velocities have been specified. We also consider that a family of reaction channels is a group of reaction channels having the same initial positions and colinear initial velocities (i.e. at the beginning of the reaction they differ in energy only). Transition state theory and the statistical theory of unimolecular reaction rates [30,31] provide expressions for the reaction rate constant that is specific to a particular reaction channel that are then averaged over the system to obtain an overall rate constant expression. A useful relationship equates the average adsorbate lifetime or reaction time (not the half-life) in a reaction that obeys first order kinetics to the reciprocal of the reaction rate constant [30]:

$$\langle \tau \rangle = k^{-1} \qquad 16)$$

We have calculated adsorbate lifetimes (and reaction times) from six dimensional molecular dynamics simulations and have constructed an Arrhenius plot with these data in order to demonstrate what one obtains from these data. There is, however, an element of arbitrariness in the operational definition of 'adsorbate lifetime' or 'reaction time'. In order to compute this quantity within the molecular dynamics formalism, one must decide when the life of the reactant particle begins and when it ends, or, when a reaction begins and ends. It is customary to speak in terms of a 'critical surface' that separates reactant states from product states on which the transition state exists [30]. When reactants cross this critical surface (in phase space) it is considered (arbitrarily) that these reactants no longer exist [30]. This convention would have the reactant lifespan end the moment the particle passes through the transition state. Another convention is to take the reactant lifetime to be the reaction time or the time required for a reactant to travel from the equilibrium reactant geometry to the region of the minimum in the potential on the product side. The way in which one defines this time quantity for use in equation 37 has an impact on the Arrhenius activation barriers that one obtains for particular reaction channels.

Figure 1 is an Arrhenius plot of the reciprocal of the time-to-transition-state obtained from our six dimensional molecular dynamics simulations vs the initial H_2 kinetic energy. In these simulations the γ and i parameters of equation 36 are 35 and 2, respectively. The DMP parameters for atomic hydrogen were obtained by fitting the DMP to the binding energies for the ontop, hollow and bridging site predicted by the BOC model. The dependence of the perpendicular reference distance on the position of the H atom over the surface was also obtained by fitting the predictions of the BOC model to an analytic form similar to equation 24. The initial velocities of the H atoms in the plane of the metal surface were equal and opposite in sign so that the H - H molecule initially stretches and rotates in a plane parallel to the metal surface. The initial perpendicular velocity was directed toward the surface. From these data the Arrhenius activation barrier is obtained by differentiating the dependent variable (ln {1/time}) with respect to 1/kT where k is the Boltamann constant. This yields an Arrhenius activation barrier of 10 kcal/mol to compare with the experimental value of 9 kcal/mol [4]. The overall reaction rate constant is obtained by averaging results such as these over all the open reaction channels.

In concluding our discussion of the BOC model we point out that the BOC model does have a foundation in the quantum mechanical description of chemisorbed specie and this fact has been discussed by Shustorovich [3,4]. Our interest in the BOC approach was sparked by images we made of some of the bonding orbitals involved in the chemisorption of radical alkane thiols on gold and silver surfaces. These images are now in the literature as figures 5-9 of reference 32. It is strikingly clear from these pictures of chemisorbed methyl sulfur radical that as the bonding interaction forms between the sulfur atom and the metal surface the bonding interaction weakens between the sulfur and the carbon which is the basis for the bond order conservation assumption. The bonding orbitals shown in reference 32 are bonding with respect to the sulfur - surface interaction and anti-bonding with respect to the sulfur - carbon bond. Orbital interactions of this kind also generate the repulsions responsible for the upright structure of the adsorbate providing that electron correlation or electrostatic attractions between the methyl group and the surface do not overcome the repulsions.

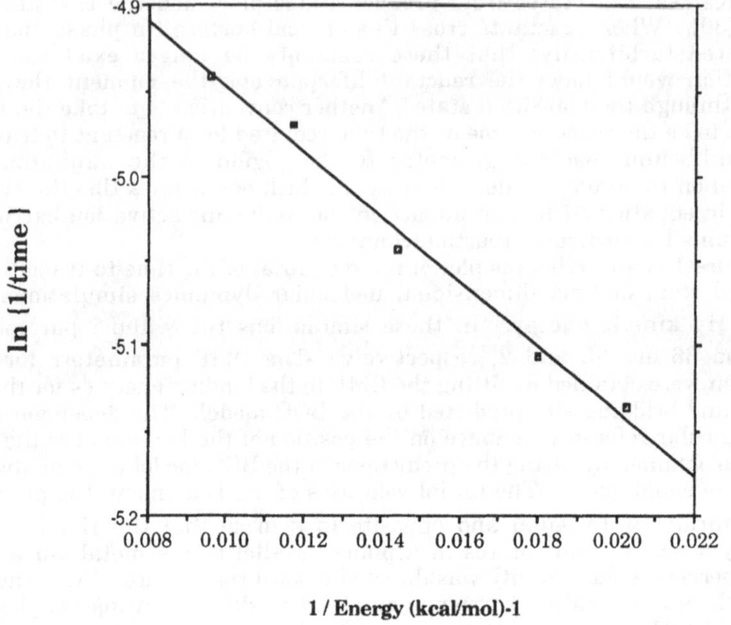

Figure 1. Arrhenius plot of six dimensional simulation data of H_2 on Ni(111).

THE ROLE AND NATURE OF THE AB INITIO CALCULATIONS

Since polyatomics are treated as pseudo-diatomics in the BOC model as described above, it cannot distinguish between a water molecule that is leaning 30 degrees from one that is chemisorbed (or physisorbed) in a perfectly upright position. Also, the energy differences between a weakly bound closed shell molecule in the ontop position and the same molecule bound in the bridging site can be so small as to be below the accuracy level of the model. Clearly, the BOC model cannot answer questions in which steric effects arising from the admolecule are important. On the other hand, it is extremely difficult to obtain chemisorption or activation energies from ab initio calculations as accurately as they can be predicted from the BOC model. However, the BOC model requires thermodynamic parameters that are often not known from experimental determinations. In this case the only alternative is to estimate them from ab initio calculations.

We have had some degree of success (and luck) in estimating the BOC parameters we needed from ab initio calculations and our experience is that one must be extremely careful. We, as others, employ a cluster model of the metal surface of interest. The more strongly bound adsorbates demand more flexibility of the cluster model of the surface with regard to the movement of electron density into and out of the chemisorption site than do weakly bound adsorbates. Cation adsorbates, such as Na^+ and K^+ cause large shifts in electron density in the model of the metal surface. Of course, some accounting of the relativistic corrections for second and third row metals must be included, most usually by the use of relativistic effective potentials. For binding energy calculations involving closed shell molecules a reasonable accounting of the electron correlation is critical. The more weakly bound the adsorbate is the more ilikely it is that systematic errors form a larger part of the ab initio results, unless much computational effort has been expended. For example the basis set superposition error in a calcuation of the chemisorption of water on a metal surface can be large enough to cause the water molecule to tend to lay on the surface rather than stand erect on the oxygen atom. The use of effective potentials for the metal atoms tends to minimize the basis set superposition error (BSSE) relative to all electron calculation arising from the metal atom basis sets because the BSSE arises largely from the core regions which are no longer involved in the calculation.

Theoreticians have long sought to model chemisorption processes with metal cluster models [33-40]. Early on Upton and Goddard [33] put forth a set of conditions that they believed clusters should obey in order for them to be useful in the modeling of infinite surface chemisorption processes. Among these conditions are the requirements that a) the ground state wavefunction should have a conduction band near the Fermi level with a significant amplitude near the chemisorption site; b) the cluster should exhibit a high density of states; c) the cluster should be highly polarizable, and, d) the cluster should posses an ionization potential similar to that of the bulk. Bauschlicher et al. [34,35,38] subsequently deduced that metal clusters having a full complement of nearest neighbors to the chemisorption site (~ 21 atoms for O chemisorption on the Ni (100) plane) make reasonable models of infinite surfaces (give converged results with respect to cluster size) for some properties such as the adsorbate - surface distance, but, that the chemisorption energy is more slowly convergent. These conclusions were based on studies of oxygen chemisorption onto clusters of one electron ECP (effective core potential) Ni atoms in which the wavefunctions for the systems studied were restricted to be of the lowest spin state.

On the basis of CSOV analyses Hermann et al. [40] realized that the orbital structure of the cluster employed in the model was more important to the stability of the chemisorption energy than previously believed. In calculations of CO chemisorption onto Cu clusters they showed that, in order to obtain a reasonable chemisorption energy, the cluster wavefunction should have high-lying orbitals of e symmetry (in C_{4v}) and that the highest-lying a_1 orbital should not be too near the Fermi level. This picture results from the bonding scheme of CO to Cu clusters [40]. Hermann et al. determined that the π type bonding is most important in this system and the presence of a doubly occupied a_1 orbital that is too close to the Fermi level generates a predominately repulsive interaction (Pauli repulsion) with the adsorbate CO molecule.

Panas et al. made a very valuable contribution [39] by stating a set of rules that the orbitals of metal clusters should obey in order to obtain a stable chemisorption energy. They demonstrated that their rules give stable results for the chemisorption energy of hydrogen and oxygen on clusters of one electron (ECP) Ni atoms of quite modest size. Panas et al. accounted for electron correlation with the contracted CI method of Siegbahn [41]. The 'Stockholm rule' for the orbital structure of the metal cluster is just that the cluster must be in a suitable 'bonding' state which is often not the ground state. Similar to the results of Hermann et al. this means that, in their Ni-O calculations [39], the cluster should not have high lying a_1 orbitals near the Fermi level. Panas et al. computed their very stable chemisorption energies by selecting such a cluster state and following that state to dissociation. This method does involve intuition regarding the bonding scheme involved in the chemisorption process of interest. Implicit in the Stockholm rule is the idea, contrary to Upton and Goddard's conclusions, that it is not important to describe the density of states, the ionization potential, or, the polarizability of the bulk with the cluster system in order to obtain stable (converged) chemisorption energies. The Stockholm rule also violates the low-spin idea of Bauschlicher. The underlying ideas are that clusters that have similar orbital structures yield similar binding energies even when the size of the cluster is relatively small, and, that the bulk metal is flexible enough to present a favorable bonding opportunity to the adsorbate. So, a cluster that presents this favorable bonding situation (through its orbital structure) to the adsorbate should be a reasonable model for the bulk so long as the cluster is large enough to accomodate the movement of electron density that accompanies bonding.

The situation is a bit different for the case of the (nonreactive) chemisorption/physisorption of closed shell molecules. These systems are essentially van der Waals complexes in the usual case. In these systems there are no formal chemical bonds (i.e. pairing of electrons) between the admolecule and the metal cluster so the above arguments regarding the Stockholm rules do not strictly apply. For example, there are no strong bonding interactions for Pauli repulsions to hinder. However, we have seen cases in which the chemisorption/physisorption energy can differ by as much as a factor of two (from, say 5 to 10 kca/mol) depending on the particular electronic state of the cluster model of the surface and the correlation method. Certainly differences in binding energies of this magnitude need to be considered in comparison to the inherent accuracy level of the methods employed.

In our analysis of the chemisorption of alkane thiols on gold described below we needed to estimate from ab initio calculations the heat of chemisorption of atomic hydrogen and sulfur on the gold (111) surface, since these quantities are not known from experimental data at present. Once these quantities were in hand we could proceed with the BOC analysis of certain elementary reactions involved. We employ the model potential ECP method of Huzinaga et al. [42] because this method preserves the nodal structure of the metal atom valence orbitals. The nodal structure of the metal atom orbitals may not have much effect at the Hartree-Fock level, but, may well have an effect on the electron correlation description. For the chemisorption of methane thiol we have employed a cluster model of the Au(111) surface that has nine top layer RECP Au atoms each with 11 quantum mechanical electrons. The second layer of the cluster model contained three RECP Au atoms each with 11 quantum mechanical electrons. Figure 2 shows the $HSCH_3$ admolecule in its chemisorption equilibrium position.

In our cluster model the orbital occupancy was chosen in the spirit of the Stockholm prepared cluster rules [39]. This means that a molecular orbital occupancy was chosen for the gold cluster that gave a reasonably good 'fit' of the frontier orbitals of the cluster to the ground state of the $HSCH_3$ adsorbate. As mentioned above this is not as critical for closed shell admolecules as it is for open shell specie. The positions of the metal atoms in the cluster model were held fixed with a nearest neighbor distance of 5.45 bohr taken from the bulk. In the 11 electron Au atoms the electron density up to and including the 5p electron density was replaced by the relativistic effective core potential. The parameters of the RECP were determined by fitting to the orbital shapes and energies of the 5d and 6s atomic Au orbitals obtained from relativistic Hartree-Fock calculations. These Hartree-Fock calculations include the Darwin and mass-velocity relativistic corrections according to the prescription given by Almlöf et al. [43]. We neglect the spin-orbit coupling. Our RECP parameters and basis sets are available from the author upon request. The projection operator or 'killing operator' [44] for gold employed in this work is described by the basis set of Gropen [45]. The valence space of atomic gold contains the 5d and 6s electron density and so the RECP atom has no p type basis functions. We include in our RECP Au basis set an uncontracted p function, centered on the nucleus, the exponent of which was optimized for the Au_{12} SH model system at the Hartree-Fock level. We observed that this set of p functions made significant contributions to the Au(111) - sulfur bond and to the orbitals involving primarily the gold atoms (in the A_2 space of C_{3v}). The sulfur and hydrogen basis sets employed in this work were the DZ and DZP basis sets. In these calculations the S-C bond distance as optimized and was found to be very nearly the same as our previous work [32]. We did not allow the SH or CH distances to vary nor was the SCH angle varied from the Hartree-Fock equilibrium values [32]. Since these are weakly bound systems the molecular structures are not expected to change much from one binding site to another and in $HSCH_3$ the SC bond length was optimized only for the ontop configuration.

In order to estimate the chemisorption energy of atomic hydrogen on the Ag/Au(111) surfaces we dissociated the H_2 molecule on large cluster models of the surfaces and employed a thermodynamic cycle to calculate the actual atomic binding energies. The reactions in the thermodynamic cycle are (the 'g' and 'ad' in parentheses indicate gas phase and surface adsorbed species, respectively):

1) $H_2(g) \longrightarrow 2\,H\,(g)$

2) $2\,H\,(g) \longrightarrow 2H\,(ad)$

3) $2\,H\,(ad) \longrightarrow H_2\,(ad)$

4) $H_2\,(ad) \longrightarrow H_2\,(g)$

It is the enthalpy of reaction 2 above that we are seeking. The enthalpy of reaction 1 is well known from experiment to be 104 kcal/mol. We calculated the dissociation energy of the H_2 molecule on the surface to get the enthalpy of reaction 3. The enthalpy for reaction 4 we also obtained from ab initio calculations. The sum of the four reaction enthalpies must add to zero. Therefore the enthalpy of reaction 2 is obtained as the negative of the sum of the other three.

The advantage in this procedure is that one obtains the atomic heat of chemisorption without having to do an open shell ab initio calculation or a calculation that requires the separation of a bonding pair of electrons. The dissociation of the H_2 molecule on a large (closed shell) model of the metal surface does not involve the separation of any bonding pairs of electrons. As the H-H bond breaks the both H - surface bonds are forming and never does one consider an 'unpairing' of any bonding electron pair. Therefore, one can expect that the contribution to the reaction energy will contain much less contributions from electron correlation. Indeed, we found this to be the case. We also found that, in order to obtain results that are converged with respect to the size of the metal cluster one has to employ a relatively large model of the surface. One reason is that if the dissociation products are close together then this constitutes a locally high coverage situation and, presumably, one is intending to compute the zero coverage limit. Another reason is that two adsorbates (the dissociation products) demand more flexibility from the wavefunction of the cluster model of the surface than does a single adsorbate. This requirement of a larger cluster model of the surface can be compensated by the fact that the results can be reasonably accurate even at the Hartree-Fock level. We obtained the 56 kcal/mol for the heat of adsorption of H on Ag(111) to compare with an experimental value of 52 kcal/mol. This value served as a measure of the quality of the predictions of this procedure. Our value for the heat of chemisorption of atomic hydrogen on the gold (111) surface obtained in this way is 46 kcal/mol [].

We obtained the value for the heat of adsorption of atomic sulfur on gold in a different and also indirect way. We determined the chemisorption energy of $HSCH_3$ on Au(111) at the level of RECP Hartree-Fock + MP2 correlation and used this value in the BOC expression for the same quantity (equation 7 for the ontop position). Solving for the Q_{0S} paramter allows one to obtain the atomic heat of chemisorption from equation 3. Our value for the atomic heat of chemisorption of sulfur on Au(111) is 80 kcal/mol. This value can be checked by using it to compute the heat of chemisorption for the SCH_3 radical and comparing the result with the experimental determination of 44 kcal/mol [46]. Our value of 80 kcal/mol also yields 44 kcal/mol for the heat of chemisorption of SCH_3.

Figure 2 shows the methane thiol in its equilibrium configuration. In accord with the predictions of the BOC model the ab initio calculations predict the ontop binding site to the be most stable. As the H atom is pulled off the sulfur atom the lowest energy position for the SCH_3 fragment shifts from the ontop site to toward the hollow site. The energy difference we obtained between the equilibrium structures of the chemisorbed $HSCH_3$ structure and the dissociation products (H + SCH_3) was 11.3 kcal/mol favoring the undissociated methane thiol.

The behavior of the reactive sticking coefficients for the series $HS(CH_2)_n$ CH_3 has been the topic of some discussion in the literature [47]. It has been observed that the methane thiol ($n = 0$) has a much lower reactive sticking probability than other members of the series [47]. It has been estimated that the molecular chemisorption energy increases by 1.9 kcal/mol for every methylene group, and, Dubois and co-workers have proposed a stabilization of the transition state (in the dissociation reaction coordinate) that is also dependent on the methylene number, n. With the BOC expressions for the activation barrier and our ab initio values for the sulfur atom chemisorption energy we have computed the activation barriers as a function of the methylene number. Table 4 contains our computed activation barriers for the extraction of the sulfur hydrogen from the alkane thiols.

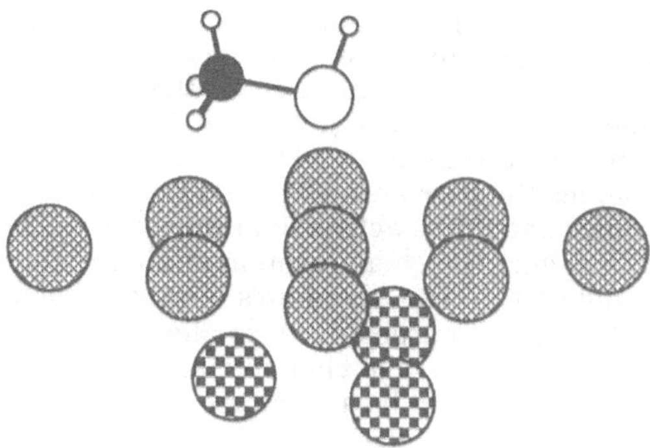

Figure 2. The ab initio equilibrium geometry of methane thiol on gold.

Table 4. BOC-MP intrinsic activation barriers (kcal/mol) for the SH bond cleavage reaction of alkane thiols on gold as a function of methylene number.

n	$E^{f}_{HS,s}$	$E^{f}_{HS,g}$
0	16.9	4.2
1	17.1	2.5
2	17.4	0.9
3	17.6	-0.8
4	17.8	-2.5
5	18.0	-4.2

The barriers are given relative to gas phase reactants as well as surface adsorbed reactants. (The barriers relative to the gas phase reactants are obtained from those relative to the surface adsorbed reactants by subtracting the molecular heat of chemisorption.) When viewed from the position of the gas phase reactants, our activation barrier, our activation barriers would agree with the description given by Dubois [47]. However, the true size of the energy hill as given by the activation barriers relative to the surface adsorbed reactants actually increases a little it for each methylene in the molecule. Our picture from the point of view of the surface adsorbed reactants is that the thermal energy liberated in the chemisorption process is available to aid the reactant in overcoming the (slightly higher) barrier to reaction.

One might reasonably ask why doesn't the S-C bond cleave rather than the H-S bond, since the S-C bond is acutally weaker than the H-S bond. We have discussed this earlier [7] and obtained a value of 17 kcal/mol for the cleavage of the S-C bond in methane thiol on Au(111). This value compares with the value of 16.9 kcal/mol for the barrier to H-S bond cleavage. This suggests that both reactions channels should be active simultaneously. The end product in many experiments involving the chemisorption of alkane thiols on gold is the self-assembled monolayer. The dynamics and thermodynamics of the formation of the end product is probably what allows one reaction channel to dominate. In the S-C bond cleavage channels atomic sulfur is often an end product [7] and its presence would be a signiture.

REFERENCES

1. J.F. Roth in Catalysis 1987, Studies in Surface Science and Catalysis (J.W. Ward, ed.), Vol. 38, p. 925, Elsevier, Amsterdam, 1988.
2. D.W. Goodman and J.E. Houston, Science 236 (1987) 403.
3. E. Shustorovich, Surf. Sci. Rep. 6 (1986) 1.
4. E. Shustorovich, Adv. Cat. 37 (1990) 101.
5. H.L. Sellers, J. Phys. Chem., 98 (1994) 968. This reference contains a misleading statement below eq. 18. The statement $F_{12} = (F_{11} F_{22})^{1/2}$ is valid for only special cases.
6. H.S. Johnston, Gas Phase Reaction Rate Theory, Ronald Press: New York, 1966.
7. H.L. Sellers, Surf. Sci. 294 (1993) 99.
8. P. Paredes Olivera, E.M. Patrito and H.L. Sellers, unpublished results.
9. Y.-F. Wang and R. Pollard, Surf. Sci., 302 (1994) 223.
10. Shustorovich, E.; Bell, A.T. *Surf. Sci.* 289 (1993) 127.
11. A.T. Bell and E. Shustorovich, J. Catal. 121 (1990) 1.
12. E. Shustorovich and A.T. Bell, Surf. Sci. 253 (1991) 386.
13. E. Shustorovich, Catal. Lett. 7 (1990) 107.
14. E. Shustorovich and A.T. Bell, Surf. Sci. 278 (1991) 359.
15. E. Shustorovich and A.T. Bell, Surf. Sci. 259 (1991) L791.
16. P. Paredes Olivera, E.M. Patrito and H.L. Sellers, Surf. Sci. in press.
17. D.L. Freeman and J.D. Doll, J. Chem. Phys. 78 (1983) 6002.
18. D.L. Freeman and J.D. Doll, J. CHem. Phys. 79 (1983) 2343.
19. E.M. Patrito, P. Paredes Olivera and H.,L. Sellers, Surf. Sci. 306 (1994) 447.
20. H.L. Sellers, J. Chem. Phys. 99 (1993) 650.
21. H.L. Sellers, J. Chem. Phys. 98 (1993) 627.
22. H.L. Sellers, Surf. Sci. 310 (1994) 281.
23. H.L. Sellers, J. Chem. Phys. submitted.
24. U. Nielsen, D. Halstead, S. Holloway and J.K. Nørskov, J. Chem. Phys., 93 (1990) 2879.
25. C.-M. Chiang and B. Jackson, J. Chem. Phys., 87 (1987) 5497.

26. D. Halstead and S. Holloway, J. Chem. Phys., 93 (1990) 2859.
27. R.C. Mowrey and B.I. Dunlap, Int. J. Quant. Chem. Symp. 25 (1991) 641.
28. C.-Y. Lee and A.E. Depristo, J. Chem. Phys., 87 (1987) 1401.
29. M.R. Hand and S. Holloway, J. Chem. Phys., 91 (1989) 7209.
30. W. Forst, Theory of Unimolecular Reactions, Academic Press, New York, 1973.
31. J. Troe, Chemical Kinetics, (Physical chemistry, series two; v. 9) (International review of science); Butterworths: Boston, 1976.
32. H.L. Sellers, A. Ulman, Y. Shnidman and J.E. Eilers, J. Am. Chem. Soc. 115 (1993) 9389.
33. T.H. Upton and W.A. Goddard, CRC critical reviews in solid state and materials sciences (CRC Press, Boca Raton, 1981).
34. C.W. Bauschlicher Jr., P.S. Bagus and H.F. Schaefer III, IBM J. Res. Dev. 22 (1978) 213.
35. P.S. Bagus, H.F. Schaefer III and C.W. Bauschlicher Jr., J. Chem. Phys. 78 (1983) 1390.
36. M.R.A. Blomberg and P.E.M Siegbahn, J. Chem. Phys. 78 (1983) 986, 5682.
37. J.N. Allison and W.A. Goddard, Surface Sci. 110 (1981) 1615.
38. C.W. Bauschlicher Jr., Chem. Phys. Lett. 129 (1986) 586.
39. I. Panas, J. Schule, P.E.M. Siegbahn and U. Wahlgren, Chem. Phys. Lett. 149 (1988) 265.
40. K. Hermann, P.S. Bagus and C.J. Nelin, Phys. Rev. B 35 (1987) 9467.
41. P.E.M. Siegbahn, Intern. J. Quant. Chem. 23 (1983) 1869.
42. S. Huzinaga, M. Klubokowski and Y. Sakai, J. Phys. Chem. 88 (1984) 4880; J. Andzelm, S. Huzinaga, M. Klubokowski and E. Radzio, Mol. Phys. 52 (1984) 1495; S. Huzinaga, L. Seijo, Z. Barandiaran and M. Klubowski, J. Chem. Phys. 86 (1987) 2132.
43. J. Almlöf, K. Faegri and H.H. Grelland, Chem. Phys. Lett. 114 (1986) 53.
44. H.L. Sellers, Chem. Phys. Lett. 178 (1991) 351.
45. O. Gropen, J. Comp. Chem. 8 (1987) 982.
46. L.H. Dubois and R.G. Nuzzo, Ann. Rev. Phys. Chem. 43 (1992) 437.
47. L.H. Dubois, B.R. Zegarski and R.G. Nuzzo, J. Chem. Phys. 98 (1993) 678.
48. G. Ert in The Nature of the Surface Chemical Bond, North Holland (T.N. Rhodin and G. Ertl, eds.), Amsterdam, 1979.
49. P. Feulner and D. Menzel, Surf. Sci. 154 (1985) 465.
50. J.T. Yates Jr., P.A. Thiel and W.H. Weinberg, Surf. Sci. 84 (1979) 427.
51. J.R. Engstrom, W. Tsai and W.H. Weinberg, J. Chem. Phys. 87 (1987) 3104.
52. G.E. Gdowski, J.A. Fair and R.J. Madix, Surf. Sci. 127 (1983) 541.
53. X.-L. Zhou J.M. White and B.E. Koel, Scur. Sci. 218 (1989) 201.
54. A.G. Sault, R.J. Madix and C.T. Campbell, Surf. Sci. 169 (1986) 347.
55. T. Engel, H. Niehus and E. Bauer, Suirf. Sci. 52 (1975) 237.
56. L. Surnev, G. Rangelov and G. Bliznakov, Surf. Sci. 159 (1985) 299.
57. G.B. Fisher and S.J. Schmeig, J. Vac. Sci. Technol. A1 (1983) 1064.
58. J.J. Vajo, W. Tsai and W.H. Weinberg, J. Phys. Chem. 89 (1985) 3243.
59. L.C. Isett and J.M. Blakely, Surf. Sci 43 (1974) 493.
60. C.H. Bartholomew, P.K. Agrawal and J.R. Katzer, Advan. Catal. 31 (1982) 170.
61. D. Brennan, D.O. Hayward and B.M.W. Trapnell, Proc. Roy. Soc. London, Ser. A 265 (1960) 81.
62. W.F. Egelhoff, Jr., J. Vac. Sci. Tech. A 5 (1987) 700.
63. E. Giamello, B. Fubini, P. Lauro and A Bossi, J. Catal. 87 (1984) 443.
64. C.T. Campbell, G. Ertl, H. Kuipers and J. Segner, Surf. Sci. 107 (1981) 220.
65. C.T. Campbell, Surf. Sci. 157 (1985) 43.
66. G.B. Fisher and J.L. Gland, Surf. Sci. 94 (1980) 446.

67. P.A. Thiel and T.E. Madey, Surf. Sci. Rep. 7 (1987) 211, 262.
68. C. Klauber, M.D. Alvey and J.T. Yates, Jr., Surf. Sci. 154 (1985) 139.
69. R.B. Hall, J. Phys. Chem. 91 (1987) 1007.
70. B.A. Sexton and A.E. Hughes, Surf. Sci. 140 (1984) 227.
71. K. Christmann and J.E. Demuth, J. Chem. Phys. 76 (1982) 6308.
72. F. Solimosi, A. Berko and T.L. Tarnoczi, Surf. Sci. 141 (1984) 533.
73. S. Brass and G. Ehrlich, J. Chem. Phys. 87 (1987) 4285.
74. G. Wedler and H. Ruhmann, Surf. Sci 121 (1982) 464.
75. F.M. Hoffmann, Surf. Sci. Rep. 3 (1983) 107.
76. C.T. Campbell and W.H. Weinberg, Chem. Phys. Lett. 179 (1991) 53.

COMPUTER SIMULATIONS OF EXCITABLE REACTION-DIFFUSION SYSTEMS

Mark R. Hoffmann and Sean P. Müller

Department of Chemistry
University of North Dakota
Grand Forks, ND 58202

INTRODUCTION

Since the pioneering studies on the Belousov-Zhabotinsky (BZ) reagent, spatio-temporal self-organization has been recognized and studied in a variety of systems. The topic has been the subject of several reviews; a recent monograph[1] and compendium[2] are excellent general references. Besides chemical reaction systems,[3-12] fluids,[13,14] lasers,[15] surface catalytic systems,[16] and, especially, biological systems[17-27] are known to exhibit bifurcation behavior. In this paper, we explore aspects of the kinetics of the changes in the concentrations of -S-H and -S-S groups in neural membrane proteins.[25-27] This system was most recently studied by Sevcikova and Marek[27] who focused on the specific kinetic parameters describing the firing of a giant squid axon. The results reported herein address two points: first, whether a semi-implicit finite difference technique (such as Crank-Nicholson) is necessary to obtain a good representation of the reaction diffusion system or whether a less expensive, fully explicit, technique suffices.[28,29] Second, the SH kinetic model is demonstrated to lead to behavior other than a traveling wave when the kinetic parameters are varied.

The SH kinetic model[25-27] consists of two coupled chemical species, denoted X and Y; we consider a system in which both chemical species are capable of diffusion. The system is prepared such that the concentrations of both species are at equilibrium for kinetic parameters corresponding to a giant squid axon. We restrict attention in this study to one spatial dimension. When a system described by the SH kinetic model is excited by a wave or other stimulus (as, *e.g.*, applied at a boundary) beyond a characteristic intensity value, the system undergoes a so-called excursion or excitation cycle. The time evolution

Theoretical and Computational Approaches to Interface Phenomena
Edited by H.L. Sellers and J.T. Golab, Plenum Press, New York, 1994

of the system can be depicted in one of several ways: a concentration profile showing the entire system at one instance of time, a concentration *versus* time plot at one particular point in space, or a plot of the concentrations of both chemical species at a specific point in space at various times. This last representation is commonly referred to as a stroboscopic map if time is discrete and as a phase portrait if time is continuous. We shall usually use this last portrayal of the system under investigation. A stroboscopic map illustrating an X, Y excursion is shown as the large curve in Figure 1.

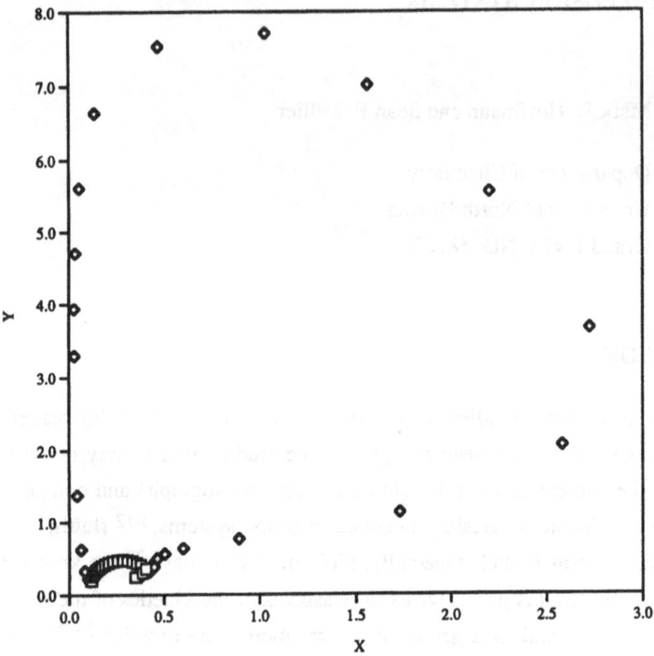

Figure 1. Local stroboscopic map near left boundary (i = 3). □ represents a subthreshold pulse of A_f = 0.450 and ◇ represents a superthreshold pulse of A_f = 0.490.

The excitation increases the concentration of species X until there is enough X to begin runaway production of species Y. The concentration of species Y increases rapidly, and eventually begins depleting X until the concentration of X is sufficiently reduced to allow the system to relax to initial concentrations of species X and Y. The minimum (or threshold) amplitude pulse, A_{th}, that is capable of initiating an excursion cycle depends on the shape of the pulse and on its time duration. If the amplitude of the stimulus greatly exceeds the threshold value, the phase portrait would resemble a larger version of the curve exhibiting excursion in Figure 1. However, if the intensity of the stimulus was below the threshold value, then no excursion would be occur, (near) linear kinetics would develop, and the stroboscopic map would look like the small curve in Figure 1.

The mathematical model that represents the above described system is defined by the following equations:

$$\frac{\partial X}{\partial t} = D_X \frac{\partial^2 X}{\partial z^2} + f(X,Y) \quad (1)$$

$$\frac{\partial Y}{\partial t} = D_Y \frac{\partial^2 Y}{\partial z^2} + g(X,Y) \quad (2)$$

where D_X and D_Y are the diffusion coefficients of species X and Y, respectively. The functions f(X, Y) and g(X, Y) denote the nonlinear reaction kinetics of each species. In the SH kinetic model,[25-27] the functions f(X, Y) and g(X, Y) are defined as:

$$f(X,Y) = \frac{\alpha v_0 + X^\gamma}{1 + X^\gamma} - X(1 + Y) \quad (3)$$

$$g(X,Y) = X(\beta + Y) - \delta Y \quad (4)$$

where α, β, γ, δ, and v_0 are dimensionless kinetic constants restricted to positive values.

To initiate the passage of waves in the system, Sevcikova and Marek[27] introduced a time-dependent combination of Direchlet and Neumann boundary conditions.[28,29] Neumann boundary conditions, *i.e.*, zero flux gradients at the boundary, are used for both the left and right boundaries, and for both chemical species, when no new material is being introduced into the system.

$$\left.\frac{du}{dz}\right|_{z=c} = 0 \quad (5)$$

where $u \in \{X, Y\}$ and $c \in \{0, L\}$. Material is introduced periodically into the system through specification of Direchlet boundary conditions for species X at the left boundary. Direchlet boundary conditions specify the function value at the boundary, *e.g.*,

$$X(0,t) = A_f \quad (6)$$

where A_f is the amplitude of the stimulus.

In Section II of this paper, two algorithms, the fully explicit Forward Time Center Space (FTCS) and the semi-implicit Crank-Nicholson (CN), are described. A novel, partially linearized CN method is introduced. In the first subsection of Section III, the FTCS and CN methods are assessed by comparing their predicted threshold value of the stimulus, A_{th}. In the second subsection, the SH model's kinetic parameters were varied to determine how each parameter affects the behavior exhibited by the system. A final section contains concluding remarks.

METHODOLOGY

The reaction diffusion system was studied using two numerical techniques. One study used the Forward Time Center Space (FTCS) explicit method,[28,29] whereas the second study used a Crank-Nicholson finite difference scheme for solving the coupled partial differential equations.[28,29] The FTCS method has the advantage of being conceptually simpler and easier to program. We investigate the accuracy of the method as a function of time step for the highly nonlinear PDE's encountered in the SH kinetic model. The FTCS method uses function information (*e.g.*, concentration) at a point's nearest neighbors to calculate spatial derivatives with which to propagate the function value of the point forward in time; *e.g.*, for a simple, nonreactive, system, the FTCS equations are:

$$\frac{u_j^{n+1} - u_j^n}{\Delta t} = D_u \left(\frac{u_{j+1}^n - 2u_j^n + u_{j-1}^n}{(\Delta z)^2} \right) \tag{7}$$

In Eq. (7), Δz and Δt represent the grid spacings in space and time, respectively; the superscript refers to discretized time and the subscript labels discretized space. In all the studies reported herein, 128 identical space intervals were used; *i.e.*, j = 0 to 128. Press *et al.*[28] provide a clear graphical representation of the procedure.

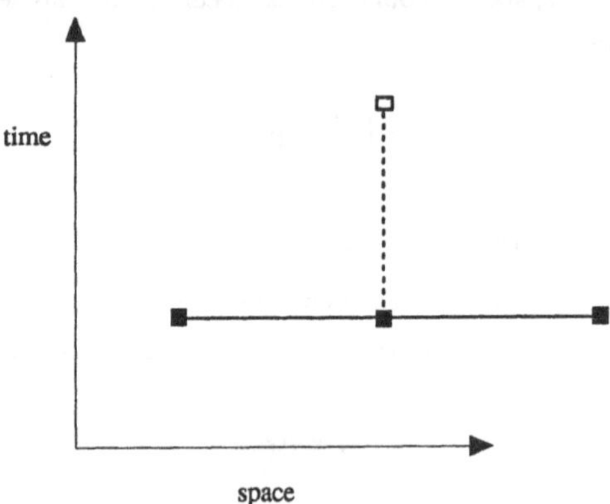

Figure 2. Graphical Representation of Forward Time Center Space method (from Ref. 28).

Extension of the FTCS method to the model's partial differential equations is straightforward; *i.e.*,

$$u_j^{n+1} - u_j^n = \alpha_u \left(u_{j+1}^n - 2u_j^n + u_{j-1}^n \right) + \Delta t\, h(X^n, Y^n) \tag{8a}$$

where

$$\alpha_u = \frac{\Delta t\, D_u}{(\Delta z)^2} \tag{8b}$$

In Eq. (8), $h(X^n,Y^n)$ is introduced to represent the kinetic term associated with species u; *i.e.*, h represents either f or g.

The Crank-Nicholson (CN) method for solving partial differential equations averages a fully explicit prediction with a fully implicit prediction to obtain a result that is second-order accurate in space and time;[28] *i.e.*, for a nonreactive system,

$$\frac{u_j^{n+1} - u_j^n}{\Delta t} = \frac{D_u}{2}\left\{\left(\frac{u_{j+1}^{n+1} - 2u_j^{n+1} + u_{j-1}^{n+1}}{(\Delta z)^2}\right) + \left(\frac{u_{j+1}^{n} - 2u_j^{n} + u_{j-1}^{n}}{(\Delta z)^2}\right)\right\} \tag{9}$$

As may be seen from Eq. (9), the CN method uses concentration information about a point's present and predicted future values and information about it's nearest neighbors' present and predicted future values. The CN method is visualized as in Figure 3.

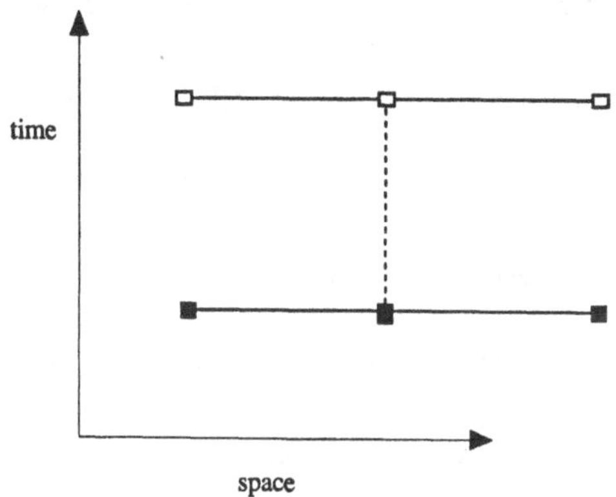

Figure 3. Graphical Representation of Crank-Nicholson method (from Ref. 28).

Because of the averaging of present and predicted future values, the CN method is stable for large time steps; in contrast, explicit methods are prone to numerical instabilities.

The Crank-Nicholson solution to a reaction diffusion equation may be defined by:

$$\frac{\partial u}{\partial t} = D_u\left\{\frac{1}{2}\left(\frac{\partial^2 u^n}{\partial z^2} + \frac{\partial^2 u^{n+1}}{\partial z^2}\right)\right\} + \frac{1}{2}\left\{h^n(X,Y) + h^{n+1}(X,Y)\right\} \tag{10}$$

In Eq. (10), $h^n(X,Y) = h(X^n,Y^n)$. The nonlinear kinetics of the present concentrations can be done straightforwards (*e.g.*, explicit method); however, the nonlinear kinetics of the future concentrations <u>cannot</u> be obtained by the solution of a set of simultaneous linear equations. We investigate the utility of using a linearized version of the kinetics to obtain a credible <u>estimate</u> of future values while retaining the full nonlinear kinetics of the present concentrations.

In the absence of reactions, the CN method obtains future values as the solution to a set of simultaneous linear equations

$$u_{j-1}^{n+1} - \frac{2(\alpha_u + 1)}{\alpha_u} u_j^{n+1} + u_{j+1}^{n+1} = -u_{j-1}^{n} + \frac{2(\alpha_u - 1)}{\alpha_u} u_j^{n} - u_{j+1}^{n} \qquad (11)$$

A similar expression for a system with reactions can be obtained with approximations. Consider first that Eq. (10) can be rearranged to:

$$u_j^{n+1} - u_j^n =$$
$$\frac{\alpha_u}{2}\left\{\left(u_{j+1}^{n+1} - 2u_j^{n+1} + u_{j-1}^{n+1}\right) + \left(u_{j+1}^{n} - 2u_j^{n} + u_{j-1}^{n}\right)\right\} + \frac{\Delta t}{2}\left(h_j^{n+1} + h_j^n\right) \qquad (12)$$

The future value kinetic term can be expressed using a Taylor expansion,

$$h(X^{n+1}, Y^{n+1}) = h(X^n, Y^n) + \frac{\partial h}{\partial X}\left(X^{n+1} - X^n\right) + \frac{\partial h}{\partial Y}\left(Y^{n+1} - Y^n\right) + \ldots \qquad (13)$$

Substituting the first order truncation of Eq. (13) into Eq. (12) and rearranging gives a set of simultaneous linear equations for u (which, however, depend parametrically on both present and future values of the other chemical species, v),

$$\frac{-\alpha_u}{2} u_{j-1}^{n+1} + \left(1 + \alpha_u - \frac{\Delta t}{2}\frac{\partial h}{\partial u}\right) u_j^{n+1} + \frac{-\alpha_u}{2} u_{j+1}^{n+1} = b'^{\,n}_j\left(u_j^n, v_j^n, v_j^{n+1}\right) \qquad (14)$$

Eq. (14) leads straightforwardly to our working equations,

$$u_{j-1}^{n+1} - \frac{2(\alpha_u + 1) - \Delta t\,(\partial h/\partial u)}{\alpha_u} u_j^{n+1} + u_{j+1}^{n+1} = b_j^n \qquad (15a)$$

The right hand sides of the simultaneous equations are given by,

$$b_j^n = -u_{j-1}^n + \frac{2(\alpha_u - 1) + \Delta t\,(\partial h/\partial u)}{\alpha_u} u_j^n - u_{j+1}^n$$
$$- \frac{2\Delta t}{\alpha_u} h(X^n, Y^n) - \frac{\Delta t}{\alpha_u}\frac{\partial h}{\partial v}\left(v_j^{n+1} - v_j^n\right) \qquad (15b)$$

Since both present and future values of the other chemical species are required for evaluation of the right hand side, further approximations are necessary. As we discuss below, the appropriate approximation to make for v_j^{n+1} varies.

We examine first the simultaneous equations for the time evolution of X. The requisite derivatives of the kinetic terms are

$$\frac{\partial f}{\partial X} = \frac{\alpha\gamma\left(X_j^n\right)^{\gamma-1}}{1+\left(X_j^n\right)^{\gamma}}\left[1 - \frac{v_0 + \left(X_j^n\right)^{\gamma}}{1+\left(X_j^n\right)^{\gamma}}\right] - \left(1 + Y_j^n\right) \tag{16a}$$

and

$$\frac{\partial f}{\partial Y} = -X_j^n \tag{16b}$$

The evolution equations decouple if we substitute an explicit estimate of Y^{n+1} to evolve X, and then go back and calculate Y^{n+1} accurately using Crank-Nicholson. The explicit or FTCS estimate of Y follows immediately from Eqs. (8) and (4).

The derivatives necessary for calculating the time evolution of Y are as follows.

$$\frac{\partial g}{\partial X} = \beta + Y_j^n \tag{17a}$$

and

$$\frac{\partial g}{\partial Y} = X_j^n - \delta \tag{17b}$$

Since the evolution of the X species for time *n* has already been performed, the X^{n+1} are available. Hence the final term in Eq. (15b), for Y time-evolution, is calculated with these updated values.

RESULTS AND DISCUSSION

Comparison of Explicit and Crank-Nicholson

A variety of pulse heights and durations were investigated to qualitatively assess the abilities of either the FTCS algorithm or the Crank-Nicholson algorithm to describe reactive diffusion behavior. As typified by Figure 4, the time evolutions obtained in all numerical experiments were remarkably similar. Furthermore, our results are also in qualitative agreement with those of Sevcikova and Marek.[27] The specific calculations shown in Figure 4 correspond to a time step of 0.0025 for both FTCS and Crank-Nicholson; the pulse amplitude was 0.474.

As can be seen from Figure 4, the Crank-Nicholson results are retarded relative to the FTCS predictions; both methods trace essentially the same path. Because the concentration scales are so large in diagrams such as Figure 4, details of the start of the discrepency

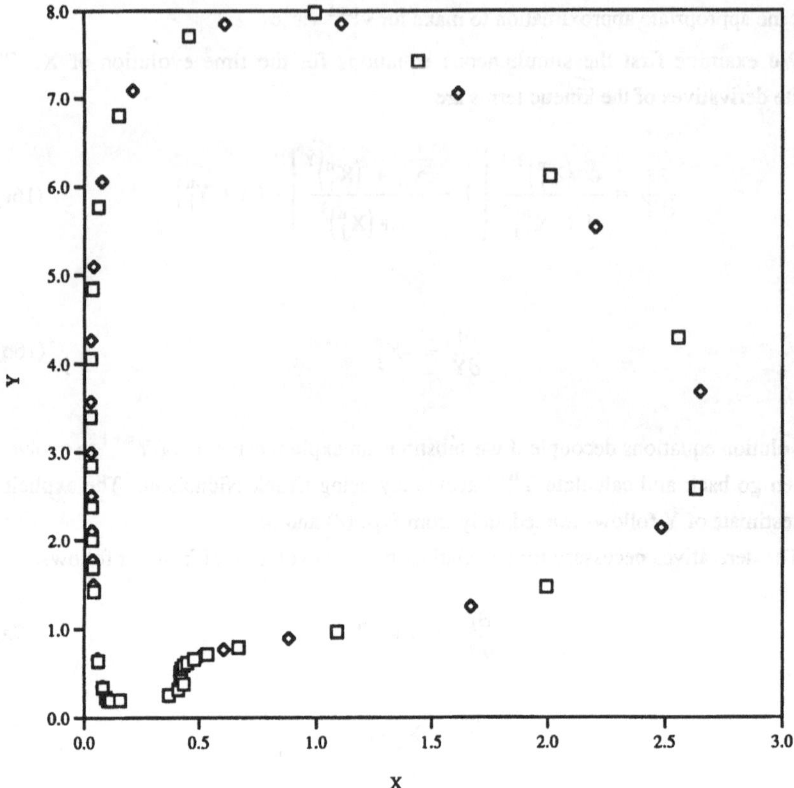

Figure 4. Local stroboscopic map near left boundary (i = 3). □ represents solution by FTCS explicit method and ◇ represents solution by Crank-Nicholson.

between FTCS and Crank-Nicholson are obscured. We reproduce the lower left-hand corner of Figure 4 in larger scale (*cf.* Figure 5). The FTCS and Crank-Nicholson methods are indistinguishable, to the accuracy of the plot, up to the time that the concentration build-up antecedent to excursion occurs (*ca.* t = 1 - 1.8). Since the excursion is highly nonlinear, small time differences in reaching a critical value are amplified and an increasing time delay develops between the numerical techniques. As the system recovers to equilibrium (*ca.* t ≥7.2), the techniques once again give the same results.

After completion of the initial qualitative studies, the threshold value of the pulse stimulus A_{th} was found for each algorithm. A series of numerical experiments, each with an initial square pulse of duration 0.6, were performed with a variety of time steps. The limit of infinitesimal time steps was then obtained with a least squares fit of calculated threshold *versus* time step. Both algorithms converged to the same limit of A_{th} = 0.472. Our value is close to, but not within, the range reported by Sevcikova and Marek;[27] *i.e.*, A_{th} = ⟨0.482, 0.4837⟩. A possible source of the discrepancy is some ambiguity in the reported threshold pulse.

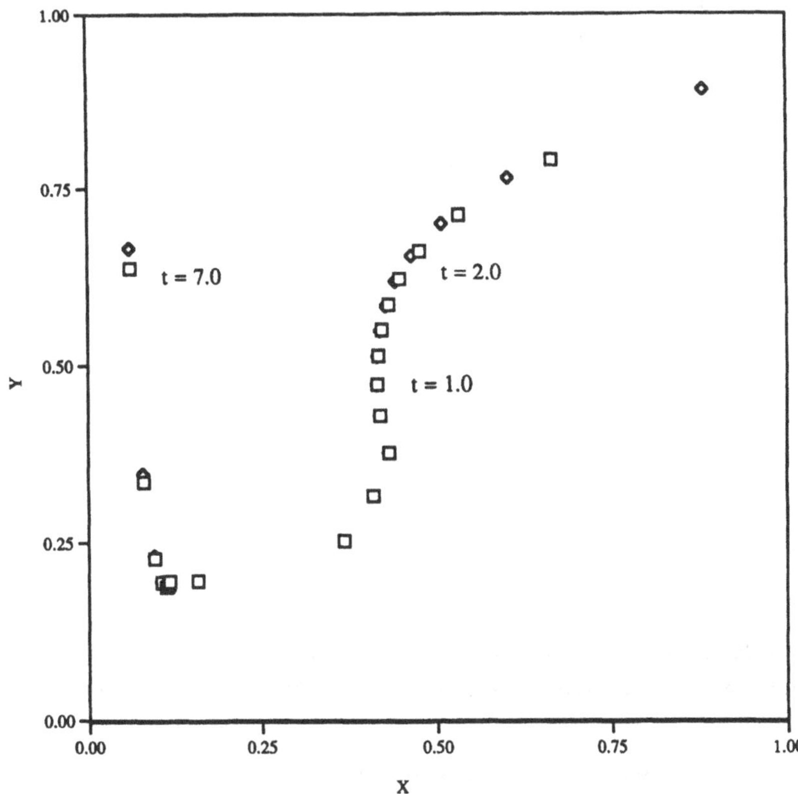

Figure 5. Enlarged view of local stroboscopic map near left boundary (i = 3). See Figure 4.

The predicted threshold stimulus was tested using the following empirical relationship for the finite-size error in the FTCS model:

$$A_{th} = 0.472094 - 0.1065574\,(\Delta t) \tag{18}$$

This equation was used to predict the threshold amplitude for a time step of 0.000025. Two additional simulations were then performed, one with a stimulus amplitude 0.000004 above and one with the stimulus amplitude 0.000004 below the predicted threshold value. Figure 6 illustrates the results of these simulations. The experiment that had a stimulus amplitude larger than the threshold value displayed an excursion cycle, while the experiment having a stimulus amplitude lower than the predicted threshold value did not display an excursion cycle. Since the stimulus amplitudes are so close to those defining the separatrix, the amount of time until bifurcation is comparatively long (*e.g.*, compare Figure 6 with Figures 1 and 5). These experiments strongly corroborate the empirical relation between error in predicted threshold and time-step size and, furthermore, lend credence to our prediction that an initial square pulse of duration 0.6 has a threshold height of 0.472.

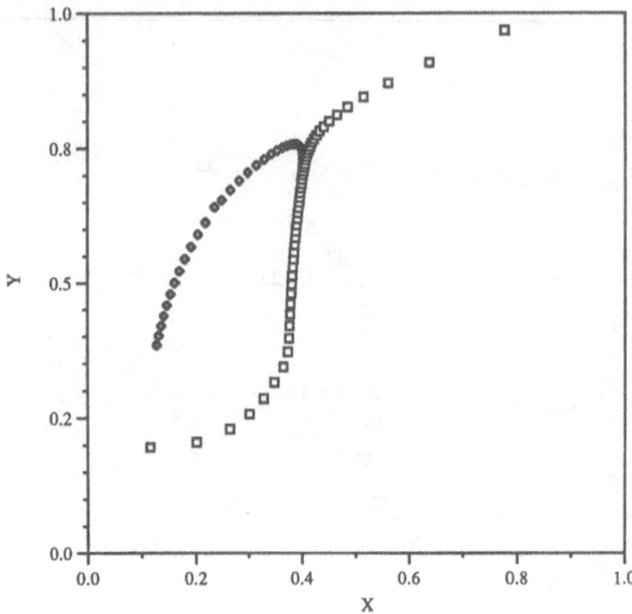

Figure 6. Local stroboscopic map near left boundary (i = 6). □ represents a superthreshold pulse of A_f = 0.472095 and ◇ represents a subthreshold pulse of A_f = 0.472087. Time intervals of 0.1 displayed.

The Forward Time Center Space algorithm (FTCS) and the Crank-Nicholson algorithms (CN) were both found capable of faithfully simulating the SH model. Though the CN algorithm does offer the advantage of using a larger time step than does the FTCS algorithm, the substantially larger computational effort to achieve the same accuracy (not reported herein) recommends against the routine use of Crank-Nicholson for these types of reactive diffusion problems.

Variation of Kinetic Parameters

After the value of the threshold amplitude of the stimulus was determined, the model's kinetic parameters (α, β, γ, δ, v_0) were varied to determine how each parameter affected the behavior of the system. This was done by varying one of the kinetic parameters and keeping all the other kinetic parameters constant. Also, the stimulus amplitude was held constant at a superthreshold level of $A_f = 0.474$ (*i.e.*, $A_{th} = 0.4718$ for a time step of $\Delta t = 0.0025$). All the kinetic simulations were conducted using the FTCS algorithm and then verified by running the simulations using the CN algorithm.

When the kinetic parameters were varied, the model displayed three main types of behavior: (1) conventional subthreshold and superthreshold behavior, (2) relaxation to new

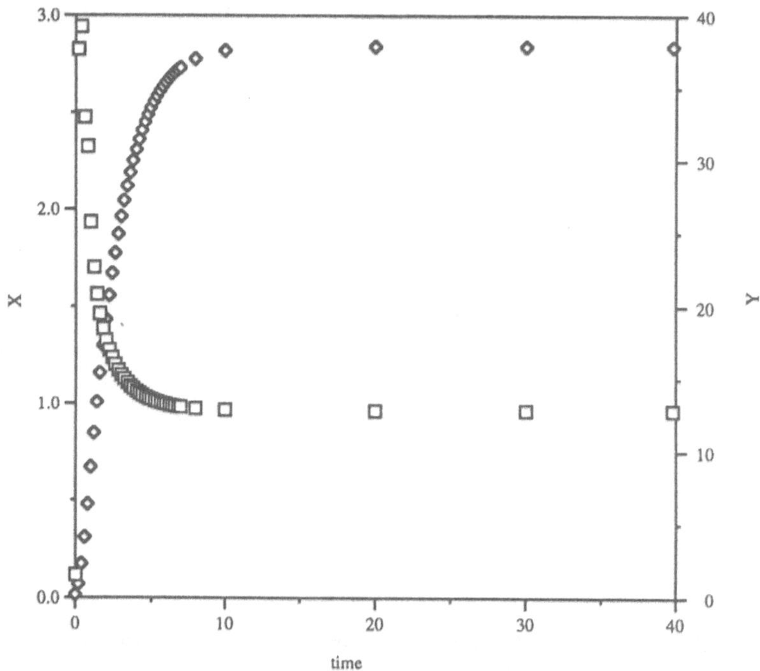

Figure 7. Local concentration near left boundary (i = 6), for $v_0 = 5.0$. □ represents concentration of X and ◇ represents concentration of Y.

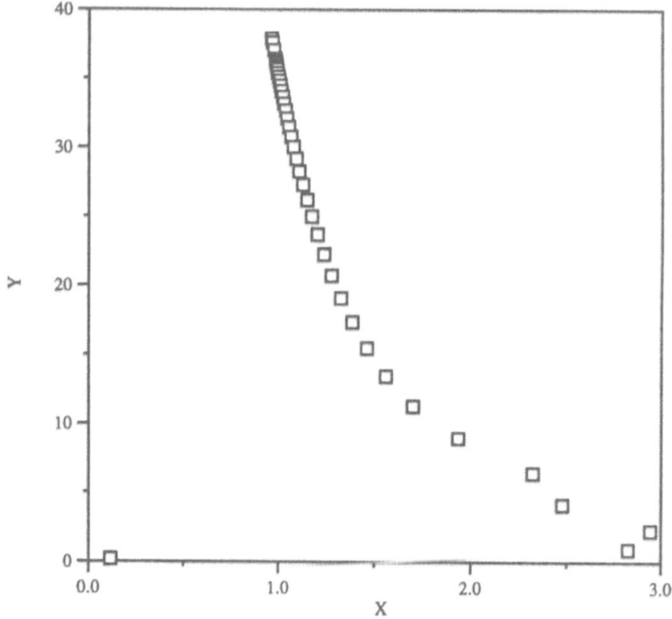

Figure 8. Local stroboscopic map near left boundary (i = 6), for $v_0 = 5.0$.

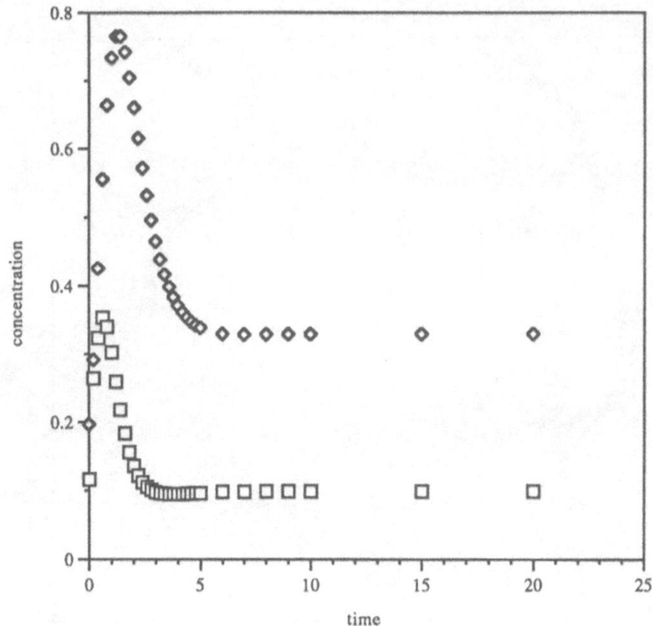

Figure 9. Local concentration near left boundary (i = 6), for $\beta = 3.0$. □ represents concentration of X and ◇ represents concentration of Y.

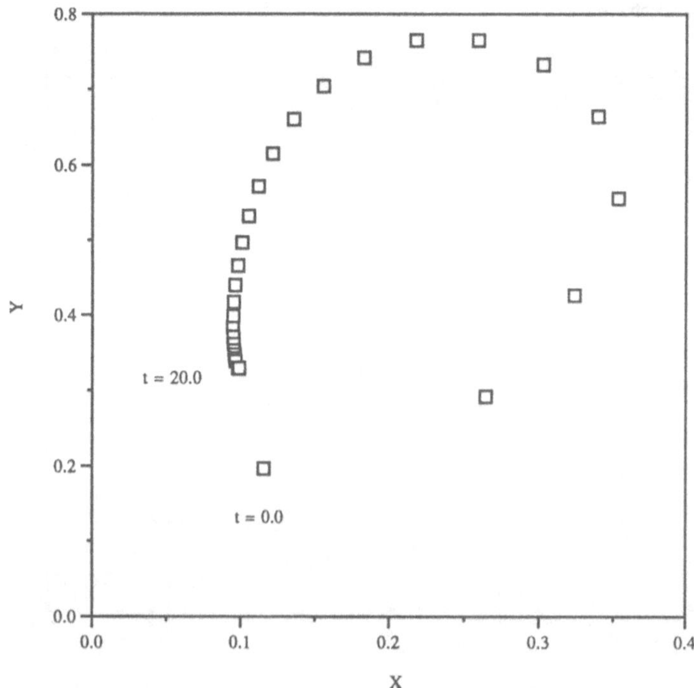

Figure 10. Local stroboscopic map near left boundary (i = 6), for $\beta = 3.0$.

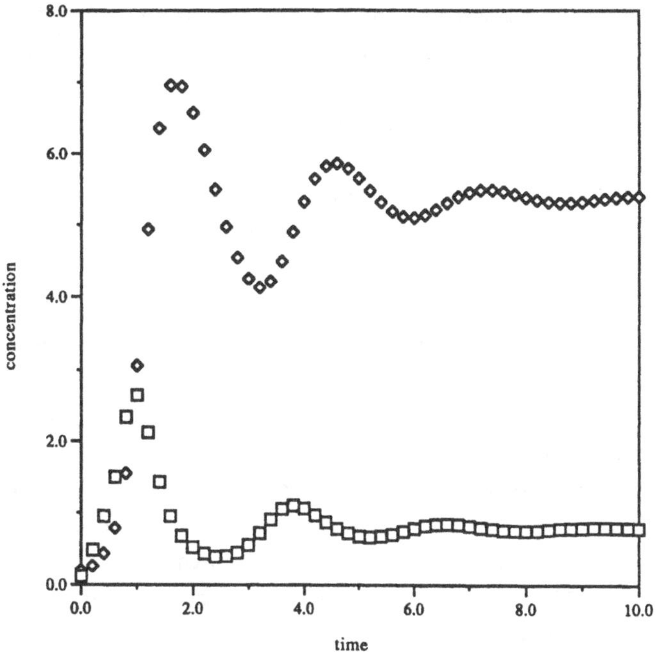

Figure 11. Local concentration near left boundary (i = 6), for $\gamma = 1.5$. □ represents concentration of X and ◇ represents concentration of Y.

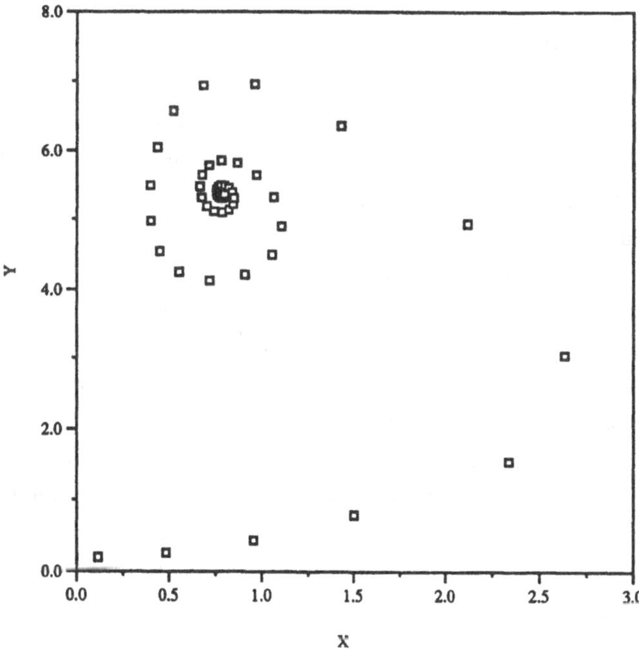

Figure 12. Local stroboscopic map near left boundary (i = 6), for $\gamma = 1.5$.

equilibrium concentrations for species X and Y, and, most remarkably, (3) generation of an infinite chemical wave train from the single stimulus pulse. Only in a very few of these simulations, in which the kinetic parameters were varied, did the model display the characteristic superthreshold behavior one would expect for a stimulus amplitude greater than the threshold amplitude. In particular, the model displayed superthreshold behavior only at values of the kinetic parameter near the normal value of that particular kinetic parameter ($\alpha = 12.0$, $\beta = 1.5$, $\gamma = 3.0$, $\delta = 1.0$, and $v_0 = 0.01$). Besides superthreshold behavior, in a few cases the model exhibited subthreshold behavior, *i.e.*, there was no excursion cycle and the system relaxed to its original equilibrium concentrations of species X and Y. Hence, we find that the threshold amplitude appears to be a fairly sensitive function of the kinetic parameters.

Although subthreshold and the expected superthreshold behavior were observed when the kinetic parameters were varied, it was more common to see the system relax to new equilibrium concentrations of species X and Y. When the system did relax to new equilibrium concentrations of the two species, the system was observed doing so in one of three ways. First, the system could relax directly to the new equilibrium concentrations of species X and Y (*cf.* Figures 7 and 8). Second, the system could display subthreshold behavior but instead of relaxing to the original concentrations of species X and Y the model relaxed to a new set of equilibrium concentrations (*cf.* Figures 9 and 10). These two mechanisms were the most widely observed of the behaviors. Finally, and more interestingly, the system could undergo a partial excursion and then exhibit damped oscillation about its new equilibrium value (*cf.* Figures 11 and 12).

All three of these types of behavior can be explained by examining the reaction diffusion equations and the forcing equations containing the kinetic parameters. At equilibrium, the change in concentration with respect to time of either species (*i.e.*, $\frac{\partial X}{\partial t}$, $\frac{\partial Y}{\partial t}$) equals 0. This can be used along with the values for the kinetic parameters to predict the equilibrium points for each set of kinetic parameters. Thus, by changing the kinetic parameters, the equilibrium concentration(s) (sometimes there are more than one set of equilibrium values) of each species changed and the system relaxed to those new values. The oscillating pattern of behavior might be understood by saying that since the system starts at concentration values for X and Y so far displaced from equilibrium that in their approach to the new equilibrium they shoot past the equilibrium; finally the system stabilizes at the new equilibrium concentrations. A more rigorous analysis of the damped oscillatory approach to equilibrium is beyond the scope of the present paper.

Finally, the most surprising discovery was that under certain conditions a single stimulus pulse could generate an infinite chemical wave train. In contrast to the damped oscillatory behavior just described, it appears that no appreciable decrease in concentration variations occurs over a very long time scale. For almost all the kinetic parameters there was a region of values that would produce an infinite wave train after being stimulated by a single pulse of a superthreshold amplitude (*cf.* Figures 13 and 14).

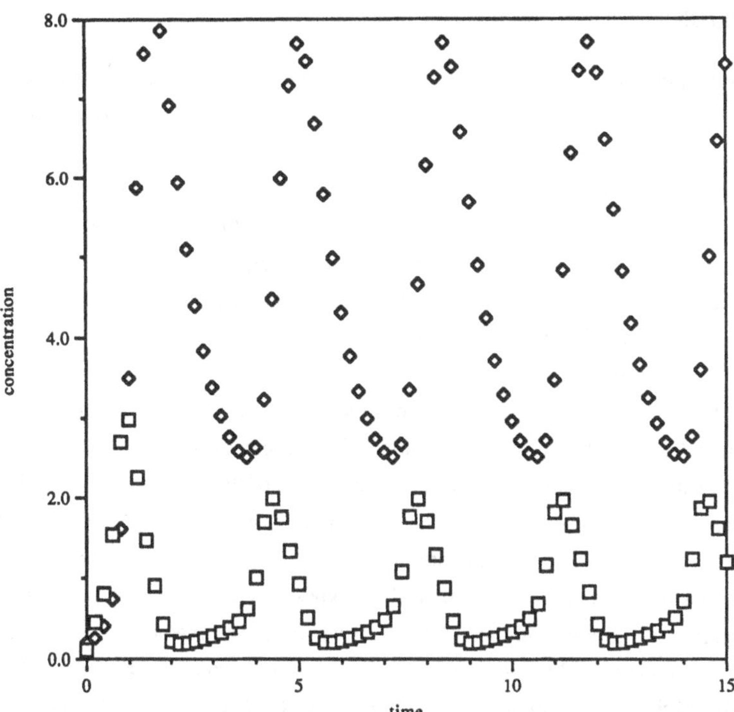

Figure 13. Local concentration near left boundary (i = 6), for $\nu_0 = 0.1$. □ represents concentration of X and ◇ represents concentration of Y.

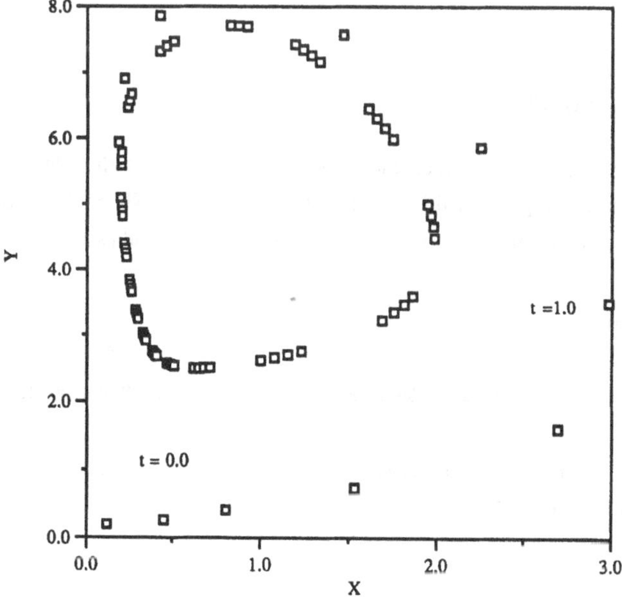

Figure 14. Local stroboscopic map near left boundary (i = 6), for $\nu_0 = 0.1$.

CONCLUSIONS

In conclusion, this project further investigated the reaction diffusion system examined by Sevcikova and Marek.[26,27] Two algorithms, Forward Time Center Space and Crank-Nicholson, were written and used to model the system. The threshold amplitude, A_{th}, for a specific time duration square pulse stimulus was found using each algorithm. To do this, the limit of infinitesimal time steps was based on an assumption of linear errors with step size. Using a least squares fit method, both models converged to the same limit of A_{th} = 0.472; our result is in slight quantitative disagreement with Sevcikova and Marek reported value of $A_{th} = \langle 0.482, 0.4837 \rangle$. Some ambiguity in their application of the pulse may account for the discrepancy.

Next, the model's kinetic parameters were varied to determine how each parameter affected the behavior exhibited by the system. Three types of behaviors were observed, conventional super- and subthreshold behavior, relaxation of the system to new equilibrium concentrations, and the generation of an infinite chemical wave train by a single stimulus pulse. This last phenomenon raises some interesting questions. For instance, if it was physically possible to obtain and maintain these types of kinetic conditions, would it be possible to store information on these "chemical waves"? We intend to investigate both damped and undamped oscillatory behavior further in the context of this kinetic model.

Finally, each of the computational techniques yielded semi-quantitatively the same results; in the limit of infinitesimal time step sizes the agreement became quantitative. Since the FTCS algorithm is computationally simpler (*e.g.*, local data structure *versus* global) and appears to converge at least as quickly (*i.e.*, based on CPU time) as does the Crank-Nicholson method, we conclude that use of the CN method for this type of problem is <u>not</u> warranted.

ACKNOWLEDGMENTS

S.M. gratefully acknowledges the financial assistance of the National Science Foundation for partial support of this research through ASEND/ND-EPSCoR REU and of the National Astronautic and Space Administration for a Space Studies Scholarship.

REFERENCES

1. H.L. Swinney and V.I. Krinsky, eds., "Waves and Patterns in Chemical and Biological Media," MIT, Cambridge (1992).
2. P. Gray and S.K. Scott, "Chemical Oscillations and Instabilities: Non-linear Chemical Kinetics," Clarendon, Oxford (1990).
3. A.T. Winfree, *Science* 175:634 (1972); 181:937 (1973).

4. J.L. Hudson and J.C. Mankin, *J. Chem. Phys.* 74:6171 (1981).

5. K. Tomita and I. Tsuda, *Prog. Theor. Phys.* 64:1138 (1980).

6. H.L. Swinney, *Physica D* 7:3 (1984).

7. R.J. Field and M. Burger, eds., "Oscillations and Traveling Waves in Chemical Systems," Wiley, New York (1985).

8. S.C. Müller, T. Plesser and B. Hess, *Science* 230:661 (1985); *Physica D* 24:87 (1987).

9. A.T. Winfree, "When Time Breaks Down," Princeton University, Princeton (1987).

10. J. Ross, S.C. Müller and C. Vidal, *Science* 240:460 (1988).

11. J.J. Tyson and J.P. Keener, *Physica D* 32:327 (1988).

12. W.Y. Tam, W. Horsthemke, Z. Noszticzius and H. Swinney, *J. Chem. Phys.* 88:3395 (1988).

13. H.L. Swinney and J.P. Golub, *Phys. Today* 31(8):41 (1978).

14. A. Libchaber, C. Larouche and S. Fauve, *J. Phys. Lett.* 43:L211 (1982).

15. K. Ikeda, H. Daido and O. Akimoto, *Phys. Rev. Lett.* 45:709 (1980).

16. M.P. Cox, G. Ertl and R. Imbihl, *Phys. Rev. Lett.* 54:1725 (1985).

17. V.I. Krinsky, *Pharmac. Theor. B* 3:539 (1978).

18. N. Shibata, P. Chen, E.G. Dixon, P.D. Wolf, N.D. Daniely, W.M. Smith and R.D. Ideker, *Am. J. Physiol.* 255:H891 (1988).

19. J.M. Davidenko, P.F. Kent, D.R. Chialvo, D.C. Michaels and J. Jalife, *Proc. Natl. Acad. Sci. (USA)* 87:8785 (1990).

20. J.M. Davidenko, P. Kent and J. Jalife, *Physica D* 49:182 (1991).

21. A.T. Winfree, *Physica D* 49:125 (1991).

22. M. Shibata and J. Bures, *J. Neurophysiol.* 38:158 (1985).

22. M.A. Allessie, F.I.M. Bonke and F.J.C. Schopman, *Circ. Res.* 39:168 (1976); 41:9 (1977).

23. A.L. Wit, S.M. Dillon, J. Coromilas, E.A. Saltman and B. Waldecker, *in:* "Mathematical Approaches to Cardiac Arrythmias," J. Jalife, ed., Ann. NY Acad. Sci., New York (1990).

24. P.C. Newell and F.M. Ross, *J. Gen. Microbiol.* 128:2715 (1982).

25. E.E. Selkov, *Biofizika* 15:1065 (1970).

26. H. Sevcikova and M. Marek, *Physica D* 26:61 (1986).

27. H. Sevcikova and M. Marek, *Physica D* 49:114 (1991).

28. W.H. Press, B.P. Flannery, S.A. Teukolsky and W.T. Vetterling, "Numerical Recipes," Cambridge University, New York (1986).

29 D. Britz, "Digital Simulation in Electrochemistry (Second Edition)," Springer-Verlag, New York (1988).

4. [illegible] Hudson and [illegible] Mankin, *J. Chem. Phys.* 74:6171 (1981).
5. K. Tomita and I. Tsuda, *Prog. Theor. Phys.* 64:1138 (1980).
6. H.L. Swinney, *Physica D* 7:3 (1983).
7. R.J. Field and M. Burger, eds., "Oscillations and Traveling Waves in Chemical Systems," Wiley, New York (1985).
8. J.C. Miller, T. Plesser, [illegible] B. Hess, *Science* 230:[illegible] (1985); *Physica D* [illegible] (1987).
9. A.T. Winfree, "[illegible]," Princeton University Press, Princeton (1987).
10. L. [illegible], [illegible] and [illegible], *Science* [illegible] (1983).
11. [illegible] and [illegible], [illegible] (1979).
12. W. [illegible], W. [illegible], [illegible] and H. Swinney, *J. Chem. Phys.* [illegible] (1986).
13. H.L. Swinney and J.P. Gollub, *Phys. Today* 31(8):41 (1978).
14. [illegible], C. [illegible] and [illegible], *J. Phys. Lett.* [illegible] (1982).
15. [illegible], [illegible] and O. [illegible], *Phys. Rev. Lett.* 45:[illegible] (1980).
16. M.P. [illegible], [illegible] and R. [illegible], *Phys. Rev. Lett.* [illegible] (1983).
17. [illegible], *Nature* [illegible] (1978).
18. [illegible]

MOLECULAR DYNAMICS COMPUTER SIMULATIONS OF CHARGED METAL ELECTRODE-AQUEOUS ELECTROLYTE INTERFACES

Michael R. Philpott and James N. Glosli†

IBM Almaden Research Center
650 Harry Road, San Jose CA 95120-6099
†Lawrence Livermore National Laboratory
Livermore CA 94550

INTRODUCTION

When two different substances are joined material flows across the interface (sometimes almost imperceptibly) until the chemical potentials of the component species are equalized. When the sustances are solid or liquid and some of the chemical species are charged then the interface developes a net electrical polarization due to the formation of an electric double layer. The existence of electric double layers was first recognized by von Helmholtz [1] who studied them in the last century. In many chemical and biological systems the electric double layer exerts a profound effect on function. For example in aqueous electrolyte solution the electric field of a charged object (electrode surface or an ion) is completely shielded by the movement of ions of opposite sign toward the surface until charge balance is achieved. The distribution of ions around charged objects is described simply by the classical theories of Gouy[2-4] and Chapman[5] for flat surfaces and by the theory of Debye and Hückel[6] for spherical ions.

The main goal of this program of study is to give a molecular basis for understanding the structure and dynamics of electric double layers at charged metal-aqueous electrolyte interface. The aim is to unify current separate descriptions of surface adsorption and solution behavior, and ultimately to include a detailed treatment of the surface crystalography and electronic properties of the metal. A key element in this effort is the correct treatment of electrostatic interactions among ions, polar neutrals and their images in the metal. In our work this is achieved with the use of the fast multipole method of Greengard and Rokhlin[7-10]. The fast multipole method was specifically designed for efficient computation of long range coulomb interactions. In our simulations we have used it to in the calculation of all electrical forces, including direct space and image interactions.

Because we calculate the electrostatic interactions accurately, without the use of finite cut-offs, this work also has important implications concerning the way in which ions and water distributions in biological[11] and clay suspensions[12] systems should be calculated.

Four previous publications summarize the work completed prior to that described in this report[13-16]. Through systematic computer simulations we have shown that a relatively simple model suffices to describe the adsorption of halide anions and alkali metal cations on neutral and charged metal surfaces. These calculations show qualitatively many of the features known to occur experimentally at electrochemical interfaces in the thermodynamically stable region of the electric double layer. Most notable are the existence of: highly oriented water layer next to the charged metal, contact adsorption layer of the larger ions, and a diffuse region of strongly hydrated ions. The contact adsorbed ions comprise the compact layer in electrochemical interfaces. In our model, contact adsorbed ions are physisorbed because there is no provision to describe covalent bonds. Typical electric fields found in our calculations correspond to $5x10^9$ V/m, and these arise from image charge densities on the electrode of about $0.1e/nm^2$. One by product of our simulations is to point up some of the limitations in current electrochemical concepts. Consider for example the outer Helmholtz plane which defines the position of closest approach of hydrated ions to the elctrode surface. This plane is shown in numerous textbook and review article illustrations[17, 18]. Our simulations show that the plane really corresponds to a narrow zone where ions of the diffuse layer ions penetrate with increasing difficulty the closer they are to the electrode surface. In this zone some mixing with contact adsorbed species can occur which is not possible in the old picture.

In the last fifteen years there have been many simulations of water and electrolyte solutions near surfaces. Some of these studies have contributed greatly to our understanding of electrochemical interfaces. For completeness some of this work is summarized here. Films of pure water between uncharged dielectric walls[13, 19-21], and charged dielectric walls[13, 14, 22, 23]. Some of this work is noteworthy because of a predicted phase transition[22, 23]. There have been numerous calculations reported for uncharged metal[24-27] walls[28-32], including one for jellium[30] and several for corrugated platinum surfaces[27-29, 31, 32] predicting that water adsorbs weakly at top sites with oxygen down on Pt(111) and Pt(100). There have also been some calculations for electrolyte solutions between uncharged and charged dielectric walls[13, 14, 33, 34] emphasizing spatial distributions and hydration shell structure. There have been studies for electrolytes between uncharged metal walls[27, 35, 36]. The work of Rose and Benjamin[36] is particularly interesting because umbrella sampling was used to calculate the free energy of adsorption. Finally we mention the studies of water between charged metal walls[37], and electrolytes between charged metal walls[37]. In much but not all of the work just summarized, the long range coulomb interactions were treated in an approximate way. The commonest approximation was to cut off all interactions beyond a certain radius like 0.10 nm. Some workers used the Ewald method or a planar modification of the method to compute the sum of long range fields correctly. Curiously many of the other summation methods, like the planewise method of de Wette [38] seem not to have been used at all. The reason why it is important to calculate the long range fields accurately is to capture the macroscopic part. The dipole component of the field is conditionally convergent and the sum must be performed in a manner consistent with the physical boundaries. This is a rather old problem which has a partial mapping onto the problem of calculating the macroscopic electromagnetic field inside a sample of arbitrary shape. Space and time (!) prevent us from pursuing this connection here.

CAPILLARY THERMODYNAMICS

Our present understanding of electric double layers and their importance in surface electrochemistry has its origin in the dropping mercury electrode experiments of Grahame[39, 40]. The basic experiment consists of measuring the radius of a sessile mercury drop as a function of electrode potential and electrolyte composition. One can then use the electrocapillary equations of Lippmann[41] to relate radius of the drop to the interfacial surface tension and so in turn to the charge on the electrode. Thermodynamic arguments are then used deduce the surface excess concentration of adsorbed cations, anions and neutral organics on the electrode surface[17, 18, 39, 40, 42, 43].

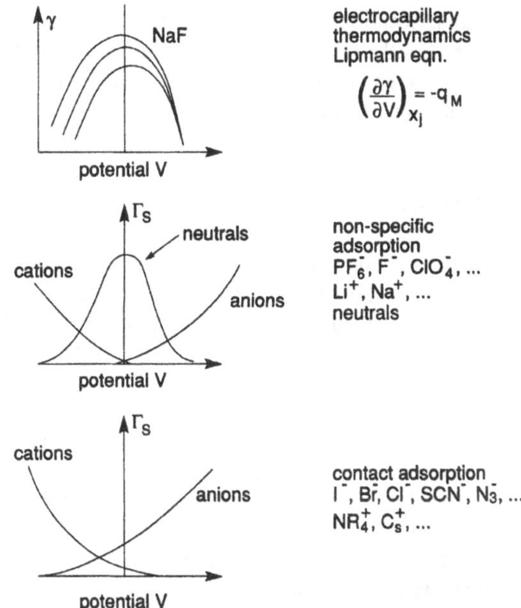

Figure 1. Schematic diagram depicting thermodynamics of adsorption: (a) plot of surface tension γ vs electrode potential V for different electrolyte compositions, (b) surface concentration Γ_S vs potential V for neutrals and strongly hydrated ions, (c) Γ_S vs potential V for contact adsorbing ions.

Figure 1 summarizes the important aspects of thermodynamic studies. For example at the top of Figure 1 the surface tension γ is depicted versus electrode potential for aqueous NaF at different concentrations. The Lippmann equation

$$\left(\frac{\partial\gamma}{\partial V}\right)_{TPN_j} = -q_M \qquad [1]$$

can be used to determine the charge on the mercury drop directly, and in turn the differential capacitance can be calculated. Contact adsorbed ions give a large contribution to the capacitance compared to ions in the diffuse layer, and this has be used to separate their contributions in mixed solutions. The middle Figure 1 shows highly schematic curves for the adsorption of neutral organics and strongly hydrated ions on noble metals. The organics adsorb most strongly when the charge on the electrode is small because the water layer nearest the metal is then least strongly bound and more easily displaced by the organic adsorbate. The final part Figure 1c shows the adsorption isotherms for larger ions, in particular it depicts behaviour like that observed for iodide and the pseudohalides CN^- and thiocyanate SCN^-. For more discussion see the review paper by Anson[44]. For solid electrodes less direct methods are used to obtain the relevant data for a thermodynamic analysis. If the differential capacitance[45, 46]

$$C(V) = \left(\frac{\partial q_M}{\partial V} \right)_{TPN_j} \quad [2]$$

can be measured in a way that inner layer and diffuse layer contributions can be separated then a variety of integration techniques can be used to determine the charge on the electrode. From these quantities the other important thermodynamic variables such as the surface concentrations Γ_S (depicted schematically in middle and bottom of Figure 1) can be obtained using the thermodynamic analysis starting from Eqn (1).

THE TRADITIONAL MODEL

On the basis of experimental methods like differential capacitance, chronocoulombmetry, ellipsometry, and UV-visible spectroscopy, wielded with consummate skill a detailed picture of the electric double layer adjacent to metal surfaces has been devised[18, 47]. We will refer to this picture as the 'traditional model'. The main features of this model are shown schematically in Figure 2 for that part of the double layer close to the electrode surface. Drawings similar to this one can be found in many textbooks and review articles on interfacial electrochemistry[48-50]. The metal is flat and carries charge (negative shown), the aqueous subphase is divided into two parts, called diffuse and compact regions. Additionally next to the charged surface there is a highly oriented layer of water shown in Figure 2 with protons oriented towards the surface. The anion is shown adsorbed in contact (physisorbed) and the strongly hydrated cations are shown no closer than two water molecules. The inner Helmholtz plane (IHP) is defined as the plane through the nuclei of the contact adsorbed anions and a similar plane through cations at their distance of closest approach is called the outer Helmholtz plane (OHP). Beyond the OHP the distribution of ions is assumed to be described by the Gouy-Chapman theory, which in it's simplest form assumes the ions are charged point-like objects and the solvent is a dielectric continuum with appropriate bulk properties. Close to the electrode the 'traditional model' calculates system properties based on static distributions or uses lattice statistics. The diffuse region in this picture[18] starts two solvent molecules from a flat electrode surface and stretches out several nanometers into the bulk electrolyte. The electrostatics and ionic distributions in this diffuse part were first described by the Gouy-Chapman theory[2-5] which predates even the Debye-Hückel[6] model of ionic atmospheres in bulk electrolytes. The ions in the diffuse layer screen the net charge of metal and any ions occupying in the compact part of the double layer. Traditionally the structure of the compact region is thought of as being rather static and resembling a parallel plate capacitor with a gap of atomic separations (0.1 - 0.2 nm). The flat surface model dates from times before the ability to make useable single crystal electrode surfaces, the advent of synchrotron sources and facilities to do surface X-ray analysis. Even so quite a considerable effort has been directed toward developing better model of the metal side. In this regard the work of Halley et al[51-53] and Schmickler[54] is particularly significant. These groups have developed theories based on jellium electrode models of the charged metal surface that attempts to capture important physics causing features in measured capacitance vs. potential curves. Focussing more on the electrolyte subphase Henderson and coworkers[55] have developed a correlation function approach and an analysis that shows that in their very reasonable model there is no sharp division between the diffuse and inner Helmholtz planes. This is in contrast to the ideas conveyed by the pictures found in many textbooks and Figure 2.

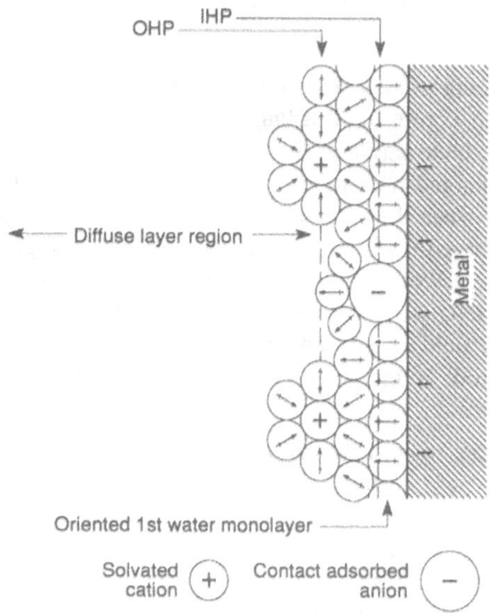

Figure 2. Schematic diagram depicting the 'traditional model' of the electric double layer found in many review articles and introductory texts on electrochemistry.

It must be pointed out that many of molecular dynamics studies performed to date have not recognized the importance of solving the electrostatics problem accurately. Intense long range fields exist at the interface because of the large dipole normal to the surface formed by the ions and their electrical images are all in phase. This problem spills over into other areas where electrostatics is important, for example most studies of water around biological objects are wrong in the way in which the dielectric polarization of water is calculated because long range correlations are lost when interactions are cut off at a finite distance. Workers in this field often resort to using distance dependent dielectric functions to shield ionized groups attached to the object's surface.

The experimental picture of electrochemical surfaces is currently undergoing rapid change due to numerous advanced in situ and ex situ UHV surface science probes of the electrochemical interface. The new synchrotron based X-ray surface crystal structure probes like grazing incidence X-ray scattering (GIXS), X-ray standing wave (XSW), surface extended X-ray absorption fine structure (SEXAFS) techniques have allowed surface geometries to be measured for the first time[56-59]. Recent studies by Hubbard and coworkers[46] reveal changes in charge state of the ion as the surface concentration is increased, changes that have analogies in UHV surface science studies of the metalization of semiconductors These studies show contact adsorption to be a complex process that can evolve into chemisorption at high coverages even if the initial step is physisorption. Local probes like scanning tunneling microscopy (STM) and atomic force microscopy (AFM) give images with local atomic scale features. The next few years will likely see major revisions in our experimental understanding of the less dynamic part of the electrochemical double layers.

MODEL FOR THE IMMERSED ELECTRODE

Consider a system consisting of two electrodes immersed in aqueous electrolyte solution as shown schematically in Figure 3. Reading from left to right there are three regions: the anode on the left, the bulk region at the center, and the cathode region on the right. If the electrodes are uncharged then in all three regions the electrolyte phase would be electrically neutral. Now when the externally applied (battery) potential is altered so that the electrodes become charged (the case shown schematically in Figure 3) the electrolyte responses by screening the electric field and the three regions acquire different net charge. One with excess anions (left) screens the positive charge on the left electrode, a bulk region in which the electric field is zero, and one with an excess of cations (right). When the charge on each electrode is included with the adjacent 'zone' of electrolyte the net charge in each region is zero.

These simple considerations suggest we can try to model an immersed electrode with it's adjacent screening region and do not have to model the whole cell as was done in two previous publications[13, 14]. This approach is useful because it reduces the number of water molecules in the calculation, however it imposes a constraint in the form of charge neutrality and requires us to choose a 'good' boundary to separate the bulk region and immersed region. We can test the boundary by scaling the size of the system to capture some bulk-like behavior. Scaling by a factor of ten suggests that four layers of water is a minimum sized system.

Our immersed electrode model therefore consists of a layer of electrolyte between two walls. The wall on the left carries no charge it is simply a restraining wall and ideally should allow a continuous transition to the bulk electrolyte region. The complete system of electroltye and electrode (always chosen to be on the right hand side in our calculations) is neutral, unless we deliberately choose to stress the system with an uncompensated charge on the right hand metal electrode. In practice we have performed both types of calculation for reasons described in more detail later. Later we will show an important result that water behaves in an uncompensated field as if the excess ionic charge were higher.

Since there is integer charge in the aqueous phase q_{Aq}, and the image charge on the metal q_{im} satisfies the equation $q_{im} + q_{aq} = 0$. In our calculations we also consider cases (to be discussed at length later) where there is additional charge q_{uc} on the electrode with the restriction that $q_T = q_{im} + q_{uc} = ne$ where $n = 0, \pm 1, \pm 2, \dots$.

In Figure 3 if one thinks of the vertical dash line as symbolizing the restraining wall then in reality the vertical line will be very close to the corresponding electrode for a macroscopic sized cell. Integral electrode charge is an essential constraint in this immersed electrode model. We stress again that the main advantage of the model is that only about half the number of water molecules are needed to simulate a system with two metal electrodes. For the restraining barrier we choose a 9-3 potential. The origin for this restraining potential is at 1.862 nm from the image plane of the metal. We refer to this restraining potential surface as the dielectric surface, and as already mentioned it's only function is to limit the extent of the fluid phase and thereby make the calculations more tractable. In all the calculations reported here the simulation cell was a cube with edge 1.862nm. The cube was periodically replicated in the xy directions parallel to the electrode surface plane. Again we mention that we have performed some scaled up calculations on cells with larger edge length by up to a factor 2 and found very similar features to those described for the smaller cells.

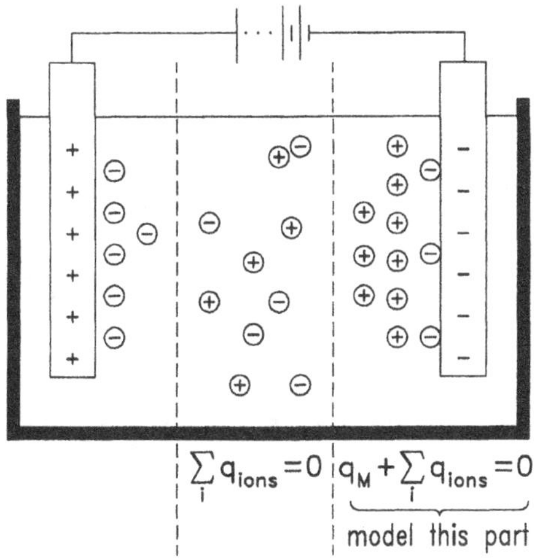

Figure 3. Immersed Electrode Model. Schematic diagram of the electrochemical cell showing the immersed electrode configuration on the right side. Vertical broken lines symbolize the transition region from diffuse layer to bulk electrolyte where the solution is electrically neutral.

MODEL FOR WATER, IONS AND THE METAL SURFACE

In all the calculations reported here we have used the parameters of the Stillinger[60, 61] ST2 water model and the extensive interaction parameter set for alkali metal ions and halide ions developed by Heinzinger and coworkers[62]. Figure 4 shows a schematic of the ST2 water model, a simple ion and the smoothly truncated Lennard-Jones potential. The ST2 water molecule model consists of a central oxygen atom (O_ST2 or O for short) surrounded by two hydrogen atoms (H_ST2 or H for short) and two massless point charges (PC_ST2 or PC for short) in a rigid tetrahedral arrangement (bond angle = $\cos^{-1}(1/\sqrt{3}\,)$). The O-H and O-PC bond lengths were 0.10 nm and 0.08 nm respectively. This small difference in bond lengths means that the water_ST2 model and its electrostatic image (i.e., $q \rightarrow -q$) behave similarly. The only Lennard-Jones 'atom' in ST2 model is the oxygen atom. The hydrogen H and point charges PC interact with their surroundings (i.e., other atoms and surfaces) by Coulomb interactions only. Their charges are q_H= 0.23570$|e|$ and q_{PC}= $-q_H$. The O atom carries no charge. The alkali metal ion and halide ion were treated as non-polarizable Lennard-Jones atoms with central point mass and charge. Figure 4 shows an ion schematically next to the water model. The atom-atom interaction parameters are taken from Heinzinger's review[28, 62].

Next we describe the interaction between water and ions and the metal and restraining wall. The metal was represented by two linearly superimposed potentials. Pauli repulsion and dispersive attractive interactions were modelled by a 9-3 potential, and the interaction of charges with the conduction electrons by a classical image potential. In the calculations described here the image plane and origin plane of the 9-3 potential were coincident. This was tantamount to choosing the image plane and the nuclear plane of the metal surface to be coincident. This was acceptable in our scheme because the Lennard-Jones core parameters σ are all large and the 'thickness' of the repulsive wall is also large (ca. 0.247 nm).

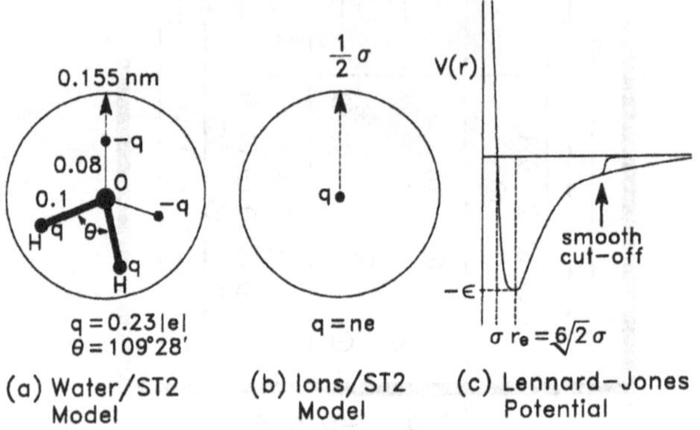

Figure 4. The Ion and Water Models. Schematic diagram summarizing the key features of the model for ST2 water and the ions. Water and ions are treated as non polarizable Lennard-Jones atoms with embedded charges. Shown on the left is a schematic drawing of the Lennard-Jones potential with smooth cut-off.

CALCULATION OF THE ELECTROSTATIC INTERACTIONS

Electrostatic sums are conditionally convergent and great care must be exercised in using cut-offs and different boundary conditions. The Coulomb field computation grows as N^2 (N number of charged particles) unless special measures are adopted. We use the fast multipole method (fmm) devised by Greengard and Roklin[7], in which the cpu time grows as N. The cross over point in efficiency for direct sum versus fmm can be as small as N = 1000 (about 250 ST2 water molecules). The original papers of Greengard should be consulted for details of this clever algorithm. Some discussion of the use of fmm in electrochemical simulations has been given by Glosli and Philpott[15].

TOTAL POTENTIAL ENERGY OF THE SYSTEM

The Coulomb interaction between molecules was represented as sum of $1/r$ interactions between atomic point charges. For the ST2 water model these interactions result in H-bonds that are too strong when the PC and H atoms are close. These interactions were softened for small molecular separation in the way described by Stillinger[60] and Lee et al[19] by introducing a switching function S that modifies specified atomic coulomb interactions at small separations. The short range part of the intermolecular interaction was modeled by Lennard-Jones potential between atom pairs on each molecule. All molecule-molecule Lennard-Jones type interactions were cut-off in a smooth fashion at a molecular separation $R = 0.68$ nm by a truncation function T. The atoms of each molecule also interacted with the surfaces at $z = \pm z_o$ where $z_o = 0.931$ nm. Both surfaces were treated as flat featureless plates with a uniform electric charge density of σ on the metal plate at $+z_o$ if there is an uncompensated charge on the metal, otherwise $\sigma = 0$ This gave rise to a uniform electric field, $E = 4\pi K\sigma$, in the z-direction where K the electrostatic coupling constant had the value 138.936 kJ.nm/(mole.e^2) in the units of this paper. The complete interaction energy U was given by the following formula

$$U = \sum_{\substack{\beta \in A_j \\ i<j}} \left(\left\{ \frac{K q_\alpha q_\beta}{r_{\alpha\beta}} \left[S(R_{ij}, R_L^{ij}, R_U^{ij}) - 1 \right] + 4\varepsilon_{\alpha\beta} \left[\left(\frac{\sigma_{\alpha\beta}}{r_{\alpha\beta}} \right)^{12} - \left(\frac{\sigma_{\alpha\beta}}{r_{\alpha\beta}} \right)^{6} \right] \right\} T(R_{ij}) + \frac{K q_\alpha q_\beta}{r_{\alpha\beta}} \right) \tag{3}$$

$$+ \sum_{\alpha} \left\{ -q_\alpha E z_\alpha + \left(\frac{A_\alpha}{(z_\alpha + z_o)^9} - \frac{B_\alpha}{(z_\alpha + z_o)^3} \right) + \left(\frac{A_\alpha}{(z_\alpha - z_o)^9} - \frac{B_\alpha}{(z_\alpha - z_o)^3} \right) \right\}$$

where i and j were molecular indices, and α and β were atomic indices. The symbol A_i represented the set of all atoms of molecular i. The symbol R_{ij} was the distance between the center of mass of molecules i and j. The symbol $r_{\alpha\beta}$ was the distance between atoms α and β. For small R we followed the practice of modifying the the coulomb energy between ST2 molecules and ions by the switching function $S(R, R_L, R_U)$ given by,

$$S(R, R_L, R_U) = \begin{cases} 0 & R < R_L \\ \dfrac{(R - R_L)^2 (3R_U - 2R - R_L)}{(R_U - R_L)^3} & R_L < R < R_U \\ 1 & R_U < R \end{cases} \tag{4}$$

The values of R_L and R_U were dependent on the types of the molecular species that were interacting. As mentioned above the tails of the Lennard-Jones pair interactions were cut off by the truncation function T. The form of T was given by,

$$T(R) = \begin{cases} 1 & R < R_L^T \\ \left(1 - \left(\dfrac{R - R_L^T}{R_U^T - R_L^T} \right)^m \right)^n & R_L^T < R < R_U^T \\ 0 & R_U^T < R \end{cases} \tag{5}$$

The same truncation function has been applied to all non Coulombic molecular interactions, with R_L^T=0.63 nm and R_U^T=0.68 nm. The integers m and n controlled the smoothness of the truncation function at R_L^T and R_U^T respectively. In this calculation $n = m = 2$ which insured that energy has continous first spatial derivatives.

All the atom-atom and atom-surface interaction parameters are given in Table I. For example we see that the (ε, σ) pairs are (0.3164, 0.3100), (0.1490, 0.2370) and (0.4080, 0.5400) for O_ST2, Li ion and I ion respectively. The units of the well depth ε are kJ/mole and the van der Waals σ is in nm. The usual combining rules were enforced for unlike species, namely: $\varepsilon_{AB} = (\varepsilon_{AA}\varepsilon_{BB})^{1/2}$ and $\sigma_{AB} = \frac{1}{2}(\sigma_{AA} + \sigma_{BB})$. The st2 model switching function (see later) interval ends R_L^{IJ} and R_U^{IJ} both vanish except for st2/st2 water pairs, where $R_L^{ST2,ST2}$=0.20160 nm and $R_U^{ST2,ST2}$=0.31287 nm.

The atom-surface parameters describing interaction with nonconduction electrons were chosen to be the same as those used by Lee at al[19], A=17.447x10^{-6} kJ(nm)^6/mole and B=76.144x10^{-3} kJ(nm)^3/mole for O, I ion and Li ion. The A and B parameters for H_st2 and PC_st2 were set to zero. The potential corresponding to these parameters describe a graphite-like surface. Real metals would have much larger ε ours were deliberately chosen to be small so as to permit coulombic interactions to dominate the physics.

Table I. The interaction parameters (q, ε, σ, A, B) and mass (m) for all atoms used in the simulations, where $q_o = e$, $\varepsilon_o = 1KJ/mole$, $\sigma_o = 1nm$, $A_o = 17.447 \times 10^{-6} KJ \cdot nm^6/mole$, $B_o = 76.144 \times 10^{-3} KJ \cdot nm^3/mole$, and $m_o = 1AMU$. The ε and σ for unlike atom pairs are formed using the combination rules $\varepsilon_{AB} = \sqrt{(\varepsilon_{AA}\varepsilon_{BB})}$ and $\sigma_{AB} = (\sigma_{AA} + \sigma_{BB})/2$.

	q/q_o	$\varepsilon/\varepsilon_o$	σ/σ_o	m/m_o	A/A_o	B/B_o
O_ST2	0.0000	0.316	0.310	16.0	1	1
H_ST2	0.2357	0.000	0.000	1.000	0	0
PC_ST2	– 0.2357	0.000	0.0	0.0	0	0
Li	1.0000	0.149	0.237	6.9	1	1
F	– 1.0000	0.050	0.400	19.0	1	1
Cl	– 1.0000	0.168	0.486	35.5	1	1
Br	– 1.0000	0.270	0.504	79.9	1	1
I	– 1.0000	0.408	0.540	129.9	1	1

In the equations of motion bond lengths and angles were explicitly constrained by a quaternion formulation of the rigid body equations of motion[63]. The equations of motion were expressed as a set of first order differential equations and a fourth order multi-step numerical scheme with a 2 fs time step was used in the integration. At each time step a small scaling correction was made to the quaternions and velocities to correct for global drift. Also the global center of mass velocities in the x and y directions was set to zero at each time step by shifting the molecular translational velocities.

In the analysis of configurations the first 100 ps were used to equilibrate the system and subsequent configurations were used in compute properties like density profiles. There were exceptions where it was evident that the system had not equilibrated. In practice it was found this occurred frequently in three ion systems, as in the case of a cation in the presence of two coadsorbed iodide ions. All the simulations were run to 1000 ps or longer, so that generally 900 or more configurations one ps apart were used to calculate averages. Typical density plots were derived from bining configurations stored every ps and with bin widths of 0.005 nm or larger.

EQUATIONS OF MOTION

All molecules, in this study, were assumed to be a rigid collection of point atoms, so that all bond lengths and bond angles within a molecule were fixed. To evolve a collection of these molecules a quaternion formulation of the rigid body equations of motion was used[63-65] The center of mass position ($\mathbf{R}_i$) and velocity ($\mathbf{V}_i$) was used to describe the translational degrees of freedoms of molecule i. The orientational motion of the molecule was described by the quaternion $\mathbf{q}_i = (q_i^0, q_i^1, q_i^2, q_i^3)$ and the rotational velocity (ω_i^p), as measured in the body frame of the molecule. The one exception to this was for monatomic molecules, in which case the orientational degrees of freedom were not needed.

The discussion of the equations of motion begin by considering the potential energy U. From Eqn 3, it can be seen that the potential energy can be treated as a scalar function of the variables ($\mathbf{R}_1$,..., $\mathbf{R}_N$; $\mathbf{r}_1$,..., $\mathbf{r}_n$), where N is the number of molecules and n is the number of atoms in the entire system.

$$U = U(\mathbf{R}_1,..., \mathbf{R}_N; \mathbf{r}_1,..., \mathbf{r}_n) \tag{6}$$

Of course not all these variables are independent. However for the purposes of the following two definitions they are treated as independent variables.

$$\mathbf{f}_\alpha \equiv -\nabla_{\mathbf{r}_\alpha} U, \qquad \mathbf{F}_i \equiv -\nabla_{\mathbf{R}_i} U. \tag{7}$$

The total force $\mathbf{F}_i^T$ and torque τ_i, can be expressed in terms of $\mathbf{f}_\alpha$ and $\mathbf{F}_i$ as

$$\mathbf{F}_i^T = \mathbf{F}_i + \sum_{\alpha \in A_i} \mathbf{f}_\alpha \qquad \tau_i = \sum_{\alpha \in A_i} (\mathbf{r}_\alpha - \mathbf{R}_i) \times \mathbf{f}_\alpha, \tag{8}$$

The translational motion of the molecule is described by the first order ordinary differential equation,

$$\frac{d\mathbf{R}_i}{dt} = \mathbf{V}_i \qquad M_i \frac{d\mathbf{V}_i}{dt} = \mathbf{F}_i^T. \tag{9}$$

For the rotational motion it is convenient to work in the body axis of the molecule, where the moment of inertia $\hat{I}_i$ is diagonal and time independent. It is useful to define the operator $\hat{Q}_i$

$$\hat{Q}_i = \begin{bmatrix} -q_i^1 & -q_i^2 & -q_i^3 \\ q_i^0 & -q_i^3 & q_i^2 \\ q_i^3 & q_i^0 & -q_i^1 \\ -q_i^2 & q_i^1 & q_i^0 \end{bmatrix} \tag{10}$$

and the body frame rotational force

$$\mathbf{F}_i^R = \tau_i^b - \omega_i^b \times (\hat{I}_i^b \omega_i^b). \qquad [11]$$

Using these quantities above, the following first order ordinary differential equations can be written to describe the rotational motion

$$\frac{d\mathbf{q}_i}{dt} = \hat{Q}_i \omega_i^b \qquad \hat{I}_i^b \frac{d\omega_i^b}{dt} = \mathbf{F}_i^R. \qquad [12]$$

This set of equations conserve the total energy for time independent potentials (U) . A constant temperature ensemble may be simulated by introducing a velocity dependent term in the acceleration terms to constrain the total kinetic energy. The constant temperature equations of motion are written as

$$\frac{d\mathbf{R}_i}{dt} = \mathbf{V}_i, \qquad M_i \frac{d\mathbf{V}_i}{dt} = \mathbf{F}_i^T - \gamma \mathbf{V}_i$$

$$\frac{d\mathbf{q}_i}{dt} = \hat{Q}_i \omega_i^b, \qquad (\hat{I}_i)^b \frac{d\omega_i^b}{dt} = \mathbf{F}_i^R - \gamma \omega_i^b, \qquad [13]$$

$$\gamma = (\sum_i \mathbf{V}_i \mathbf{F}_i^T + \omega_i^b \mathbf{F}_i^b)/(2K)$$

This choice of γ ensures that the total kinetic energy of the systems is constant.

INTRODUCTION TO THE RESULTS

The purpose of this section is to briefly introduce and explain some of the terms and concepts used in the description and discussion of the results. We define inner layer, inner surface field and the external or as we will also refer to it the 'uncompensated' field.

In our calculations the inner layer is defined to be all ions and molecules in contact with the electrode. Recall that the surface is represented by a 9-3 potential that acts on the center of the molecules and ions. This means that in our model the inner most layer, i.e., the first layer, has a simpler structure because steric effects due to dimension perpendicular to the surface are suppressed. In a more realistic model Pauli repulsion would ensure that the center of water was closer to the surface that the center of an iodide ion. One consequence of our model is that some hydrophobic effects at the surface are accentuated, and some steric effects based on volume are decreased in importance.

In equilibrium states the charge on the flat electrode is just the net image charge, which is equal in magnitude but opposite in sign to the sum of the charges on the ions. In this case there is no field in the bulk because the fields from the ions and their images cancel. When the charge on the electrode is only the image charge, then the potential drop occurs between the metal surface and the ions. There is essentially no electric field beyond the ions. We call the field generated by the ions and their images the inner surface field. Since the water molecules have large permanent dipole moments those molecules between the ions and the metal will try to align with the field.

It is useful from the theoretical point of view to stress the system by placing extra uncompensated charge on the electrode. In this case there is a field across the sample, just as occurs when we have water without ions between charged plates. We refer to this field as the external, applied or uncompensated electric field. In general then we can think of the total charge on the electrode as the sum of the image charge and the extra uncompensated charge. In passing we emphasize again that in electrochemical systems it is not possible to have uncompensated charge (i.e., unshielded) on an electrode immersed in electrolyte. However, it is possible in principle to place uncompensated charge on an emersed electrode, so that there are physically realizable states resembling those of the immersed model with uncompensated charge. However the uncompensated charge densities reachable in practice are about 100 times smaller than the ones used in this paper. In our calculations with uncompensated charge we have deliberately chosen charge densities equal in magnitude to the image charge density in order to better understand the effect of the surface field on the inner layer of water.

Finally we state again that the inner surface field is localized close to the metal, in contrast to the field of uncompensated charge which extends across the entire sample. At the surface the total surface field consists of the inner field and any applied field due to uncompensated charge.

POSITIVELY CHARGED METAL IN WATER WITHOUT IONS

Water is the common primary chemical species in all the calculations described in this paper and it is important to know its behaviour in electric fields without the additional structure changing effects of ions. In this simulation 158 ST2 model water molecules were confined between the metal surface at $z_0 = 0.931$ nm and the 'dielectric' wall at $z = -z_0$. Gap between surfaces is $\Delta z = 1.862$ nm = L the edge length of the simulation cell. The water film thickness corresponds to about 4.5 layers of water. For comparison we note that in the calculations of Lee, McCammon and Rossky[19] there were 216 water ST2 molecules equivalent to about six layers. The image plane of the metal was at z = 0.931 nm the right hand surface. Left confining boundary is a dielectric surface (with dielectric constant $\varepsilon=1$) with no image field. The repulsive part of the 9-3 potentials on both sides of the box began at $|z|=0.682$ nm.

There are no ions in this sample. The electric field across the system comes from (uncompensated) positive charge density of $q_T = q_{uc} = +1$ on plate with area L^2, or equivalently an charge density of $0.29e/(nm)^2$. The electric field in vacuum due to this charge is about 5.2 GV/m . Density profiles in the z direction were obtained by averaging 900 configurations over the xy plane. These profiles are shown in Figure 5. The total surface field is just the applied field. Since the field attracts negative charge to the metal the water orients with point charges PC_ST2 directed towards the metal. In comparing Figure 5 profiles with similar ones obtained by reversing the direction of the external field (not shown here) we note there is slightly more structure in the negative field than in zero field (not shown here) or the positive field (Figure 5). The structure is most prominent on the metal side because the O-H bond is longer than PC-O bond and so the proton-proton image interaction being larger pulls the water molecule closer to the metal.

In the PC_ST2 density profile in Figure 5 a distinct shoulder occurs at about 0.7 nm clearly inside the repulsive region of the 9-3 potential. This is permitted because the wall potential acts the center of the water molecule not the component atoms. That the system is strongly polarized as shown by the z component of the dipole density $\rho_p(z)$., and the microscopic charge density $\rho_q(z)$. There is a very pronounced oscillation at the surface in both profiles. Given the existence of oscillations in $\rho_{wat}(z)$ these are not unexpected. What is not clear is whether they result from the superposition of two wall effects. This has been checked with larger N simulations where they occur to the same degree implying the oscillation is intrinsic to one surface. The microscopic charge density has a small value in the interior region of the film and a large oscillation centered at approximately 0.7 nm. The dipole polarization shows large negative deviations at both surfaces, and a uniform positive value across the interior. The behaviour of the dipole and charge densities is that consistent with a high dielectric response material.

All the simple rigid water models display dielectric saturation at high fields because reorientation is the possible relaxation process, and when the molecules are all highly oriented the dielectric is saturated. In the fields in the calculations used by Brodsky et al[22, 23] water was highly oriented and close to saturation, since their system like the one studied here polarizes by molecular reorientation. We have also repeated our calculations in fields up to 1.5 times stronger and find that the interior polarization increases linearly, implying that dielectric saturation has not occurred. However the response of molecules near the surface is different and the dipolar density did not increase linearly with the external field. This is consistent with the presence of more highly oriented and more densely packed water molecules next to the surfaces.

The question of dielectric response of thin films and near surfaces in larger systems is of great interest. If we assume linear response of the electric polarization to an external field then the dielectric constant is given by

$$\frac{\varepsilon_0}{\bar{\varepsilon}} = 1 - \frac{1}{2zq_T} \int_{-z}^{z} \bar{P}_z(z')\, dz' \qquad [14]$$

In the integrand $\overline{P}_z$ is the dipole density parallel to the surface normal (z) averaged over the xy plane. The electrode charge is q_T, and the integration limit is bounded $z < L/2$. The central portion of the dipole density is linear in the applied field, and appears to provide a convenient 'quick' and unambiguous way to calculate the dielectric 'constant' $\overline{\varepsilon}_{zz}$. However in practise this is not easily realized because to check convergence one must evaluate the integral for very large system to avoid interference from the walls.

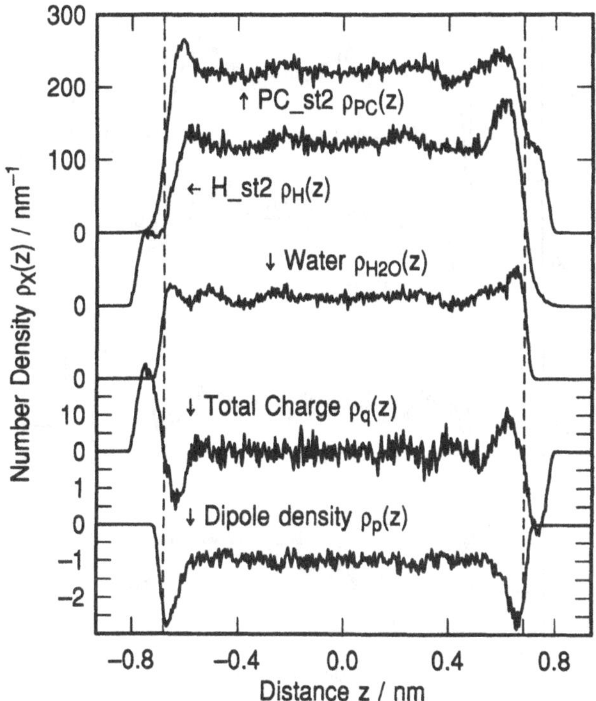

Figure 5. Water without any ions between charged plates with $q_T = q_{uc} = +1$. Component density profile plots for 158 ST2 water molecules adjacent to charged metal electrode on right side. The total electric charge density on the electrode is one positive electron on the simulation plate of area L^2, where L = 1.862 nm. 0.29 e/(nm)^2 The vacuum electric field is approximately 5 GV/m.

FLUORIDE SOLUTIONS. STRONG ELECTRIC FIELD EFFECTS

In this section we discuss several simulations performed with one and two fluorides in the cell (effective ionic concentrations of 0.35 M and 0.70 M respectively) containing 157 and 156 water molecules respectively. We will show that at the higher concentration there is more dense oriented layer of surface water and that the inner surface electric field drives the effect.

Figure 6 shows some representative results for one fluoride ion in the form of density profiles for the fluoride ion and the atomic components of water. In this calculation $q_T = q_{im} = +1$ and $q_{uc} = 0$. The surface field arises from the negatively charged fluoride ion

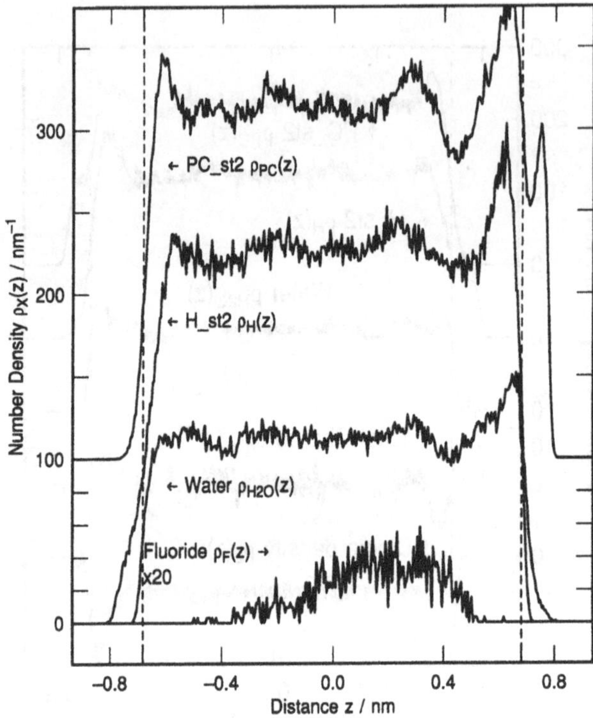

Figure 6. Adsorption of fluoride anions. Density profiles for one fluoride F^- and 157 ST2 water molecules between a metal electrode and the dielectric boundary. Charge on the metal $q_T = q_{im} = +1$. Image plane at z = 0.931 nm . Repulsive portion of the wall potentials begin at |z| = 0.682 nm. Note oriented water layer forming near the metal (rhs) with point charge (PC top curve) pointing at metal.

and its image. Because the ion is distributed uniformly in the xy plane the surface electric field is similar to that inside a capacitor (with a 1.4 nm gap). Features in the water density profiles near the metal surface resemble those already seen for water without ions in Figure 5.

Features on the left at the dielectric boundary do not compare well, in fact they resemble water in zero field (not shown here). These results show that in the single ion case there is no field at the dielectric and a high field at the metal surface. This indicates that the inner field felt by surface waters is similar to the externally applied field in the absence of ions. Closer comparison shows that at the metal side the water peaks are uniformly stronger. In particular the PC_ST2 density profile shows a distinct peak near 0.75 nm, to be compared with the shoulder at 0.7 nm in Figure 5. The extra height and structure may be due to water in the solvation sheath of the fluoride ion at its position of closest approach. Recall that the fluoride ion is strongly hydrated and shows no propensity to contact adsorb. Fluoride remains hydrated even at the point of closest approach (ca. 0.55 nm) to the metal. However at this distance the solvation shell is splayed against the electrode. The distribution is broad (-0.4 nm to 0.5 nm, peak near 0.2 nm) covering almost all accessible configuration space consistent with the ion being strongly hydrated and a permanent resident of the diffuse layer. Note also that the fluoride distribution is diffuse across the entire water film, and there being no hint of a sharp plane or barrier (equivalent to an outer Helmholtz plane OHP in the traditional model) across which the ion is prevented from passing. In these simulations the fluoride behaves more like a strongly hydrated ion in the Grahame model of the electric double layer. Figure 7 shows density profiles for the case of two fluoride ions in the simulation cell with 156 water molecules. The total microscopic charge density is also shown. Charge on the metal $q_T = q_{im} = +2$. The density profile for two fluorides resembles that of a single fluoride except it is shifted further from the metal (range is from -0.5 to 0.5 nm with the peak near 0.0 nm). The electric field due to metal charge and ions is zero near the dielectric boundary as seen by comparing water component profiles with previous two figures. The inner surface field has a much higher value as judged by sharper more intense peaks near the metal surface. The sharp water peak near 0.65 nm has twice the interior density value. Some of this water has one PC pointing at the electrode. Also note that the fluoride density profile is shifted away from the metal surface, consistent with the notion that the compacted layer of surface water hinders the approach of strongly hydrated ions. This effect is consistent with the concept of the OHP being approximately two water molecules distant from the electrode. The position of the first H_ST2 peak, and the near surface oscillation in microscopic charge confirm the overall picture as described.

Next we show the importance of the long range interfacial dipole field by repeating the calculation for one fluoride shown in Figure 6 with an uncompensated field equivalent to a charge of one positive electron on the plate area L^2. The charge on the metal is given by $q_T = q_{im} + q_{uc} = +2$, with $q_{im} = q_{uc} = +1$. Figure 8 shows the density profiles for one fluoride F⁻ and 157 ST2 water molecules between a charged metal electrode and a dielectric boundary. Note the similarities in the density profiles for water and its components in the range z > 0 nm in Figures 7 and 8. In this region the density profile almost

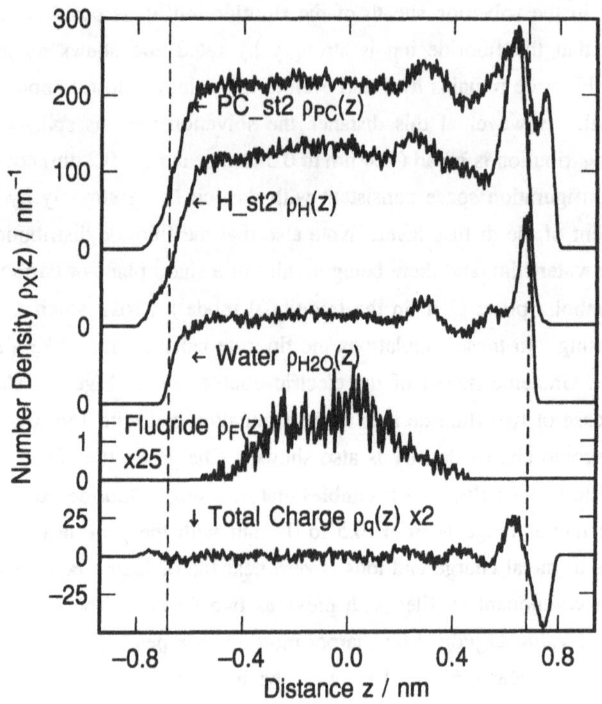

Figure 7. Adsorption of anions continued. Density profiles for two fluoride F⁻ and 156 ST2 water molecules between a metal electrode and the dielectric boundary. Charge on the metal $q_T = q_{im} = +2$. Image plane at z = 0.931 nm. Repulsive portion of the wall potentials begin at |z| = 0.682 nm. Note the high density oriented water peak near 0.65 nm and the shift in the fluoride distribution away from the metal compared to Figure 6.

superimpose. Since we have shown that the water profiles are sensitive to the electric field it suggests that the inner fields are almost the same in the two cases. Furthermore these calculations show that it is the field in the inner layer and not steric effects due to ions that drives the structural changes.

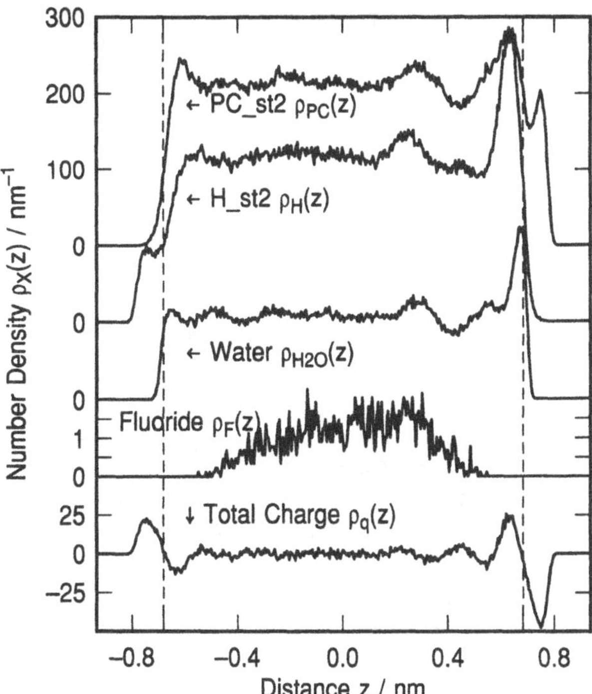

Figure 8. Adsorption of anions continued. Surface electric field effects. The charge on the metal is given by $q_T = q_{im} + q_{uc} = +2$, with $q_{im} = q_{uc} = +1$. Density profiles for one fluoride F^- and 157 ST2 water molecules between a charged metal electrode and a dielectric boundary. Note similarity of water and component density profiles in Figure 7 for z > 0 nm. Image plane at z = 0.931 nm. Repulsive portion of the wall potentials begin at |z| = 0.682 nm.

ELECTRIC FIELD EFFECTS. CASE STUDY OF SODIUM FLUORIDE

In this section we consider the effect of interchanging anions and cations with similar strengths of hydration. We have chosen NaF solutions to explore this case, and have chosen to work also with an uncompensated field ($q_{uc} = \pm 1$) to mimic higher effective concentrations than actually used in the calculation.

Figure 9 shows the density profiles for a simulation with one fluoride F^-, two sodium ions Na^+ and 155 ST2 water molecules between a charged metal electrode and a dielectric boundary. The charge on the metal was $q_T = q_{im} + q_{uc} = -2$, with $q_{im} = q_{uc} = -1$. Note that even though the field has opposite sign there is a superficial similarity of water density profile with those shown in Figure 7 and 8.

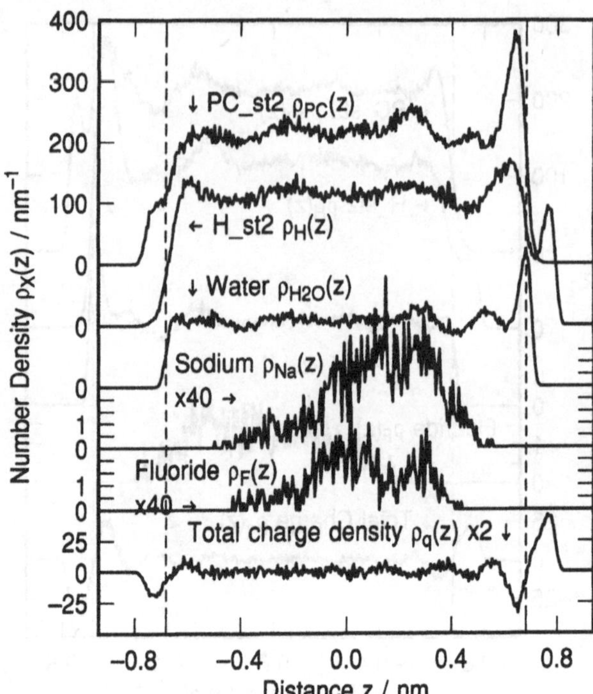

Figure 9. Approximate equivalence interchanged anions and cations, part 1. The charge on the metal was $q_T = q_{im} + q_{uc} = -2$, with $q_{im} = q_{uc} = -1$. Note similarity of water and component density profiles in Figure 7. Density profiles for one fluoride F^-, two sodium ions Na^+ and 155 ST2 water molecules between a charged metal electrode and a dielectric boundary. Image plane at z = 0.931 nm . Repulsive portion of the wall potentials begin at |z| = 0.682 nm.

Figure 10 displays the profiles for two fluoride F^-, one sodium ions Na^+ and 155 ST2 water molecules The charge on the metal was now positive $q_T = q_{im} + q_{uc} = 2$, with $q_{im} = q_{uc} = +1$. Note similarity of water and component density profiles in Figure 7. Note also that the density profiles for Na and F in Figures 9 and 10 when interchanged are almost the same. These calculation extend to mixed electrolytes the conclusion reached in the last section that the water structure in the inner layer is the result of surface electric fields.

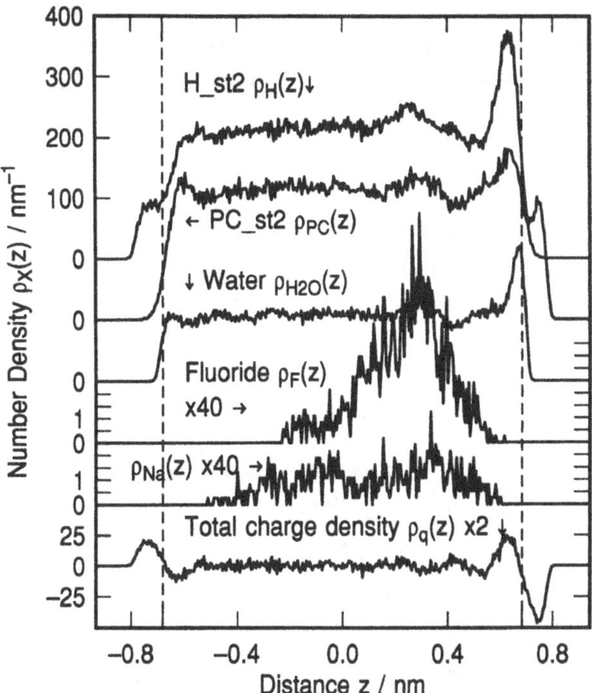

Figure 10. Approximate equivalence interchanged anions and cations, part 2. Note similarity of water and component density profiles in Figure 7. The charge on the metal was positive $q_T = q_{im} + q_{uc} = 2$, with $q_{im} = q_{uc} = +1$. Density profiles for two fluoride F^-, one sodium ions Na^+ and 155 ST2 water molecules between a charged metal electrode and a dielectric boundary. Image plane at z = 0.931 nm . Repulsive portion of the wall potentials begin at |z| = 0.682 nm.

SMALL CATION COADSORBED WITH IODIDE IONS

In this set of simulations we explore another important aspect of adsorption on metal electrodes, namely the ability of strong contact adsorbers like iodide ions I^- to adsorb on positively charged electrodes in sufficient excess to change the sign of the charge at the interface as observed by an ion in the diffuse layer. In this case cations are attracted out of the diffuse layer region to compensate the excess negative ion charge at the interface. The lithium ion was chosen as cation. We have performed these calculations without and with uncompensated charge.

In the first case the charge on the metal was $q_T = q_{im} = +1$, and in the second case $q_T = q_{im} + q_{uc} = 2$. Figure 11 displays the density profiles for all components of the system and the microscopic charge density. The simulation time was 1000 ps with the first 100 ps discarded for equilibration, and then configurations were stored every 1.0 ps.

Note that both iodide ions adsorbed in one sharply peaked distribution mostly inside the onset of the repulsive wall region, but with a small tail out to smaller z positions. The iodide distribution in zero field is characteristically different from the case when an attractive field exists there being a longer tail into the electrolyte indicating a more weakly bound state. The single Li ion occupies a (possibly weakly bimodal) diffuse-like distribution between -0.6 nm and 0.6 nm. The water profiles hint that the inner surface electric field across the first water layer is weak because the iodide charge is so close to the metal, i.e., the waters are outside the main part of the capacitor is this model. Additionally the large size of the iodides tends to exclude some water from the surface.

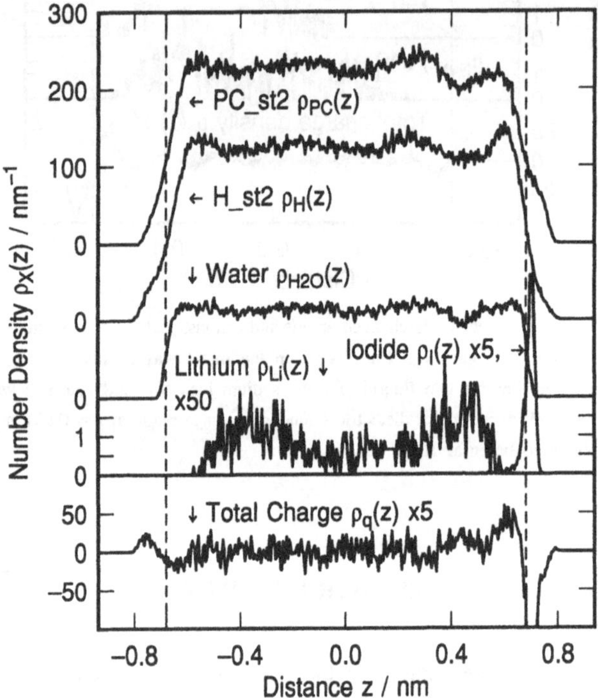

Figure 11. Coadsorption of ions. The charge on the metal was $q_T = q_{im} = +1$. Interface becomes effectively negatively charged after the adsorption of two iodide ions, thereby attracting the positive Li ion. Density profiles for two I^-, one lithium cation Li^+ and 155 ST2 water molecules using the immersed electrode model.

In the second case the charge on the metal was $q_T = q_{im} + q_{uc} = +2$ due to the presence of the extra charge $q_{uc} = +1$. The net charge on the electrode repells the lithium ion. Figure 12 displays the density profiles for all components of the system and the charge density. The simulation time was 1000 ps with the first 100 ps discarded for equilibration, configurations were stored every 0.5 ps. Note that both iodide ions were adsorbed in a sharply peaked distribution that resembles the single adsorbed iodide distribution calculated by Glosli and Philpott[16]. The single Li ion occupies diffuse-like distribution between -0.6 nm and 0.3 nm and peaks around -0.1 nm. The region between 0.0 and 0.3 nm defines a region of reduced probability of penetrating into inner layer and contacting either the wall or the adsorbed iodide ions.

It is noticeable that the water structure near the metal surface though less well defined than in the case of adsorbed fluoride (see Figures 7, 8 or 10) shows more structure than in Figure 11. Overall the structure resembles water in the field of a positively charged electrode as shown in Figure 5. Some artifacts are expected due to some surface PC reaching the high field region between the iodide layer and the metal.

A final word of caution. The statistics for lithium ions in the simulations described in this section are not as good as in previous calculations. The presence of two iodides on the surface creates a rough surface and the calculations should be run several nanoseconds to permit the positive ion to explore all configuration space.

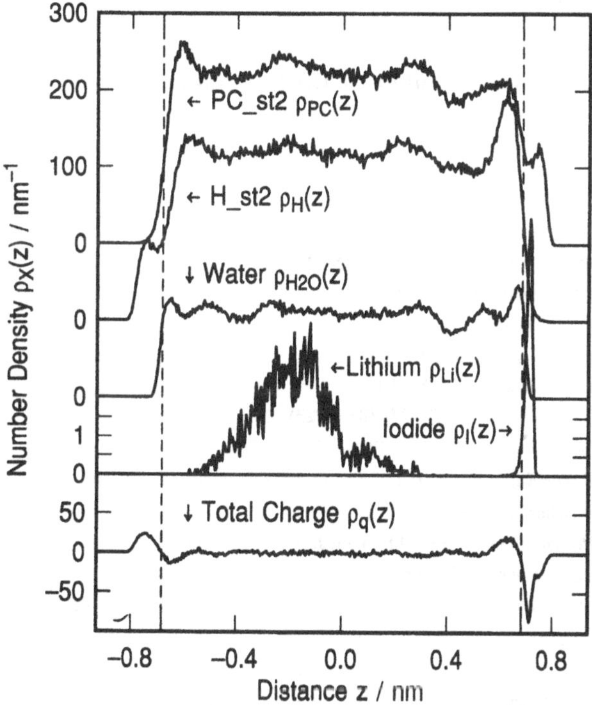

Figure 12. Coadsorption of ions continued. Density profiles for two I^-, one Li^+ and 155 ST2 water molecules next to immersed electrode. The charge on the metal was $q_T = q_{im} + q_{uc} = +2$.

CONCLUSIONS

In this paper we have shown how a simple model suffices to mimic many phenomena familiar from experiments on electric double layers at the electrolyte-metal interface. There was emphasis on electrostatic interactions and understanding the role played by the inner surface field in driving structural changes in the surface water layer. This is expected to be important for all polar systems. A key element of the calculations was the use of the fast multipole method to accurately and efficiently calculate coulomb interactions so that macroscopic electric fields were computed correctly. Among the phenomena studied were: an oriented boundary layer of water at the electrode when it is charged, penetration of nominally diffuse layer species like hydrated fluoride into the inner layer, and attraction of cations to a positively charged electrode induced by contact adsorption of large ions like iodide.

Finally we point out that the techniques described in this report with minor modifications can be extended to more complex systems, for example: microelectrodes, polymer coated electrodes, biological membranes, globular protein surfaces, and clay surfaces.

ACKNOWLEDGEMENTS

This research was supported in part by the Office of Naval Research.

REFERENCES

1 H. L. von Helmholtz, Ann. Physik **89**, 211 (1853).
2 G. Gouy, Ann. chim. phys. **8**, 291 (1906).
3 G. Gouy, J. Physique **9**, 457 (1910).
4 G. Gouy, Ann. phys. **7**, 129 (1917).
5 D. L. Chapman, Phil. Mag. **25**, 508 (1913).
6 P. Debye and E. Hückel, Physik. Z. **24**, 185 (1923).
7 L. Greengard and V. Rokhlin, J. Comp. Phys. **73**, 325-348 (1987).
8 L. F. Greengard, *The Rapid Evaluation of Potential Fields in Particle Systems*. (The MIT Press, Cambridge, Massachusetts, 1987).
9 J. Carrier, L. Greengard, and V. Rokhlin, Siam J. Sci. Stat. Comput. **9**, 669 (1988).
10 L. Greengard and V. Rokhlin, Chemica Scripta **29A**, 139-144 (1989).
11 R. B. Gennis, *Biomembranes. Molecular Structure and Function*. (Springer-Verlag, New York, 1989), pp. 235-269.
12 H. van Olphen, *Clay Colloid Chemistry* (Kreiger, Malabar Florida, 1991).
13 J. N. Glosli and M. R. Philpott, J. Chem. Phys. **96**, 6962-6969 (1992).
14 J. N. Glosli and M. R. Philpott, J. Chem. Phys. **98**, 9995-10008 (1993).
15 J. N. Glosli and M. R. Philpott, Electrochem. Soc. Symposium Proc. **93-5**, 80-90 (1993).
16 J. N. Glosli and M. R. Philpott, Electrochem. Soc. Symposium Proc. **93-5**, 90-103 (1993).

17 J. O. Bockris and A. K. Reddy, *Modern Electrochemistry, Vol.1* (Plenum Press, New York, 1973).
18 J. O. Bockris and A. K. Reddy, *Modern Electrochemistry, Vol.2* (Plenum Press, New York, 1973).
19 C. Y. Lee, J. A. McCammon, and P. J. Rossky, J. Chem. Phys. **80**, 4448-4455 (1984).
20 J. P. Valleau and A. A. Gardner, J. Chem. Phys. **86**, 4162-4170 (1987).
21 Y. J. Rhee, J. W. Halley, J. Hautman, and A. Rahman, Phys. Rev. B **40**, 36-42 (1989).
22 A. M. Brodsky, M. Watanabe, and W. P. Reinhardt, Electrochimica Acta **36**, 1695-1697 (1991).
23 M. Watanabe, A. M. Brodsky, and W. P. Reinhardt, J. Phys. Chem. **95**, 4593 (1991).
24 N. Parsonage and D. Nicholson, J. Chem. Soc. Faraday Trans. 2 **82**, 1521-1535 (1986).
25 N. Parsonage and D. Nicholson, J. Chem. Soc. Faraday Trans. 2 **83**, 663 - 673 (1987).
26 A. A. Gardner and J. P. Valleau, J. Chem. Phys. **86**, 4171-4176 (1987).
27 E. Spohr and K. Heinzinger, Electrochimica Acta **33**, 1211-1222 (1988).
28 E. Spohr and K. Heinzinger, Ber. Bunsenges. Phys. Chem. **92**, 1358-1363 (1988).
29 K. Heinzinger and E. Spohr, Electrochimica Acta **34**, 1849-1856 (1989).
30 J. Hautman, J. W. Halley, and Y. Rhee, J. Chem. Phys. **91**, 467-472 (1989).
31 K. Foster, K. Raghavan, and M. Berkowitz, Chem. Phys. Lett. **162**, 32-388 (1989).
32 K. Raghavan, K. Foster, K. Motakabbir, and M. Berkowitz, J. Chem. Phys. **94**, 2110-2117 (1991).
33 R. Kjellander and S. Marcelja, Chemica Scripta **25**, 73-80 (1985).
34 E. Spohr and K. Heinzinger, J. Chem. Phys. **84**, 2304-2309 (1986).
35 J. Seitz-Beywl, M. Poxleitner, and K. Heinzinger, Z. Naturforsch. **46A**, 876 (1991).
36 D. A. Rose and I. Benjamin, J. Chem. Phys. **95**, 6856-6865 (1991).
37 K. Heinzinger, Pure Appl. Chem. **63**, 1733-1742 (1991).
38 G. E. Schacher and F. W. de Wette, Phys. Rev. **136A**, 78-91 (1965).
39 D. C. Grahame., Chem. Rev. **41**, 441 (1947).
40 D. C. Grahame., J. Amer. Chem. Soc. **79**, 2093 (1957).
41 G. Lippmann, Ann. Chim. Phys. (Paris) **5**, 494 (1875).
42 H. D. Hurawitz, J. Electroanal. Chem. **10**, 35 (1965).
43 E. Dutkiewicz and R. Parsons, J. Electroanal. Chem. **11**, 100 (1966).
44 F. C. Anson, Accts Chem. Res. **8**, 400-407 (1975).
45 R. Parsons, Chem. Rev. **90**, 813-826 (1990).
46 A. T. Hubbard, Chem. Rev. **88**, 633 - 656 (1988).
47 J. O. Bockris and A. Gonzalez-Martin, *Spectroscopic and Diffraction Techniques in Interfacial Electrochemistry, NATO ASI Series C* (Kluwer,Dordrecht,Holland, 1990), pp. 1-54.
48 J. W. Albery, *Electrode Kinetics* (Clarendon Press, Oxford, 1975).
49 L. Antropov, *Theoretical Electrochemistry* (Mir Publishers, Moscow, 1972).
50 J. O. Bockris and S. U. Khan, *Quantum Electrochemistry* (Plenum Press, New York, 1979).
51 J. W. Halley, B. Johnson, D. Price, and M. Schwalm, Phys. Rev. B **31 B**, 7695-7709 (1985).
52 J. W. Halley, Supperlattices and Microstructures **2**, 165-172 (1986).
53 J. W. Halley and D. Price, Phys. Rev. B **35 B**, 9095-9102 (1987).
54 W. Schmickler and D. Henderson, Prog. Surf. Sci. **22**, 323 (1986).
55 D. Henderson, *Trends in Interfacial Electrochemistry* (Reidel, Dordrecht, Holland, 1986), p. 183.
56 O. R. Melroy, M. G. Samant, G. L. Borges, J. G. Gordon, L. Blum, J. H. White, M. J. Albarelli, M. McMillan, and H. D. Abruna, Langmuir **4**, 728 (1988).
57 M. F. Toney, J. G. Gordon, and O. R. Melroy, SPIE Proceedings **1550**, 140 (1991).
58 M. F. Toney, J. N. Howard, J. Richer, G. L. Borges, J. G. Gordon, O. R. Melroy, D. G. Wiesler, D. Yee, and L. B. Sorensen, Nature (1994).
59 T. Tadjeddine, D. Guay, M. Ladouceur, and G. Tourillon, Phys. Rev. Lett. **66**, 2235 - 2238 (1991).
60 F. H. Stillinger and A. Rahman, J. Chem. Phys. **60**, 1545 (1974).
61 O. Steinhauser, Mol. Phys. **45**, 335-348 (1982).

62 K. Heinzinger, *Computer Modelling of Fluids Polymers and Solids*, (Kluwer,Dordrecht, 1990), pp. 357-404.
63 M. P. Allen and D. J. Tildesley, *Computer Simulation of Liquids* (Oxford University Press, Oxford, 1989), pp. 88-90.
64 D. J. Evans, Mol. Phys. **34**, 317-325 (1977).
65 D. J. Evans and S. Murad, Mol. Phys. **34**, 327-331 (1977).

MOLECULAR DYNAMICS COMPUTER SIMULATIONS OF AQUEOUS SOLUTION/PLATINUM INTERFACE

Max L. Berkowitz and Lalith Perera

Department of Chemistry
University of North Carolina
Chapel Hill, NC 27599

INTRODUCTION

A detailed molecular level description of structure and dynamics of water and aqueous solutions next to metallic surfaces is of fundamental importance for electrochemistry, catalysis, and corrosion studies.[1] Structural details about the water monolayer/metal interface can be obtained using different experimental techniques.[2] In the case of bulk aqueous solution/metal interface, where very little of molecular level information is available from experiment, computer simulations can play a very significant role.

The first computer simulation of water bounded by a surface was a Monte Carlo (MC) simulation,[3] where the surface was represented by a hard wall. Actually in this simulation and in most of the other following simulations,[4-8] a water lamina bounded by two surfaces was considered. In this way one can improve the statistics from the simulation, since the averages can be calculated over two interfaces. (Needless to say, the width of water lamina should be large enough to eliminate interference effects from both interfaces). The discontinuous (hard wall) potential describing the water-surface interaction,

Theoretical and Computational Approaches to Interface Phenomena
Edited by H.L. Sellers and J.T. Golab, Plenum Press, New York, 1994

which is by the way a convenient form of a potential to use in MC simulations, was supposed to represent a model for water-hydrocarbon interface. Few more simulations of such interface followed the initial work of Jonsson;[3] in some of the work the MC simulation method was used. Other simulations were performed using the Molecular Dynamics (MD) computer simulation technique. In MD simulations the water-hydrocarbon surface interaction was represented by an analytical function, such as, for example, the "9-3" potential:

$$U(z) = A/z^9 - C/z^3 \tag{1}$$

In equation (1) z is the distance between the surface and the oxygen of the water. Most notable of these simulations was the molecular dynamics (MD) simulation of Lee et.al.[6] In this work it was shown that the surfaces produce water density oscillations that extended at least 1 nm into the liquid. Lee et.al. also observed the significant orientational preference extended at least 0.7 nm into the liquid. But perhaps the most important contribution of the work was the structural interpretation of the result given by the authors. They argued that the inert wall displayed the tendency to organize adjacent water into an ice-I structure with its c axis perpendicular to the wall, since this structure allowed the water molecules to maintain the maximum number of hydrogen bonds. To create such a structure, one of the hydrogens of the water molecules adjacent to the wall had to point toward the wall. The existence of an orientational ordering of water next to the wall seems to be independent on the model of water used in the simulations. For example, Lee et. al. used the ST2[9] model of water, while Valleau and Gardner[10] obtained same qualitative results for TIPS2[11] water.

The first simulation of water-metal interface was performed by Parsonage and Nicholson.[7,8] To represent the water-surface interaction they added to the "9-3" potential, like the one given in equation (1), the electrostatic image potential. That means that the wall was considered to be "classical" metallic, i.e. the metal had an infinite dielectric constant. A similar model was considered by Gardner and Valleau,[12] who performed a MC simulation with TIPS2 water next to the "classical" metallic hard wall. As Gardner and Valleau pointed out the presence of the image forces greatly enhanced the tendency (already observed for water at the inert wall) to point one of the hydrogens toward the walls. This tendency contradicts experimental observations, which show that water tends to situate with its oxygen atom located next to the transition and noble metal surfaces, while the water hydrogens are pointing away.[2] Theoretical calculations on bonding of water to transition metal clusters support the same conclusion.[13,14] Based on one of such calculations of water on Pt cluster,[13] Spohr and Heinzinger developed an interaction potential for water-platinum atom.[15] Although obtained from a cluster calculation, nevertheless the Spohr and Heinzinger potential correctly predicts the adsorption of water on the top site of the Pt metal surface and exhibits a minimum for orientations with the oxygen atom closer to the surface than the hydrogen atoms. It also predicts the correct order of magnitude of the adsorption energy (~40 kJ/mol) and the adsorption distance. Using this potential, Spohr performed a molecular dynamics simulation of a slab of water between two Pt (100) surfaces.[16] In Spohr's simulation the Pt surfaces were explicitly represented by 550 Pt atoms and the dynamics of these atoms was included in the calculation. To avoid the calculation of the dynamics of Pt, (which is later disregarded in the analysis) we have developed a water-platinum surface potential,[17,18] which takes into account the symmetry of Pt surface and its corrugation. This potential allowed us to concentrate on the motion of water molecules only, replacing all Pt atoms by an effective external field. In addition we have used a rigid water model in our simulation while Spohr used a flexible model. Comparison of structural data from our simulation with Spohr's simulation of water at the Pt (100) interface displayed a very good agreement between these two simulations. The advantage of our approach is that it results in the substantial saving of the computational time and therefore permits to perform simulations on larger size systems and over longer time periods. These kind of simulations are necessary for the study of aqueous ionic solutions.

METHODS

Potential Energy Surface

From the outset we should make it very clear that our results depend on the quality of the potentials used in the simulations. Among the four potentials that we need for our simulations (water-water, water-surface, water-ion and ion-surface potentials) the best known to us are water-water potentials, since a large number of investigations were devoted to this subject. In our simulations we used the SPC/E model of water[19] to represent the water-water interactions, since it is known to provide a good description of bulk properties of water.[19] The water-ion potentials are also considered to be known rather well, (for some ions better than for others) although many unresolved questions and discrepancies between the experimental data and simulations still exist.[20] In our simulations we represented the ion-water interaction by a Lennard-Jones plus Coulomb terms, i.e. we used the following form for this interaction:

$$U_{wi} = (A/r^{12} - C/r^{6}) + \sum_j q_i q_j / r_{ij} \qquad (2)$$

where r is the distance between the ion center and the oxygen site on the water,

q_i the charge of the ion, q_j is the charge on the site j of water molecule, r_{ij} is the distance between the ion center and the site j of the water. The water-surface and especially the ion-surface potentials are the least known potentials. Due to the size of the problem, the calculation of these potentials using ab-initio methods is a very difficult task.[13,14,21-24] The potentials we used to represent the ion-surface and water-surface interactions were deduced from the calculations done on water-small Pt cluster[13] and ion-small Pt cluster[24] complexes. To describe the ion-Pt interaction we used the same potentials as the ones developed by Seitz-Beywl et.al.[24] The interaction of a water molecule with the Pt surface was represented in our simulations by the sum of the interactions of individual atoms in water and the surface, i.e.

$$V_{H_2O,Pt} = V_{H,Pt} + V_{H,Pt} + V_{O,Pt} \tag{3}$$

where the oxygen-Pt and hydrogen-Pt interactions are given by the following expressions:

$$V_{O,Pt} = f_0(z) + f_1(z)Q_1(x,y) + f_2Q_2(x,y) \tag{4a}$$

$$V_{H,Pt} = g_0(z) + g_1(z)Q_1(x,y) + g_2(z)Q_2(x,y) \tag{4b}$$

Here z denotes the distance along the normal to the surface.

The functional form of equations (4) is very general and can describe the interaction of water with any face of the Pt surface. To specify the interaction of water with the Pt(100) surface the following expressions for Q_1 and Q_2 are used:

$$Q_1(x,y) = \cos(2\pi x/a)\cos(2\pi y/a) \tag{5a}$$

$$Q_2(x,y) = \cos(4\pi x/a) + \cos(4\pi y/a) \tag{5b}$$

To describe the water/Pt(111) surface interaction the expressions for the Q functions have the form:

$$Q_1(x,y) = \cos(2\pi s_1) + \cos(2\pi s_2) + \cos(2\pi s_3) \tag{6a}$$

$$Q_2(x,y) = \sin(2\pi s_1) + \sin(2\pi s_2) - \sin(2\pi s_3) \quad (6b)$$

and

$$s_1 = (\sqrt{2}/a)(x - y/\sqrt{3}) \quad (7a)$$

$$s_2 = 2\sqrt{2}y/\sqrt{3}a \quad (7b)$$

$$s_3 = s_1 + s_2 \quad (7c)$$

In equations (5) and (7), a=0.392 nm, which is the Pt lattice constant. The functional form of the functions $f_0(z)$, $g_0(z)$, $f_1(z)$, $g_1(z)$ and $f_2(z)$, $g_2(z)$ and the potential parameters used in the simulations are described elsewhere.[17,18,25,26] Our potential functions, obtained through the fit to the semi-empirical calculations performed on clusters, predict the adsorption of water molecules through the oxygen on the a-top sites of the lattice. Such behavior is in agreement with the experimental conclusions.[2]

Molecular Dynamics Simulations

Three simulations are considered here: two simulations of the water/Pt interface and one simulation of the water/Pt interface with added ions. The first water/Pt simulation was performed on 512 water molecules enclosed between two Pt(100) surfaces, while the second water/Pt simulation considered 1298 water molecules enclosed between two Pt(111) surfaces. In the third simulation we considered 512 water molecules and two ions (Li^+ and I^-) enclosed between two Pt(100) walls. The metallic walls in our simulations were parallel to the xy plane. The molecular dynamics cell in the first simulation and in the third simulation (Pt(100)) was a parallelepiped with a square base, while a hexagon was used as the base of the cell for the second simulation (Pt(111)). The shape of the cell was dictated by the symmetry of the problem and the size was chosen to produce a bulk density for water in the middle of the cell. Since the interatomic distance between Pt atoms (~0.277 nm) is commensurate to the distance between water molecules in ice and the hexagonal structure of the lattice (for Pt(111) case) is capable of promoting ice-like structure in water, we had to find a unit cell which will not obstruct a possible creation of such ice-like structures at the interfaces. Therefore, we have chosen a hexagonal prism as our molecular dynamics box with the hexagon as its base. Periodic boundary conditions were used in x and y directions and the contribution of long-range forces was not included in the simulations. The potentials were constructed in such a way that one can think that the centers of the first row of Pt atoms are located at the distances $|z| = 2.11$ nm in the first and third simulations and $|z| = 1.5$ nm in the second simulation. The temperature of the simulations was 300 K.

RESULTS

Water/Pt interface

The oxygen and hydrogen density profiles obtained from the water/Pt simulations are displayed in figure 1. A quick glance at the figure suggests that the structures of water next to the Pt(100) and next to the Pt(111) are similar. The oxygen density plots show two pronounced peaks corresponding to two layers of water: the adsorbed layer and the layer hydrogen bonded to it. A weak third peak, corresponding to the third layer, can also be seen. Although being very similar to each other the plots for water/Pt(100) and water/Pt(111) interfaces display some differences. The location of the first peak of oxygen density for the water/Pt(111) interface is closer to the line of centers of the first row of Pt atoms than in the case of the water/Pt(100) interface. That means that a stronger adsorption occurs on the Pt(111) surface. Also, the width of the first peak for oxygen density in water/Pt(100) is broader than the one observed for water/Pt(111). This can be easily explained. Since the lattice structure for the Pt(111) surface is hexagonal with the spacing between Pt atoms commensurate with the distance between water molecules in ice, the Pt(111) surface promotes ice-like structures consisting of hydrogen bonded six-membered rings which are flat most of the time.[18] The Pt(100) surface is quadratic; therefore for water molecules to be adsorbed in this arrangement and still be engaged in a hydrogen bonding pattern, the adsorbed water molecules of the first layer must create a puckered structure. Another difference between the top and bottom panels of figure 1 is the appearance of just one peak in hydrogen density in the region between two layers in the case of the water/Pt(100) interface, while two small peaks are seen in this region for the water/Pt(111) interface. This is again due to a sharper structure of water next to the Pt(111) surface. The two peaks seen in the case of the water/Pt(111) interface are due to hydrogens of the first and of the second layers correspondingly. In the case of Pt(100) the single peak is due to the hydrogens which belong to both the first and the second layer.

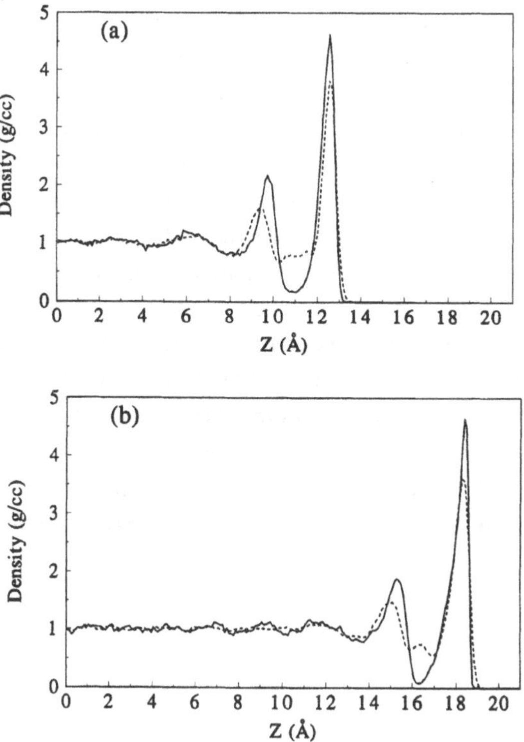

Figure 1. Oxygen (solid line) and hydrogen (dotted line) density profiles from the simulations of (a) H_2O/Pt(111) and (b) H_2O/Pt(100) interfaces. z is the distance from the center of the simulation box.

Figures 2, 3 and 4 compare the orientational structure of water obtained from our simulations. To study the orientational order of the water molecules we have divided the space between the metal surfaces into bins of 0.1 nm width. (The bin closest to the metal surface was numbered as bin number 1). The orientational distributions of angles between the water dipole vector and the normal vector pointing into the Pt surface along the z-direction were calculated in every bin. Also a distribution of angles between an OH vector and the surface normal vector was calculated. These distributions for the first two bins in both simulations are shown in figure 2. The water molecules in the first two bins belong to the adsorbed layer. As it was pointed out in the literature, the adsorbed layer is actually a bilayer.[2] We can see that our two simulations confirm the existence of the bilayer. The simulations also predict a similar orientational structure of the adsorbed layer which is independent of the face of the surface. From the orientational distributions shown in figure 2 and from the peaks location in these distributions we conclude that most of the water molecules in the first sublayer have their dipole and OH bonds (and the plane of the molecule) parallel to the surface. This conclusion is confirmed by the density plots in figure 1, which show that the maxima for the first oxygen and first hydrogen peaks occur at the same position. The distribution of dipolar orientations for the molecules of the second sublayer of the first layer is bimodal. This indicates that this sublayer is made up of "flip-up" and "flop-down" molecules ("flip-up" are molecules which have their dipoles pointing away from the surface, while "flop-down" molecules have their dipoles pointing in the direction towards the surface). The locations of the peaks of the OH orientational distributions for the second sublayer show that for the "flip-up" molecules one of their hydrogens is pointing away from the surface, while the second hydrogen is creating a tetrahedral angle with the first hydrogen. This configuration corresponds to the maximum of the dipole orientation distribution observed at $\cos\Theta \sim -0.525$. The second maximum of the dipole orientation distribution observed at $\cos\Theta \sim 0.2$ is due to the "flop-down" molecules, which have both their hydrogens pointing towards the waters of the adsorbed sublayer.

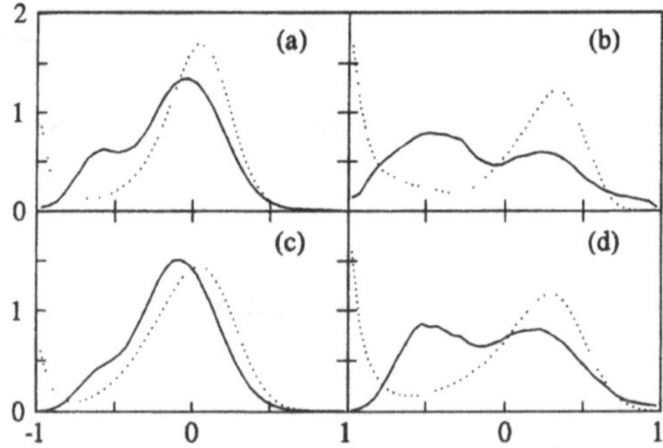

Figure 2. Distributions of dipole (solid line) and OH (dotted line) orientations as a function of $\cos\Theta$ (Θ is the angle between the dipole or OH vector and the inward normal to the surface). Figures (a) $1.25 < z < 1.35$nm and (b) $1.15 < z < 1.25$nm are for H_2O/Pt(111) and (c) $1.80 < z < 1.90$ nm and (d) $1.70 < z < 1.80$ nm are for H_2O/Pt(100).

Figures 3 and 4 display the orientational distribution in the next six bins, i.e., for Δz's of 0.1 nm which are successively farther away from the surface.

The distributions in the third bin are for the water molecules which are between the adsorbed and the second layer. Since there are only a few such molecules the distributions have large statistical uncertainties. The orientational distributions in bins 4 and 5 for the water/Pt(100) and water/Pt(111) interfaces are very similar to each other.

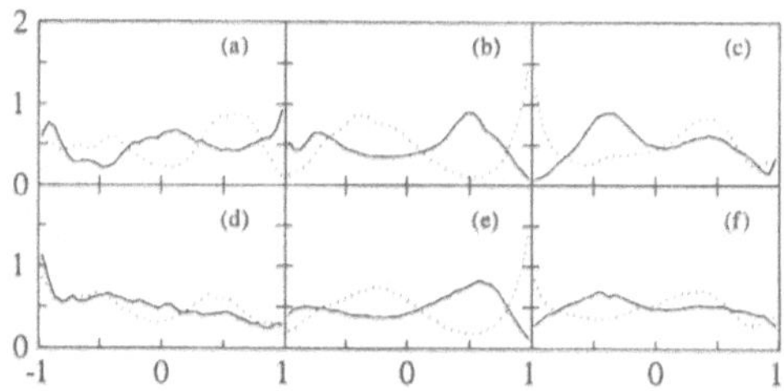

Figure 3. Distributions of dipole (solid line) and OH (dotted line) orientations for the second layer as a function of cosΘ for H_2O/Pt(111) (figures. a-c) and H_2O/Pt(100) (figures. d-f).

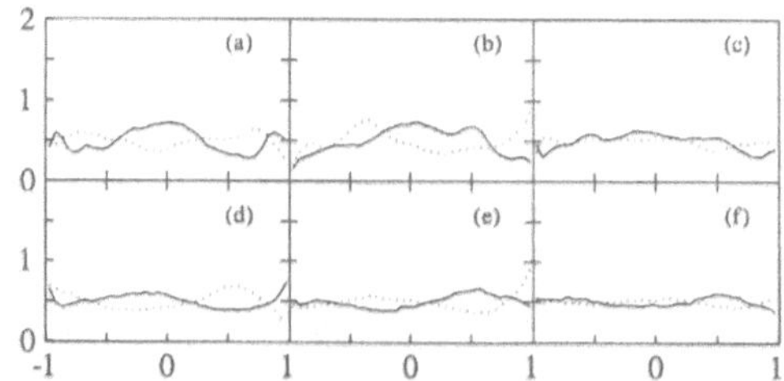

Figure 4. Distributions of dipole (solid line) and OH (dotted line) orientations for the third layer as a function of cosΘ for H_2O/Pt(111) (figs. a-c) and H_2O/Pt(100) (figs. d-f).

The positions of the peaks for these distributions and the fact that the distributions in bins 4 and 5 are mirror images of each other with respect to cosΘ =0 suggest that water displays an orientational ordering biased towards the ordering observed in ice-I, similar to the one observed by Lee et. al.[6] We also observe that the distance between the second and the third layers is

0.35nm, which is the distance between layers in ice-I structure. The distributions in bins 6, 7 and 8 describe the orientations of water molecules in the third layer (see figure. 4). The distributions for water/Pt(100) are again very similar to the distributions for the water/Pt(111) interface. Although the distributions for the third layer are broader than the distributions for the second layer, the positions of the peaks in the third layer are at the same angles where the positions of the peaks for the second layer are found. Also the shapes of the second layer distributions and third layer distributions are similar. This means that the bias towards ice-I structure is still observed in the third layer, but it is much weaker. No orientational bias was observed beyond the third peak.

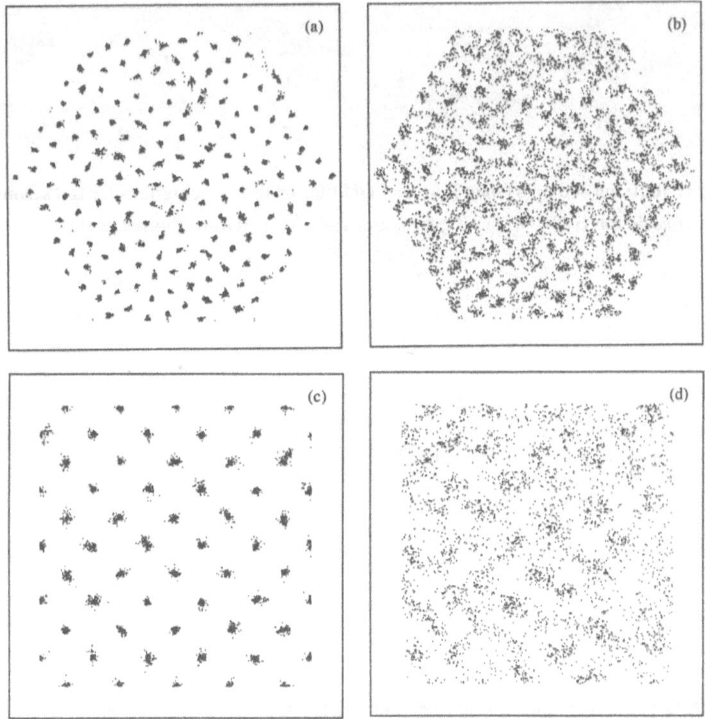

Figure 5. Projection of positions of oxygen atoms on the xy plane (plane parallel to the Pt surface) for the first two layers of water adsorbed on Pt(111) (figures. (a) and (b)) and on Pt(100) (figures. (c) and (d)).

The most illustrative expositions of the structure and dynamics of water in our simulations are the scatter plots in figures 5a-5d. In these figures a dot has been drawn every 0.25 ps at the (x,y) position of each oxygen atom. Figure 5a depicts positions of waters in the xy plane for the adsorbed layer of the water/Pt(111) interface, while figure 5b depicts the positions of the second layer of water for the same interface. Figures 5c and 5d are the corresponding figures for the water/Pt(100) interface. The molecules in the adsorbed layer repeat the structure of the underlying face centered cubic Pt surface and their motions are mostly oscillations about equilibrium positions. Such a motion is characteristic for the solid phase. We also observe that for the Pt(100) surface all available a-top absorption sites are covered by water molecules, while for the Pt(111) surface some of the a-top sites are free, since water creates ice-like hexagonal rings with a free Pt site in the center of the ring. This observed difference in the coverage of the surface sites by water molecules can have important consequences for the adsorption of the impurities and salt molecules in aqueous solutions. For the second layer we observe that the density of points is much more homogeneous thus indicating that the layer has a liquid-like character. The calculation of the apparent diffusion coefficients of the corresponding layers confirms this conclusion.[25]

One can also calculate the water contribution to the surface potential of the interface. As was shown by Wilson et.al.[27] the value of the surface potential depends on the distribution of the solvent charge density and not just on the distribution of the solvent dipole density at the interface. We calculate the surface potential by performing a straightforward integration of the Poisson's equation, which results in the following expression for the surface potential :

$$\Delta\phi = -\frac{1}{\epsilon_0}\int_0^z dz' \int_0^{z'} \rho(z'')\,dz'' \tag{8}$$

The resulting profile of the surface potential is shown on figure 6, together with the profile of charge distribution that is responsible for the surface potential.

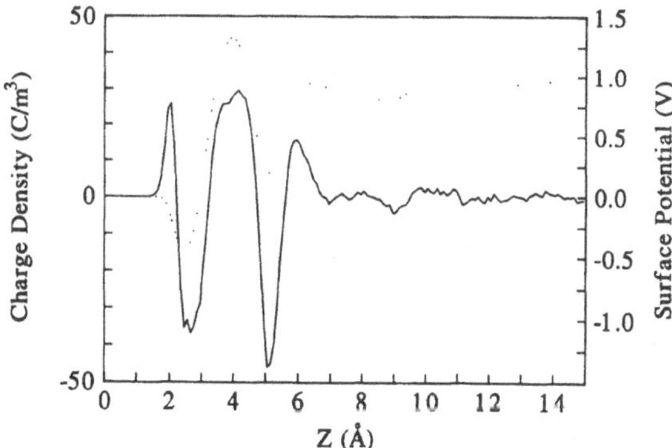

Figure 6. Surface Potential (dotted) and charge density (solid line) for the water/Pt(111) interface.

Free Energy Profiles for Li^+ and I^- Ions Approaching Pt(100) Surface

As we have shown above, molecular dynamics computer simulations on water/Pt interface can produce a rather detailed picture of the structure and dynamics of water at this interface. It is quite obvious that the next step in the investigation of this interface is to study the effects due to the addition of ions to the interface. And indeed this was done recently by Rose and Benjamin in the study of the solvation of Na^+ and Cl^- at the water/Pt(100) interface.[28] Very recently Spohr considered the solvation of the I^- ion at the same interface.[29]

Computer simulations can provide us with a molecular basis for the understanding of the structure and dynamics of electric double layers at metal/aqueous solution interface.[30,31] For example, one of the important questions that computer simulations can help in answering is what is the most stable configuration that ions achieve at the interface? According to the traditional picture, anion loses its water solvation shell and comes in contact with the metal surface, while a small cation preserves its solvation shell and therefore does not come in close contact with the metal. To see if this picture is correct one should calculate the probability $p(r)$ of finding the ion at distance r from the surface plane, or one can calculate the potential of mean force (pmf) between the surface and the ion, $W(r)$, which is directly connected to the probability $p(r)$ by a simple relationship:

$$W(r) = -kT \ lnp(r) \tag{9}$$

In equation (1) k is the Boltzmann constant and T is the temperature of the system. Here we present the results from our calculations on the pmf between the I^- ion solvated in water and Pt(100) surface, and the pmf between Li^+ solvated in water and Pt(100) surface. The two particular ions were chosen because they represent the extremes in their sizes and therefore are well suited for testing the classical picture. In addition the gas phase potentials of interaction between these ions and the Pt surface are available in the literature.[24]

To calculate the potential of mean force we first calculated the mean force acting on the constrained ion at a set of specified positions and subsequently integrated this force. In general, due to the corrugation of the surface the mean force depends on the position of the ion in the plane parallel to the surface (x,y plane) and also on the distance from the surface (z-coordinate). A complete exploration of such a force is too complicated and we therefore restricted ourselves to the study of the potential of mean force when x and y coordinates were fixed, so that the ion was always located on a line perpendicular to the surface and crossing through the hollow site of the Pt surface. (The ab initio calculation predict that both ions are absorbed at the hollow site). Heavy masses were assigned to the ions in order to constraint them at the desired positions. Initially the ions were placed in the middle region of the box, each about 1.02 nm away from the corresponding surface. Once the positions were selected, the system was equilibrated for 5 ps with a 2.0 fs time step. The mean force acting on the ion was calculated after subsequent 26 ps of the production run. To move the ions to their new positions, the heavy masses were removed and an external forces were applied along the required directions to allow the system to evolve for 1-2 ps, until the ions came to the new positions closer to the walls. The average force was checked to be close to zero (within the statistical accuracy) in x and y directions. Potentials of mean force along z-direction were evaluated from the integration of the mean forces calculated at 30 different points along the line described above. This method is basically the same as that used by Spohr in his recent calculation of the pmf for the I^-/Pt(100) surface.[29] Note that in our simulation we are able to get two pmf's simultaneously and observe that the system is also electroneutral in our simulation.

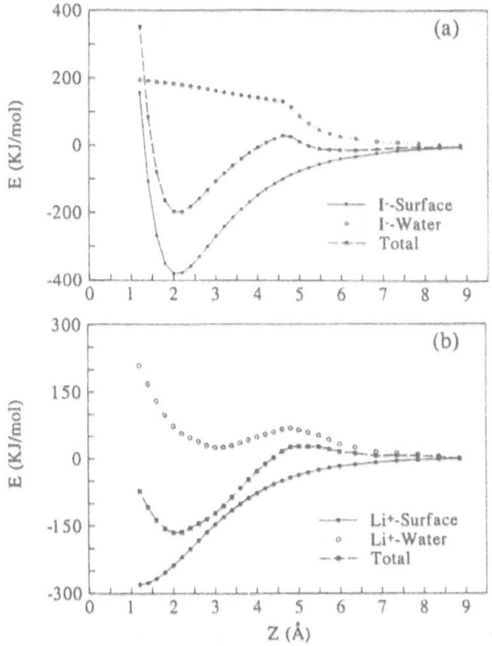

Figure 7. The potential of mean force and its components, calculated for (a) the I^- ion (b) the Li^+ ion, as a function of the distance to the ion from the surface. In these calculations we have assumed that the contibution from water into the pmf is zero beyond 0.9 nm.

The pmf's, which are obtained through the integration of the mean forces are displayed in figure 7. To understand the effect that the solvent has on the pmf, we display the corresponding contributions of the ion-surface potential and the mean force due to the water. The pmf for I^-/Pt(100) in water displays two minima: a deep one situated 0.2 nm away from the metal surface and another shallow one at 0.6 nm away from the surface. As figure 7a shows, the most probable position of I^- is in contact with the Pt(100) surface; at this position the ion is even closer to the Pt surface than the first water layer. This is not surprising, considering that I^- displays a strong attraction towards the hollow site on Pt(100) surface (the minimum of the gas phase I^-/Pt(100) potential is found for a hollow site configuration[24]). The second minimum in the I^-/Pt(100) pmf is located around the position where the second water layer is observed, while the barrier peak in pmf is located right between the first two water layers. While the global minimum in pmf is due to the interaction between the ion and the wall the second minimum and the barrier are mostly due to the water-ion interaction. The barrier in the pmf exists because a large distortion in the structure of the water has to be created prior to penetration of a large ion such as I towards the surface. This is also confirmed by observation of trajectories of water molecules that belong to the layer adjacent to the wall.[26] Even when the ion is in contact with the Pt surface, its surrounding water molecules are still not in their a-top positions due to the large size of the ion (see figure. 8).

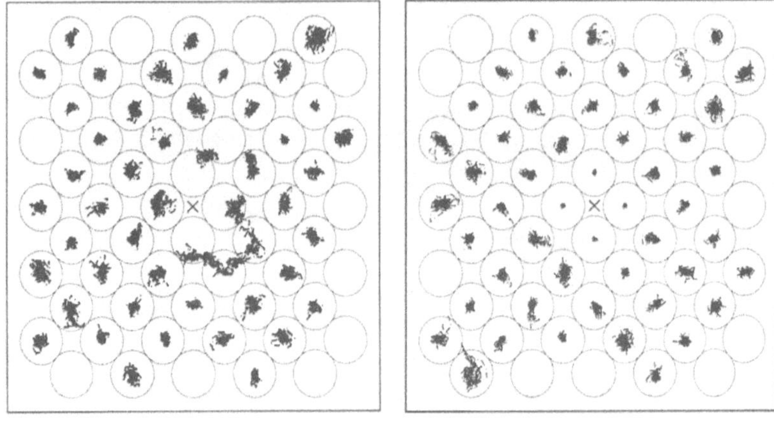

Figure 8. XY projection of the trajectories of the molecules adjacent to the wall from a 26 ps simulation in which the ions (left Iodide, right Lithium) are in the global minima of the respective potentials of mean force. The ion projections are at (0,0) and are marked by a cross.

The shape of the pmf for I^- that predicts a contact adsorption of the ion to the Pt surface is in agreement with the traditional picture, while the pmf for Li^+ ion presents a picture that is in disaccord with the traditional description.

According to this description a small ion such as Li^+ should keep its solvation shell around itself and be separated from the metal surface by roughly two water layers, one due to its own shell and another due to the adsorption to the metal surface water layer. As figure 7 shows the most probable location of Li^+ ion is not far from the surface, i.e. 0.21 nm away from it. At the same time this is not a case of a contact adsorption, since the contact adsorption for Li^+ would occur at 0.1 nm as the surface-ion potential shows.[24] The location and the value of the minimum in the pmf for Li^+ is the result of the interplay between the ion-surface and the ion-water interactions. Because the ion experiences a large attraction towards the surface (its value is 265 kJ/mol[24]) it wants to be attached to the surface. But since it wants to keep its solvation shell because of a strong ion-water interaction it can not come into close contact with the wall. As a result the ion finds itself in a slightly elevated position from the surface and is surrounded by four nearest water molecules, which are sitting in an a-top positions of the Pt surface. The compromise state is also due to the fact that the small size of Li^+ ion allows it to sit on the hollow adsorption site without disturbing the water neighbors that attach to the a-top sites. This picture is confirmed by the display (see figure. 8) of the trajectories of water molecules that belong to the adsorbed layer.

Finally to illustrate the geometry of water/ion/surface complex when the ion is next to the surface, we display the snapshots of typical configurations in figure 9.

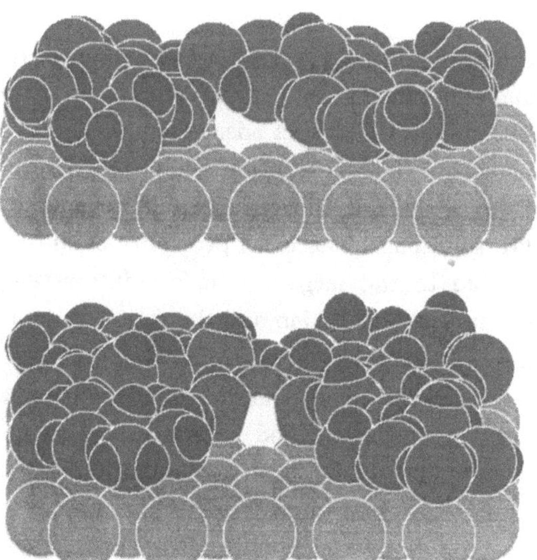

Figure 9. Snapshots of typical configurations from the Iodide/Pt/water (top) and Lithium/Pt/water (bottom) systems. Not all of the adsorbed water molecules are shown for clarity.

CONCLUSIONS

Our simulations show that water layer adsorbed on Pt(100) and Pt(111) is actually a bilayer. The adsorption is stronger on the Pt(111) since it is easier for water to satisfy the geometric constraints of this hexagonal surface. The occupation of adsorption sites on Pt(100) is complete, while it is ~80% for the Pt(111).[18] The orientational structures of the adsorbed bilayer on Pt(100) and Pt(111) are similar. They are even more similar in the second and the third layers, where the orientational structure in both cases is biased towards the one observed in ice-I with the c-axis pointing along the z-direction. As we can see, the influence of the geometry of Pt at the interface is mostly felt in the adsorbed layer, where water adjusts to the surface. Beyond the adsorption layer the differences in the structure of water are minimal. We also would like to mention here that it seems that the water model we use in our simulations predicts a very "robust" behavior of water. Even a strong perturbation on water such as the one from the regular structure of Pt is nullified after a distance of ~1 nm. The decay of surface induced perturbations in water over distance of 1 nm was also observed experimentally in studies of water in small micro-pores.[32]

It is interesting to compare the structure of water at the Pt interface with the structure of water at the water/hydrophobic wall interface.[6] The simulations show that while the structure of the water layer adjacent to the Pt surface is determined by the nature of the Pt surface, the structure of water beyond the first layer is similar to the structure of water observed at the water/hydrophobic wall interface.

From our calculations of the potential of mean force for the I^- and Li^+ ions solvated in water and approaching the Pt (100) uncharged surface we observed that the shape of pmf predicts a contact adsorption of the I ion at the surface. For the Li^+ ion although a contact adsorption is not observed, the most probable location of the ion is much closer to the surface than predicted by the traditional picture of electrolyte solutions. The close proximity of the Li^+ ion to the Pt surface is not only due to the large ion-surface interaction, but is also a consequence of the small size of the ion and the fact that its adsorption site is the hollow site, while water adsorbs at the a-top position.

We are now witnessing the appearance of many computer simulations related to the study of the electrified interfaces. In our opinion these simulations have more qualitative, at best semi-quantitative character. Before one can embark on a serious quantitative analysis of the interfaces, through the use of computer simulations, one needs to obtain reliable potentials to describe the interactions between the molecules at the interface. In addition one also has to know the effect of the approximate treatment of the long-range electrostatic forces on the values and shapes of such quantities as pmf's and surface potentials. Right now we observe that although the detailed shape of the density profiles of water and ionic pmf's may be a function of the potentials and the value of the cut-off used in our calculations, the qualitative features of the results are in agreement with other recent simulations on electrolyte solutions at metal surfaces.

ACKNOWLEDGEMENTS

Collaborations with Drs. K. Foster and K. Raghavan are gratefully acknowledged. This work was supported by a grant from the Office of Naval Research. The simulations were performed on the Cray YMP at the North Carolina Supercomputing Center and on Convex C240 at UNC.

REFERENCES

1. J. Lipkowski, and P.N. Ross. "Structure of Electrified Interfaces," VCH, New York, 1993.
2. P.A. Thiel, and T.E. Madey, *Surf. Sci. Rept.* 7:211 (1987).
3. B. Jönsson, *Chem. Phys. Lett.* 82:520 (1981).
4. M. Marchesi, *Chem. Phys. Lett.* 97:224 (1983).
5. G. Barabino, C. Gavotti, and M. Marchesi, *Chem. Phys. Lett.* 104:478 (1984).
6. C.Y. Lee, J.A. McCammon, and P. Rossky, *J. Chem. Phys.* 80:4448 (1984).
7. N.G. Parsonage, and D. Nicholson, *J. Chem. Soc. Faraday Trans. 2.* 82:1521 (1986).
8. N.G. Parsonage, and D. Nicholson, *J. Chem. Soc. Faraday Trans. 2.* 83:663 (1987).
9. F.H. Stillinger, and A. Rahman, *J. Chem. Phys.* 60:1545 (1974).
10. J.P. Valleau, and A.A. Gardner, *J. Chem. Phys.* 86:4162 (1987).
11. W.L. Jorgensen, J. Chandrasekhar, J.D. Madura, R. Impey, and M.L. Klein, *J. Chem. Phys.* 79:926 (1983).
12. A.A. Gardner, and J.P. Valleau, *J. Chem. Phys.* 86:4171 (1987).
13. S. Holloway, and K.H. Bennemann, *Surf. Sci.* 101:327 (1980).
14. M.W. Ribarsky, W.D. Luedtke, and U. Landman, *Phys. Rev. B.* 32:1430 (1985).
15. E. Spohr, and K. Heinzinger, *Ber. Bunsenges. Phys. Chem.* 92:1358 (1988).
16. E. Spohr, *J. Phys. Chem.* 93:6171 (1989).
17. K. Foster, K. Raghavan, and M. Berkowitz, *Chem. Phys. Lett.* 162:32 (1989).
18. K. Raghavan, K. Foster, K. Motakabbir, and M. Berkowitz, *J. Chem.Phys.* 94:2110 (1991).
19. H.J.C. Berendsen, J.R. Grigera, and T.P. Straatsma, *J. Phys. Chem.* 91:6269 (1987).
20. O. Ohtaki, and T. Radnai, *Chem. Rev.* 93:1157 (1993).
21. H. Sellers, and P. Sudhakar, *J. Chem. Phys.* 97:6644 (1992).
22. H. Sellers, *J. Chem. Phys.* 98:627 (1993).
23. H. Yang, and J.L. Whitten, *Surf. Sci.* 223:131 (1991).
24. J. Seitz-Beywl, M. Poxleitner, M. Probst, and K. Heinzinger, *Int. J. Quant. Chem.* 42:1141 (1992).
25. K. Raghavan, K. Foster, and M. Berkowitz, *Chem. Phys. Lett.* 177:426 (1991).
26. L. Perera, and M. Berkowitz, *J. Phys. Chem.* 97:13803 (1993).

27. M. Wilson, A. Pohorille, and L. Pratt, *J. Chem. Phys.* 90:5211 (1989).
28. D. Rose, and I. Benjamin, *J. Chem. Phys.* 95:6856 (1991).
29. E. Spohr, *Chem. Phys. Lett.* 207:214 (1993).
30. J. Glosli, and M. Philpott, *J. Chem Phys.* 96:6962 (1992).
31. J. Glosli, and M. Philpott, *J. Chem Phys.* 98:9995 (1993).
32. M. Lefleur, M. Pigeon, M. Pezolet, and J.-P. Caille, *J. Phys. Chem.* 93:1522 (1989).

DIFFUSION MECHANISMS OF FLEXIBLE MOLECULES ON METALLIC SURFACES

Marvin Silverberg

Department of Chemistry, Biochemistry, and Molecular Biology
Oregon Graduate Institute of Science & Technology
P.O. Box 91000
Portland, Oregon 97291-1000

I. INTRODUCTION

The dynamics of flexible molecules in condensed phases is a subject of intense investigation for a variety of systems. Particularly relevant to heterogeneous chemical catalysis is the behavior of large flexible molecules adsorbed to metallic surfaces. Transition metals serve as catalysts for a variety of organic reactions and so the diffusion dynamics of organic molecules on such surfaces are especially important. In addition, the inter-relationship among internal degrees of freedom and cohesive motion of the molecules constitutes a complex subject rich in intriguing phenomena. The purpose of the present paper is to delineate some of these relationships for systems of relevance to heterogeneous chemical catalysis.

Typical catalytic reactions occurring on transition metal surfaces occur via a series of elementary kinetic steps. Reactants typically adsorb to the surface from the gas or liquid phase (sometimes dissociatively). Once adsorbed, the atoms or molecules diffuse among adsorption sites. Reactions among coadsorbed species occurs upon encounter, sometimes at specific surface features, such as impurities or defects. Useful product species typically desorb. For various reactions involving alkane species, characterizing diffusion mechanism can contribute to an understanding of the overall catalytic process.

Extensive measurements of alkane diffusion on metallic surfaces have been obtained by laser induced thermal desorption.[1] Diffusion coefficients have been measured for alkanes of various chain length and for pentane isomers. The general trend observed is that long alkane chains diffuse more slowly than shorter ones and highly branched isomers diffuse more rapidly than linear ones. Due to the complexity of these systems, theories of such diffusive motion have not been extensively developed. The theory of the diffusion of small molecules and atoms on surfaces is quite advanced compared to that of large molecules.[2,3,4] A common theoretical

Theoretical and Computational Approaches to Interface Phenomena
Edited by H.L. Sellers and J.T. Golab, Plenum Press, New York, 1994

approach is to describe the diffusive motion of adsorbed atoms and molecules by kinetic hopping models. Adsorbed particles are regarded as executing sequential uncorrelated hops among neighboring adsorption sites of minimum potential energy. The absence of correlations among sequential hops results from rapid dissipation of translational energy, followed by long residence times in adsorption sites. The rapid dissipation of energy by the system makes this a strong friction description of diffusion. An increase in temperature will result in more frequently occurring uncorrelated hopping. For weak friction, there is a significant distribution of hopping lengths and correlated hopping among sites occurs, since the translational energy is not necessarily dissipated in a single hop. For example, molecular dynamics simulations of Rh atoms adsorbed on Rh(111) exhibit correlated diffusion barrier crossings and re-crossings. However, on the more highly corrugated Rh(100) surface, adsorbed Rh atoms undergo simple uncorrelated hopping among nearest-neighbor sites.[4] Molecular dynamics simulations of CO diffusion on Ni(111) demonstrate that long-distance jumps rather than short-distance hops become the dominant diffusion mechanism at high temperatures.[3]

The goal of the present paper is to determine the nature of alkane diffusion in the dilute sub-monolayer limit, on single crystal metallic surfaces, by molecular dynamics. There are reasons to suspect that hopping model descriptions are necessarily incomplete. Specifically, large molecules such as alkanes are bonded to surfaces over an extended area. The adsorption occurs at sites of minimal potential energy for the entire molecule subject to the overall constraints imposed by the molecular bonds. The center of mass of the molecule is a collective degree of freedom obtained from the position of many atoms, each interacting with a different part of the surface. Thermal excitation and de-excitation are consequently far more complex processes, since energy is partitioned among internal degrees of freedom of the molecule and transferred back and forth to the surface via all the constituent atoms. The molecule may change conformations or may be frozen into specific conformations while adsorbed and diffusing. The existence of hopping motion of some kind, including long jumps, cannot be assumed *a priori*. It must be determined whether this in fact occurs and whether it is related to conformational change. To determine what occurs to alkanes, the molecular dynamics simulations are performed with qualitatively correct intramolecular potentials. These potentials must be accurately represented since diffusive motion might display a sensitive dependence on molecular stiffness.

II. METHODOLOGY

A. General

The diffusion of alkane molecules on solid fcc(111), fcc(100) and fcc(110) surfaces was simulated by MD calculations. Large scale averaging over initial conditions was not performed but rather typical initial conditions were used for calculating representative trajectories for each examined temperature. A detailed description of the method employed has been published elsewhere.[5] A single step algorithm was employed in integration of the equations of motion[6]

$$\mathbf{r}_i(t+h) = \mathbf{r}_i(t) + \mathbf{v}_i(t)h + \frac{1}{2}\mathbf{a}_i(t)h^2$$

$$\mathbf{v}_i(t+h) = \mathbf{v}_i(t) + \frac{1}{2}[\mathbf{a}_i(t+h) + \mathbf{a}_i(t)]h \tag{1}$$

where $\mathbf{r}_i$, $\mathbf{v}_i$ and $\mathbf{a}_i$ are the position, velocity and acceleration of particle i, and h is the time increment. The time step was set to one femtosecond in all simulations. All forces are assumed pairwise additive. The total force on each particle is obtained from contributions from all particles located within an interaction radius. The cut-off distance of this interaction radius was chosen as 2.7 times the Lennard-Jones parameter σ of the particular pair of interacting atoms. No distance cut-off was used for intramolecular potentials.

B. The Surface

Instead of modeling metallic surfaces with fully detailed potentials, crystal faces of a face-centered cubic Lennard-Jones solid was used. The Lennard-Jones parameters are those of tungsten and so the metal is an fcc allotrope of tungsten. One cannot attribute the results of alkane diffusion to a specific metal surface. Rather, the aim was to examine properties of diffusing molecules on typical metal surfaces. Subtleties arising from specific metallic potentials will be addressed in future work.

C. The Adsorbate

N-pentane is modeled by a unified atom model in which methyl and methylene groups are treated as single atoms with a mass equal to the total mass of the carbon and the attached hydrogens. The radii of the potentials are correspondingly adjusted. Methyl and methylene segments can be referred to simply as carbon atoms or segments. Adjacent carbons in these molecules are bound to each other by Morse potentials with the parameters D= 109.6 kcal/mol, ω_e = 1854.71 cm^{-1} and R = 1.526 Å.[7]

Further intramolecular potentials employed are the van der Ploeg-Berendsen harmonic potential[8] for the three bending modes in the pentane,

$$V_b(\theta_i) = \frac{1}{2}k_b(\theta_i - \theta_0)^2 \tag{2}$$

is used in the modified form

$$V_b(\theta_i) = k_b(1 - \cos(\theta_i - \theta_0)) \tag{3}$$

where k_b is the force constant, θ_0 is the equilibrium bond angle, and θ_i is the instantaneous bending angle where i = 1, 2 or 3. The Rychaert-Bellemans potential[9]

$$V_d(\phi_i) = \sum_{n=0}^{5} c_n(\cos\phi_i)^n \tag{4}$$

where $i = 1$ or 2, is used to model the potentials for the two torsional modes of pentane. The force constant parameters used are those employed by Leggetter and Tildesley in their study of liquid hydrocarbon layers adsorbed on graphite.[10] The values of the parameters for these potentials are listed in Table 1. A Lennard-Jones potential with the parameters $\sigma_{CC} = 3.923$ Å and $\epsilon_{CC} = 0.14$ kcal/mol[7] exists to prevent overlap of the end segments in the pentane.

Table 1. Parameters for the intermolecular and intramolecular potentials

ϵ_{MM}	24.63 kcal/mol
σ_{MM}	2.56 Å
ϵ_{MC}	0.722 kcal/mol
σ_{MC}	2.6716 Å
ϵ_{CC}	0.14 kcal/mol
σ_{CC}	3.923 Å
D	109.5 kcal/mol
ω_e	1854.71 cm^{-1}
R	1.526 Å
k_b	124.41 kcal/mol
θ_0	109.47°
c_0	2.2176 kcal/mol
c_1	2.9051 kcal/mol
c_2	-3.1356 kcal/mol
c_3	-0.7313 kcal/mol
c_4	6.2713 kcal/mol
c_5	-7.5271 kcal/mol

The chain potentials accurately control the rate of conformational isomerization and so one can reliably account for the effect of chain dynamics on diffusion mechanisms. Lennard-Jones parameters for the interaction between molecular segments and surface metal atoms are $\sigma_{MC} = 2.6716$ Å and $\epsilon_{MC} = 0.722$ kcal/mol.[7]

D. Preparation of the System

The system is equilibrated in a two-step procedure. The surface itself is equilibrated at the desired temperature. Following this, the molecule is adsorbed and allowed time to thermally equilibrate with the surface. The complete algorithm for these steps has been detailed elsewhere.[5] After equilibration, it was found that the adsorbed molecule is thermally stable in one of three conformations. These were the trans-trans (TT), trans-cis (TC) and gauche-gauche (GG) as depicted in Fig. 1.

III. RESULTS

Molecular trajectories on the surface are characterized by several quantities. The position and conformation of the molecule were followed as functions of time. The angle of orientation of the long axis of the molecule in the plane of the surface is also

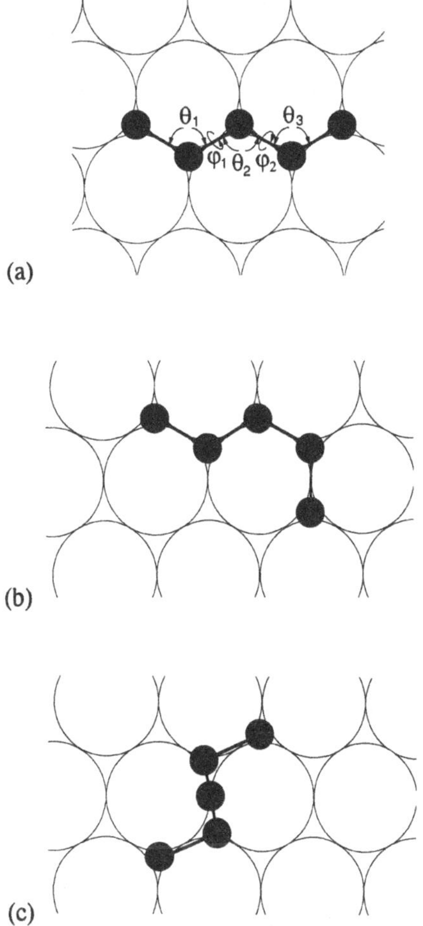

Figure 1. The three conformations of n-pentane discussed in the text as viewed from above, in their preferred adsorption sites. (a) trans-trans or TT, (b) trans-cis or TC and (c) gauche-gauche or GG. Torsion and bending angles are indicated explicitly in the TT conformation.

followed. Internal conformations of the molecule are indicated by the magnitude of the two torsion angles A_1 and A_2. In addition, the orientation of the molecule with respect to the surface can be determined from the z-component of the normal vector to the planes defined by three sequential carbons along the chain. The normal vectors are defined by the vector products and the three relevant vector components are: $B_1 = [(\mathbf{r}_3\text{-}\mathbf{r}_2)\times(\mathbf{r}_2\text{-}\mathbf{r}_1)]_z$, $B_2 = [(\mathbf{r}_4\text{-}\mathbf{r}_3)\times(\mathbf{r}_3\text{-}\mathbf{r}_2)]_z$ and $B_3 = [(\mathbf{r}_5\text{-}\mathbf{r}_4)\times(\mathbf{r}_4\text{-}\mathbf{r}_3)]_z$, where $\mathbf{r}_i$ is the position of the ith atom along the chain at time t. The qualitative descriptions presented below follow from examination of these quantities.

A. FCC(111) Surface

Trajectories of 250 picosecond duration were obtained for n-pentane in the temperature range 50 K-400 K. At 100 K, conformational change is rare on the picosecond time scale. Three trajectories were run on 100 K surfaces, beginning in the thermally equilibrated TT, TC and GG conformations, respectively. The paths of the center of mass during these trajectories are illustrated in Fig. 2. The trajectories

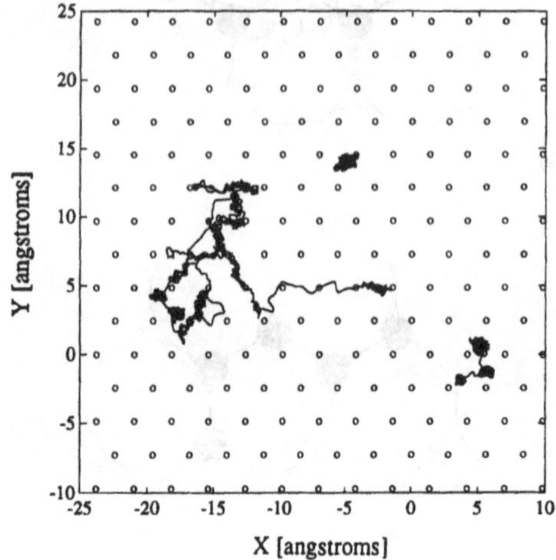

Figure 2. The path swept out by the center of mass of the n-pentane molecule during its trajectories, at 100 K, on the fcc(111) surface. The long trajectory began with the molecule in the GG conformation. The trajectory which started in the TC conformation did not hop, while the TT conformation hopped twice. The circles indicate the optimal position of the centers of the atoms of the topmost layer of the metal surface.

which start in the TT and TC conformations have all the carbons of the molecule initially located near. three-fold hollow sites and when the molecule moves, it rapidly equilibrates thermally in a neighboring set of hollow sites. During this 250 psec trajectory, the TT conformation hops twice to neighboring sets of three-fold sites. The molecule changes to the GG conformation while hopping between sites and relaxes rapidly back into the TT conformation at the new site. Several attempts occurred during which the GG conformation was adopted and the molecule relaxed into the TT conformation without hopping. The molecule does not roll over its diffusion barrier but rather slides over it transverse to the molecular axis while remaining parallel to the surface. The TC conformation is unchanged during the entire trajectory which starts in that conformation. No hops are observed. The GG conformation starts with carbon 3 elevated above a bridge site and the other carbons in three-fold sites. The molecule moves a much greater distance during the 250 psec trajectory than it does in the other two conformations at this temperature. The paths of the center of mass during these trajectories and the displacement versus time of the GG trajectory are illustrated in Fig. 2. Several long jumps of the GG occur. For a duration of about 30 psec starting near 150 psec, the molecule changes to the TT conformation, and then returns to the GG conformation. In addition, the molecule switches several times between the right-handed GG conformation and the left-handed GG conformation. The molecule remains upright in these conformations at all times. The diffusion occurs by sliding and hopping of the molecule while it remains astride rows of surface atoms with carbon 3 on top of the atoms and the others aligned beside these rows.

The adsorption potential energy of the molecule on the surface is given by

$$V_{ads} = \sum_i \sum_\alpha V_{i\alpha}(|\mathbf{r}_i - \mathbf{r}_\alpha|)\,, \tag{5}$$

where the index i is summed over atoms of the adsorbed molecule and α is summed over atoms of the metallic surface. Both the TT and TC trajectories show a maximum in the magnitude of the adsorption energy of about 16.5 kcal/mol. The maximum value of the magnitude of the adsorption energy in the GG trajectory is around 14.5 kcal/mol. This value goes to 16.5 kcal/mol during the interval of time that the molecule switches to the TT conformation and rises again when the molecule assumes the GG conformation again. This indicates that the molecule falls in to a deeper potential well upon changing to the TT conformation. The sudden jump in V_{ads} due to the conformational change is of greater magnitude than the average fluctuation in V_{ads} at this temperature. The greater mobility of the TT conformation than the TC conformation, at 100 K is associated with the greater frequency of large fluctuations in the values of V_{ads} during their respective trajectories.

Three trajectories were obtained at 200 K with the molecule initially in one of the three basic conformations. The displacement of the molecule and the path of the trajectory which begins in the TC conformation are shown in Fig. 3. Very long,

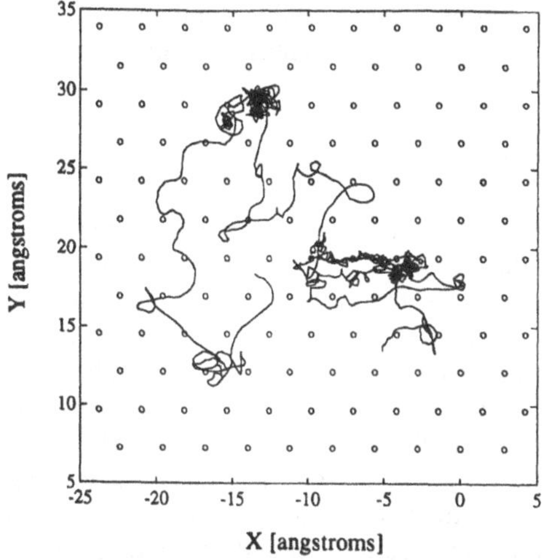

Figure 3. The path swept out by the center of mass of the n-pentane molecule during its trajectory on the fcc(111) surface, at 200 K. The trajectory began with the molecule in the TC conformation. The circles indicate the optimal position of the centers of the atoms of the topmost layer of the metal surface.

convoluted continuous motion across the surface is in evidence. There is also localization for long periods in specific sites and some hopping of the molecule among closely located adsorption sites. The molecule switches to the GG conformation for a period, passing briefly through the TT conformation. This period coincides with a period of rapid continuous sliding motion across the surface. The molecule then switches to the CT conformation, again passing through the TT conformation, and remains fixed in this latter conformation for the remainder of the trajectory. At this point the molecule slows down and moves by hopping rather than sliding. Each of these conformational changes occurs in about one picosecond. Vertical orientation with respect to the surface is not changed while the conformation is fixed: the molecule remains parallel to the surface. The absence of conformational change during tens of picoseconds of diffusive motion shows that conformational and diffusional barriers are distinct.

The trajectory which began in the TT conformation flips into the TC conformation after about 20 psec and remains in this conformation for the remaining 230 psec of the trajectory. The trajectory's overall appearance is the same as the trajectory displayed. A trajectory which began in the GG conformation at 200 K, flipped to the TT conformation after 20 psec and then switched suddenly to the TC. It remained in this conformation for about 60 psec, switched to the TT for about 10 psec and then passed to the TC until the end of the trajectory. Conformational change occurred in these cases by rotations about the terminal C-C bonds.

Trajectories run at higher and higher temperatures show longer and longer correlated slides and shorter periods of localization. In addition, conformational change becomes more rapid. At 300 K, around the desorption threshold, conformational change occurred on an average of every 20 psec, during periods of continuous sliding. Partial desorption occurs as well, during which time, transitions between TT and GG are continuous. Hopping diffusion was observed exclusively for the TT and TC conformations at 100 K and a combination of hopping and sliding for the GG conformation. As temperature increases, sliding appears in each conformation and eventually dominates over hopping.

B. FCC(100) Surface

The above pattern of diffusion dynamics is largely reproduced for the n-pentane adsorbed on the fcc(100) surface. However, only two stable conformers were observed for any length of time: the TT and the TC. Hopping diffusion was observed exclusively at low temperatures and slides dominated increasingly for elevated temperatures. An interesting quantity is the angle formed between the long axis of the molecule and the x-axis of the surface. Figure 4 depicts this quantity for three trajectories which all began in the TC conformation at 100 K, 200 K and 400 K. At the lowest temperature, only discrete changes occur in the angle formed with the surface. For most of the time the angle fluctuates around fixed values. At the intermediate temperature there are long periods of fixed angles and a few periods of gradual change of orientation. The latter correspond to the long slides of the molecule across the surface. During hopping motion the angle changes suddenly and discretely while during long slides, the angle changes continuously. For the highest temperature examined (actually in a precursor state to desorption), the orientation changes continuously throughout. Furthermore, during this trajectory the molecule slides across the surface all the time without ever being trapped in particular adsorption sites. Effectively, the molecule is rotating freely in the plane of the surface while sliding and it is rotationally constrained while hopping.

C. FCC (110) Surface

The fcc(110) surface is highly anisotropic. Figure 5 shows the center of mass position during two trajectories of duration 500 psec. The anisotropy of the surface

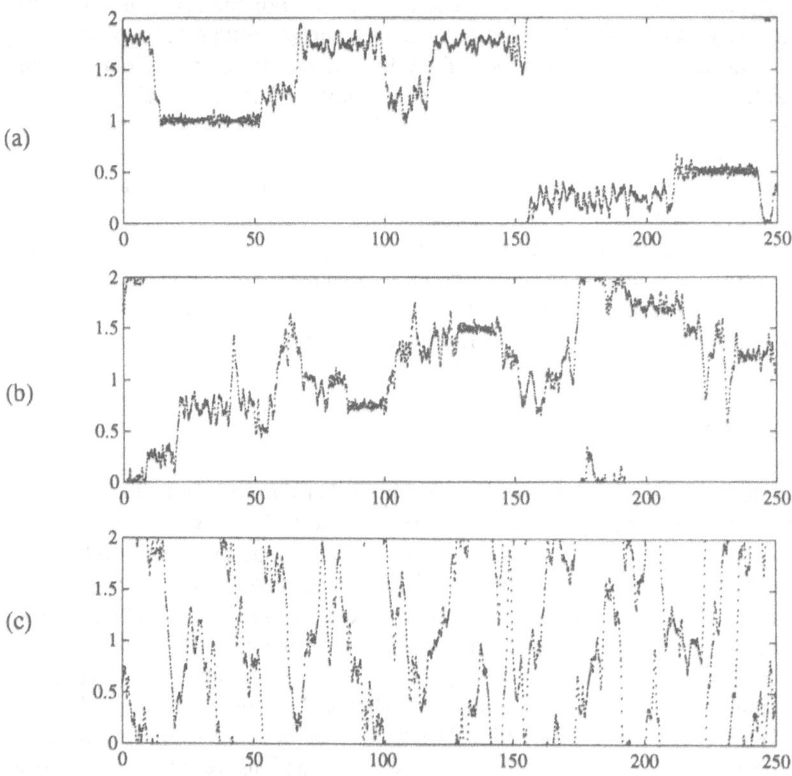

Figure 4. Angle of orientation with respect to the x-axis of the fcc(100) surface as a function of time for three trajectories at (a) 100 K, (b) 200 K, (c) 400 K.

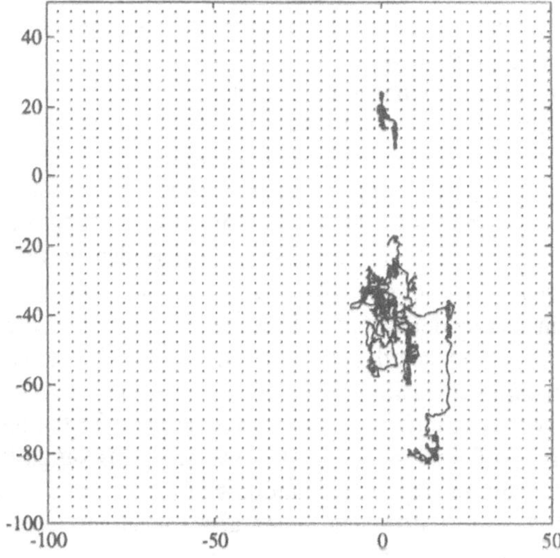

Figure 5. The paths swept out by the center of mass of the n-pentane molecule for two 500 psec trajectories on the fcc(110) surface. The shorter trajectory is at 200 K while the longer one is at 400 K. The dots indicate the optimal position of the centers of the atoms of the topmost layer of the metal surface.

is clearly reflected in the anisotropy of these two paths. For the lower temperature trajectory at 200 K, the molecule is located within a surface trough most of the time and aligned parallel to it. It adopts the TT conformation. In the coordinate system in which the troughs are parallel to the y-axis, the molecule rotates freely in the xz-plane. The molecule hops back and forth within the trough remaining in the TT conformation. When the molecule hops sideways, it switches to the TC conformation and rotates in the xy plane. At 400 K both hopping and long slides along and across the troughs are in evidence. During these slides the molecules rotate relatively freely in the xy plane. The overall motion of the molecule is again a combination of hopping and sliding. Rotational motion in the xy-plane is correlated with sliding motion and is punctuated by the periods of hopping. During the latter, the molecule is rotationally constrained in the xy-plane but resumes free rotation in the xz-plane.

IV. DISCUSSION

These studies of n-pentane have demonstrated that the diffusion of alkanes on metallic surfaces is considerably more complex than the diffusion of metal atoms or small molecules such as CO on metallic surfaces. On the atomically smooth (111) surface, the molecule displays different diffusional properties in three different conformations. There are barriers to conformational isomerization among these adsorbed states. Not only must the molecule overcome its own intrinsic conformational barriers, but it must also overcome the barriers imposed by the surface. In general, this raises the barrier height. Diffusional barriers in each of these conformers is lower than the conformational barriers. The evidence suggests that these diffusional barriers are distinct in the various conformations resulting in distinct diffusion rates. Diffusion occurs for long periods of time with the molecule locked into single conformations. The molecule in the GG state diffuses far more rapidly than in either of the other conformations; thus its diffusion barrier is lower and it experiences a lower effective friction. It slides readily at temperatures for which the other conformations hop infrequently.

On the fcc(100) surface the onset of hopping and sliding occurs at about the same temperatures for the surface stable conformations. On the highly anisotropic fcc(110) surface hopping is observed at very different rates parallel to and transverse to the surface's troughs. Hopping along troughs occurs predominantly in the TT conformation and across troughs in the TC conformation. Diffusion mechanisms of the different conformations are highly directionally dependent in this case.

This summarizes much of our qualitative understanding of the diffusion of adsorbed n-pentane in this model system. These results show that the transition states to diffusion are more complicated and more subtle than previous studies would suggest. To obtain accurate diffusion coefficients, using complete molecular dynamics simulations, proper averaging must be performed to account for the multiple transition states and the hierarchy of time scales in the system. This suggests that averaging should be done over many trajectories of hundreds of picoseconds duration to account for the effects of competitive diffusing states and transitions between them, since conformational isomerization occurs on a time scale of tens or hundreds of picoseconds or longer. Initial conditions must be carefully considered as well. A proper equilibrium distribution of the various conformational states must be obtained for accuracy. Computational models which address these issues are currently being developed for various linear and branched alkanes, and will be reported upon in future publications.

The metal adsorbate potentials used in the present study were chosen to be representative but to correspond to no particular face-centered cubic metal. Nonetheless, the above results may suggest some general characteristics of the diffusion of flexible molecules on surfaces. The weak friction behavior on fcc surfaces and the multiplicity of adsorption/diffusion states of linear alkanes should follow the pattern established here. Additional issues include: (1) the existence of multiple diffusion rates on atomically smooth surfaces has no correspondence in adsorbed atomic or molecular systems which have a single adsorption site; (2) the generality of this phenomenon in adsorbed molecular systems and its relevance to surface reaction mechanisms is an intriguing issue; and (3) more exact potentials could have a profound effect on the hierarchy of barriers and should be explored as they become available.

Acknowledgments

The author gratefully acknowledges the National Science Foundation, Division of Chemical and Thermal Systems (Grant CTS-9211333). Part of this work was performed while the author was a visiting scholar at the Institute for Advanced Studies of the Hebrew University of Jerusalem.

REFERENCES

1. (a) S.M. George, A.M. DeSantolo, and R.B. Hall, *Surf. Sci.* **159**:L425 (1985); (b) C.H. Mak, J.L. Brand, A.A. Deckert, and S.M. George, *J. Chem. Phys.* **B85**:1676 (1986); (c) C.H. Mak, B.G. Koehler, J.L. Brand, and S.M. George, *J. Chem. Phys.* **87**:2340 (1987); (d) C.H. Mak, J.L. Brand, B.G. Koehler, and S.M. George, *Surf. Sci.* **191**:108 (1987); (e) C.H. Mak, B.G. Koehler, and S.M. George, *J. Vac. Sci. Technol.* A **6**:856 (1988); (f) E.D. Westre, M.V. Arena, A.A. Deckert, J.L. Brand, and S.M. George, *J. Chem. Phys.* **94**:293 (1990).
2. (a) R. Gomer, *Rep. Prog. Phys.* **53**:917 (1990) and references therein; (b) A.G. Naumovets and Y.S. Vedula, *Surf. Sci. Rep.* **4**:365 (1985) and references therein.
3. K.D. Dobbs and D.J. Doren, *J. Chem. Phys.* **97**:3722 (1992) and references therein.
4. (a) D.E. Sanders and A.E. DePristo, *Surf. Sci.* **264**:L169 (1992), (b) J.M. Cohen and A.F. Voter, *J. Chem. Phys.* **91**:5082 (1989) and references therein.
5. M. Silverberg, *J. Chem. Phys.* **99**:9255 (1993).
6. A.L. Trayanov and M.G. Prisant, *J. Chem. Phys.* **94**:2352 (1991).
7. (a) D. Cohen and Y. Zeiri, *J. Chem. Phys.* **97**:1531 (1992); (b) D. Cohen and Y. Zeiri, *Surf. Sci.* **274**:173 (1992).
8. P. van der Ploeg and H.J.C. Berendsen, *Molec. Phys.* **49**:233 (1983).
9. J.P. Ryckaert and A. Bellemans, *Faraday Discuss. Chem. Soc.* **66**:95 (1978).
10. (a) S. Leggetter and D.J. Tildesley, *Molec. Phys.* **68**:519 (1989); (b) S. Leggetter and D.J. Tildesley, *Ber. Bunsenges. Phys. Chem.* **94**:285 (1990).

COMPUTER SIMULATION OF SOLVATION IN SUPERCRITICAL FLUIDS

Gregory S. Anderson,[1] Kristen M. Hegvik,[2] and Mark R. Hoffmann[2]

Chemical Engineering Department[1]
Chemistry Department[2]
University of North Dakota
Grand Forks, ND 58202

INTRODUCTION

Supercritical fluids have become one the fastest growing areas of interest in both research and process development. Possible applications range from extraction processes to mediums for polymer synthesis.[1,2] The interest in supercritical fluids is due to their relatively inert nature at normal conditions--thus handling and disposal is safe and inexpensive. However, working with supercritical fluids can be expensive due to the high pressures necessary for proper operating conditions and the possible corrosive effects of the supercritical fluid on equipment. Due to these concerns, experimental studies must be especially well planned to obtain the maximum amount of information from a minimum number of experiments.

The ultimate goal of this project is to develop a computational model which will quickly and accurately predict the solubility of a given solute in a supercritical solvent. The model will be able to predict solubilities at any temperature and pressure above the critical point. Having an accurate prediction of solubility will make experimental design of projects such as environmental extractions and polymer syntheses in supercritical fluids more efficient.

METHODOLOGY

In a recent review, Cochran *et.al.* identify three types of semi-empirical approximations to the Kirkwood fluctuation integral for supercritical fluids: excluded volume (EV), local composition (LC), and the hard sphere expansion conformal solution (CS).[3] We report herein an investigation of the temperature dependence of parameters in the simplest of the models: the excluded volume model. In this model, the Kirkwood fluctuation integral[3] is approximated by

$$G_{21} \approx \alpha_{21}(G_{11}+V_{11}) - V_{21} \tag{1}$$

where

$$\alpha_{21} \approx \frac{\epsilon_{21}}{\epsilon_{11}} \frac{V_{21}}{V_{11}} \tag{2}$$

V_{11} and ε_{11} are the excluded volume and the (effective) potential energy of a solvent-solvent pair. Likewise, V_{21} and ε_{21} are the excluded volume and potential energy of a solvent-solute pair. One may suspect that ε_{11} and V_{11} can be derived from the van der Waal's equations. However, our numerical experiments (not reported here) indicate that while these parameters may be related to van der Waal σ and U_{min}, nonunity numerical prefactors are involved. Initial investigations indicate that reasonable results (e.g., as measured by χ^2) are obtained for a σ_{11} approximately two to three times the distance of the van der Waal's minimum. This result is in agreement with the values used by Cochran *et al.*.[3] For the study reported here we follow Cochran and estimate

$$V_{11} \sim V_c$$

and

$$\varepsilon_{11} / k \sim T_c.$$

where V_c and T_c are the critical volume and temperature, respectively, of the pure solvent. V_{21} and α_{21} are fit according to experimental data using a computer program to perform the nonlinear least squares fit.[4] Once V_{21} and α_{21} have been fit to experimental data, ε_{21} may be extracted from the definition of α_{21}.

The solubility of a nonvolatile solute in a solvent can be derived from Kirkwood-Buff fluctuation integral theory as,[3]

$$\ln y_2 = \ln\frac{P_{sat}}{P} + \ln Z + \frac{\upsilon_2}{kT}(P-P_{sat}) + \frac{1}{kT}\int_{P_{sat}}^{P} G_{21}\,dP \tag{3}$$

where the Kirkwood fluctuation integral, G_{21}, is related to the equation of state through

$$\int \frac{kT}{\rho_2}\,d\rho_2 = \int \{G_{21}+\upsilon_2\}\,dP \tag{4}$$

where υ_2 is the molar volume of the solute in the solid phase. In the excluded volume model (cf. Eq. (1)), the pressure integral over the fluctuation integral becomes

$$\int_{P_{sat}}^{P} G_{21}\,dP \approx \alpha_{21}\int_{P_{sat}}^{P} G_{11}\,dP + (\alpha_{21}V_{11}-V_{21})(P-P_{sat}) \tag{5}$$

The first term on the right-hand-side can be evaluated in terms of the compressibility of the pure solvent,

$$\frac{1}{kT}\int_{P_{sat}}^{P} G_{11}dP=\int_{\rho_{sat}}^{\rho}\frac{1}{\rho}d\rho\,(1-Z)-\int_{\rho_{sat}}^{\rho}d\rho\,(dZ/d\rho)_T \tag{6}$$

RESULTS

Figures 1-4 illustrate our model's prediction of the solubilities of 1,4-naphthoquinone, acridine, benzoic acid, and phenanthrene in supercritical carbon dioxide. The choices of solutes studied was based on experimental data available in the literature.[5,6] The compressibility was calculated from the equation of state of Haung *et. al.*.[7] Tables 1-4 list the extracted parameters, V_{21} and ε_{21}, for each solute studied. Parameters were evaluated at each temperature allowing us to observe any trends in the values.

Table 1
CO_2 - 1,4-Naphthoquinone

Temperature (K)	Excluded Volume (m^3/g mol)	Solvent-Solute Potential (K)
343	.3521E-3	460.3
328	.3633E-3	446.6
318	.3845E-3	427.7

Table 2
CO_2 - Acridine

Temperature (K)	Excluded Volume (m^3/g mol)	Solvent-Solute Potential (K)
343	.4532E-3	392.7
328	.4372E-3	397.9
318	.4085E-3	416.0

Table 3
CO_2 - Benzoic Acid

Temperature (K)	Excluded Volume (m^3/g mol)	Solvent-Solute Potential (K)
343	.2623E-3	573.8
328	.2607E-3	570.4
318	.2630E-3	563.1

Table 4
CO_2 - Phenanthrene

Temperature (K)	Excluded Volume (m^3/g mol)	Solvent-Solute Potential (K)
343	.4373E-3	393.5
323	.4435E-3	388.0

DISCUSSION

Our current model does follow the experimental trends as displayed in the plots of solubility. However, due to the unavailability of experimental uncertainties we have not yet analyzed the statistical error, so it is difficult to draw quantitative conclusions from the predictions of V_{21} and ε_{21}. Prior to obtaining the predictions of V_{21} and ε_{21} we believed that V_{21} would be directly proportional to temperature and ε_{21} would be inversely proportional to temperature. A study of the data shows that only acridine follows our anticipated trend.

We feel our next step, following an analysis of statistical error, will be to find a temperature dependent function used to predict the model parameters. It seems unrealistic to base V_{11} and ε_{11} on the critical points without a dependence on temperature when it is obvious that V_{21} and ε_{21} are temperature dependent. The additional temperature relation should reduce the experimental error and lead us to a more accurate prediction of solubility in supercritical fluids. Additional experimental solubility data would be vital to developing a statistically meaningful general model of solubility.

ACKNOWLEDGEMENTS

G.S.A. gratefully acknowledges the Ronald E. McNair Scholarship Program.

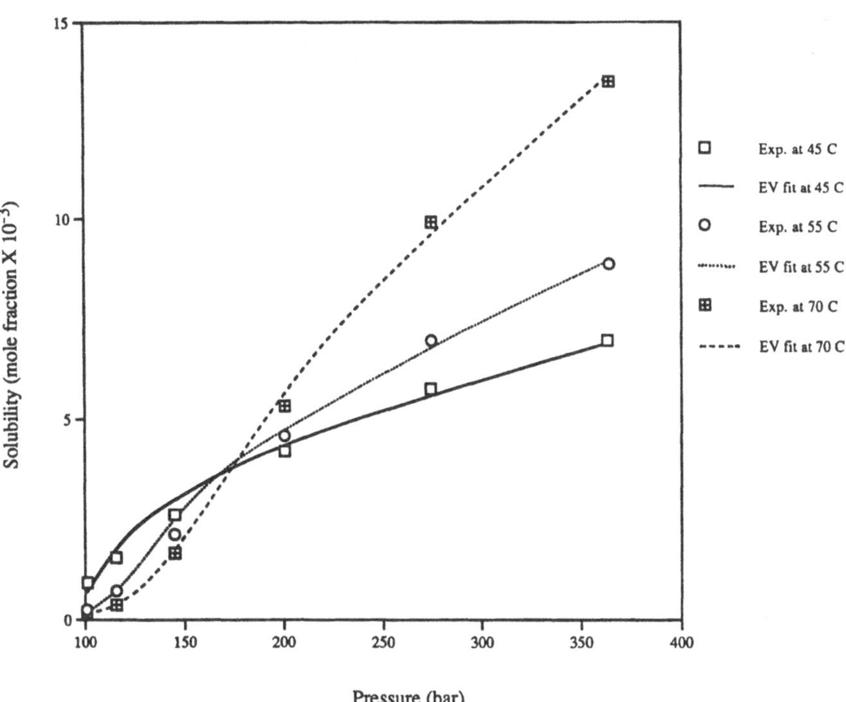

FIGURE 1: Solubility of 1,4-naphthoquinone in supercritical carbon dioxide.

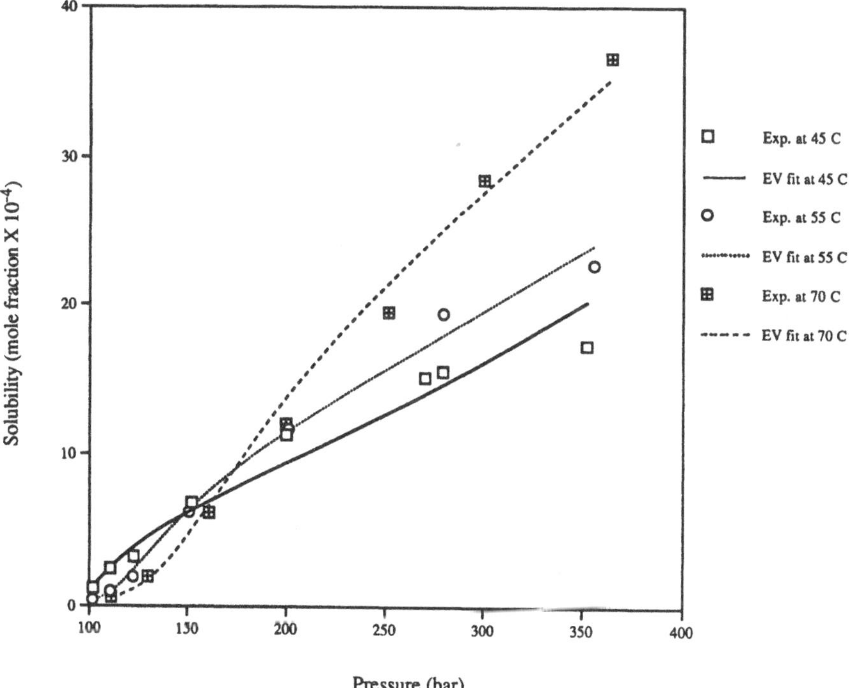

FIGURE 2: Solubility of acridine in supercritical carbon dioxide.

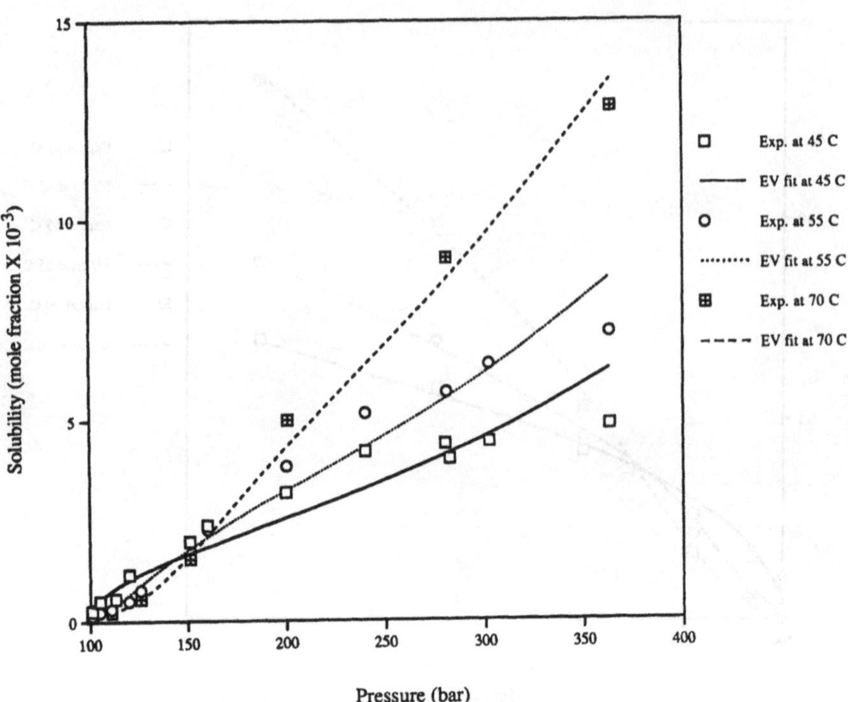

FIGURE 3: Solubility of benzoic acid in supercritical carbon dioxide.

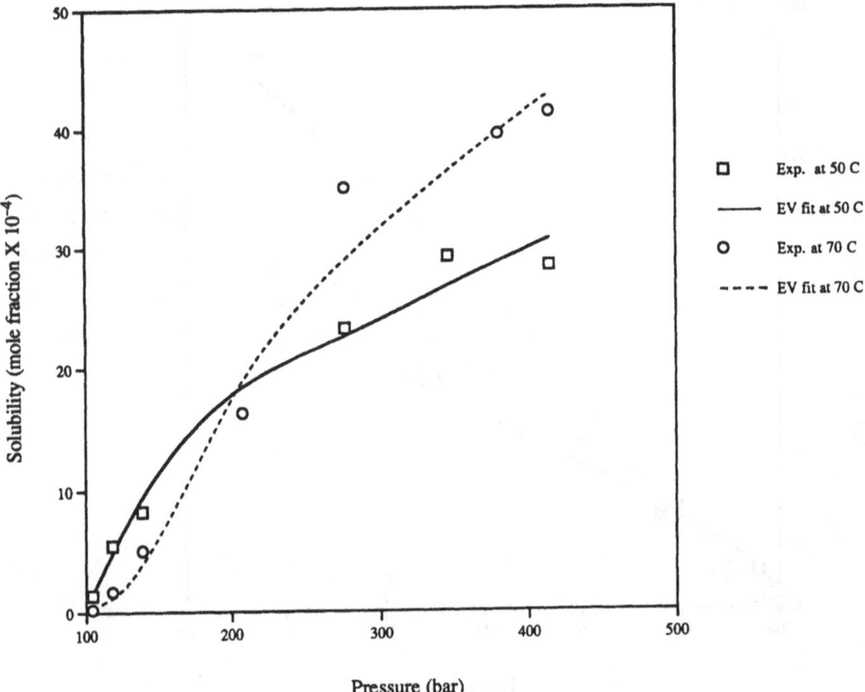

FIGURE 4: Solubility of phenanthrene in supercritical carbon dioxide.

REFERENCES

1. J.M.L. Penninger, M. Radosy, M.A. McHugh, V.J. Krukonis (editors), "Supercritical Fluid Technology," Elsevier Science Publishers, New York, NY (1985).

2. J.M. DeSimone, Synthesis of Fluoropolymers in Supercritical Carbon Dioxide, *Science*, 257:5072 (1992).

3. H.D. Cochran, L.L. Lee, D.M. Pfund, Structures and Properties of Supercritical Fluid Mixtures from Kirkwood-Buff Fluctuation Theory and Integral Equation Methods, *in*: "Advances in Thermodynamics," Vol. 2, E. Matteoli, G.A. Mansoori, ed., Taylor and Francis, New York, NY (1990).

4. W.H. Press, B.P. Flannery, S.A. Teukolsky, W.T. Vetterling, "Numerical Recipes," Cambridge University Press, New York, NY (1986).

5. W.J. Schmitt, R.C. Reid, Solubility of Monofunctional Organic Solids in Chemically Diverse Supercritical Fluids, *J. Chem. Eng. Data*, 31:2 (1986).

6. K.P. Johnston, D.H. Ziger, C.A. Eckert, Solubilities of Hydrocarbon Solids in Supercritical Fluids. The Augmented van der Waals Treatment, *Ind. Eng. Chem. Fundam.*, 21:3 (1982).

7. F. Haung, M. Li, L.L. Lee, K.E. Starling, An Accurate Equation of State for Carbon Dioxide, *J. Che. Japan*, 18:6 (1985).

STRUCTURE-FUNCTION MODELING IN BLOOD COAGULATION: INTERFACES, BIOLOGY AND CHEMISTRY

Michael N. Liebman

Bioinformatics Program
Amoco Technology Company
Mail Code F- 2
150 West Warrenville Road
Naperville, IL 60563-8460

INTRODUCTION

Interfaces in biology can occur between any two biological substances, where the interface is defined as the boundary between two regions separated in matter or in space. These interfaces can occur at the *macroscopic level*, e.g. tissue-solid: surgical implants, dental adhesives; tissue-gas: skin, lung alveoli; cell-solution: blood cell membrane-blood; at the *macromolecular level*, e.g. antibody-antigen, enzyme-substrate, receptor-ligand, enzyme-solvent; and even at the *submacromolecular* level. Within a macromolecule, the submacromolecular interfaces involve protein secondary structures and/or protein structural domains and produce the protein's tertiary structure. Thus these interfaces and their resulting influence on structure form the basis for defining the function of that macromolecule in its higher order interactions.

Interfaces in biology can also occur between academic disciplines, e.g. biophysics, biochemistry, biomathematics, where the interface represents the region of complementarity between these disciplines. Our research focuses on developing enabling technologies based on the interfaces between disciplines to understand those interfaces which exist between substances and are critical in rational protein engineering, manipulation of biological pathways, genetic disease analysis and pharmaceutical design.

The goal of our research is in developing technology to realize value from data such as the results of the Human Genome Project. As such, we address here the downstream processing and utilization fo the sequence data being collected by other research groups. The tools and methodologies which we are attempting to develop are targeted not solely at questions which might exist today, but also at those questions which cannot yet be asked. In this report we present an overview of our approach and an example of its application in analysis of the genetic disease associated with hemophilia B. We suggest that this approach is generalizable to a wide-range of problems beyond the example presented here.

BACKGROUND

We have developed a representation of the relationship between information and knowledge in biology, and recognizing that these are somewhat orthogonal, produced the

Theoretical and Computational Approaches to Interface Phenomena
Edited by H.L. Sellers and J.T. Golab, Plenum Press, New York, 1994

plot shown below as a model for the organization of information in genome modeling. By placing information as the ordinate and knowledge as the abcissa, the axes are constructed to show the hierarchical development of information databases, which would exist as vertical lines, and the increase in value which coincides with progression along the knowledge axis from genome sequence towards medical application. We have specifically note the difference in referring to function *in vitro* from function *in vivo* because of its relevance to utilizing available data to analyze specific problems, i.e. the relevance of laboratory-convenient assays in predicting function in real-time, in vivo, delivery systems. In an academic sense, this representation can be used to describe all of the biological and disciplinary interfaces listed above. In reality it is a two-dimensional representation of more scientific experimentation and results than could be hoped to be collected in one's lifetime. the value of this organization is in being able to "plot" where domain-specific databases exist, and what upstream (or downstream) linkages enable their use in predicting/evaluating other information. Does knowledge exist to enable a sequence database to be used to generate a structure or function database? The focus of our tool development is in attempting to solve problems by developing tools to enable the bridging between such data to occur and to evaluate which linkages are essential to a specific problem and potentially generalizable to larger classes of problems for which external value can be assigned. We term the construct of these linkages as providing us with a virtual database which enables the integration, at a distance, of these underlying databases.

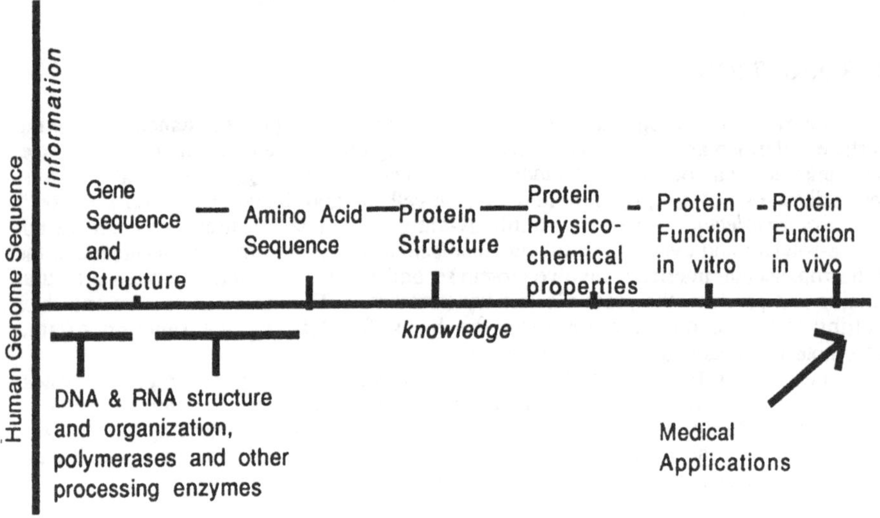

Figure 1. Model of the BioInformation Pathway for the Human Genome

CONVENTIONAL DATABASES COMPRISING THE VIRTUAL DATABASE

The conventional categorization of existing databases which are available for access and integration within biological problem-solving includes, but is not limited to:

<u>nucleic acid sequences</u>:
1. GenBank (GB)
2. EMBL

3. Backbone Database (NCBI)
4. genome sequences for model organisms
5. human chromosome databases

amino acid sequences:
1. Protein Identification Resource (PIR)
2. SwissProt (Swiss)
3. X-ray crystallographically determined protein structures (PDB)
4. natural mutant hemoglobin sequences (HMB)
5. Bio-engineered mutants (PERI)

structural databases:
1. Protein structures from x-ray crystallography (PDB)
2. Protein structures from nuclear magnetic resonance (PDB and NMR)
3. small molecule x-ray crystallographic structures (CCDB)
4. Protein secondary structure analysis: Greer-Levitt algorithm (G-L); Kabsch-Sanders algorithm (K-S).

clinical databases:
1. Hemoglobinopathies (HMB)
2. Coagulation Factor IX (FIX)

BASIS FOR REPRESENTATIONS AND DATABASE INTEGRATION

The basis for developing the representations we have used to integrate the underlying databases involves developing the concepts of inheritable characteristics which a protein exhibits. These are shown below:

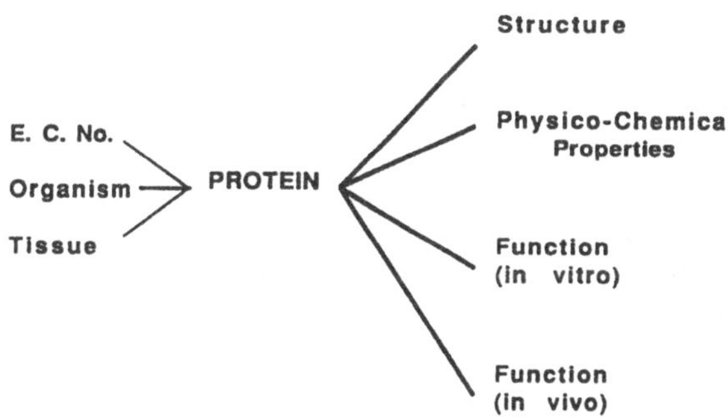

Figure 2. Protein Characteristics

Within these concepts, a set of descriptors can be listed which reflect the individual characteristics which are measurable, calculable or definable. Examples of these descriptors, within these concepts, are presented below:

structural parameters: folding domains; sequence of secondary structure (substructure); substructure composition; substructure hierarchy; cis-peptides; deviations from ideal configuration in distances or angles;

structure determination: structural resolution; structural refinement method; method of data collection; crystallization conditions, e.g. solvent, pH, temperature; presence of non-crystallographic intra- and inter-molecular symmetry;

physico-chemical parameters: molecular volume, solvent accessible surface; rugosity (surface area to volume ratio); ellipsoid of revolution parameters, e.g. axial ratio through center of mass, orientation; radius of gyration; molecular dipole moment, e.g. backbone and side-chains; packing density; bulk hydrophobicity; amphiphilicity; molecular weight; link to spectroscopic data bases;

function (in vitro): substrate specificity; inhibitor specificity; binding pocket based on substrate/inhibitor contacts; mechanism; binding constant/turnover rate; antigenic determinants;

function (in vivo): phylogenetic source of protein; organ site of activity; environmental conditions at site of activity; substrate specificity; inhibitor specificity; cofactors; binding constant/turnover rate; participation in metabolic pathway or cascade; link to nucleic acid data base sequences, e.g. exon boundaries; Enzyme Commission classification; membrane localization; post-translational modification, e.g. glycosylation, sites and type; isolatable form, e.g. zymogen, prepro-, proenzyme.

Focusing on these descriptors, the existing databases are observed to contribute data to multiple characteristics. PDB contains data on specificity and conformational response to in vitro and in vivo interactions, solvent perturbation, pH and temperature variation, site-directed mutations (natural and bio-engineered), variation in organism, evolutionary relationships, etc., and it can thus be used to address a variety of questions by appropriate selection of the constituent data. In addition, flexibility is required to adapt to new descriptors which might be developed and do not exist within the present database. Examples of concepts and their potential database sources are listed below:

Structure	Databases(examples)
primary structure	
nucleic acid	GB, EMBL
exon/introns	ATC
amino acid	PIR, Swiss
mutations	
naturally occurring	HMB, FIX, ATC
bio-engineered	PERI, ATC
super-families	PIR
secondary structure	
composition	PDB
ordering	PDB, G-L, K-S
predicted or modeled	
substructures	ATC
domains	ATC
tertiary structure	
x-ray crystallography	PDB
nuclear mag. resonance	NMR
modeled by homology	
quaternary structure	
crystal packing	ATC

Physico-chemical Properties

intrinsic	
hydrophobicity	derived
dipole moment	ATC
solvent accessibility	derived
van der Waals surface	derived
extrinsic	
molecular recognition	ATC
molecular dynamics	Traj

Function- in vitro

structural change	
x-ray crystallography	PDB
nuclear mag. resonance	NMR
Fourier transform infrared	ATC
Circular Dichroism	ATC-USDA
molecular interactions	
recognition	ATC
specificity	
kinetics	
thermodynamics	
solvent interactions	ATC
metal/ion interactions	ATC
solution vs crystal environment	ATC

Function- in vivo

biological pathways	ATC
molecular recognition	
molecular specificity	
clinical observations	HMB, FIX, ATC
chromosomal abnormalities	various, ATC
post-translational modification	

Note: ATC represents proprietary databases under development within the Bioinformatics Program, Diagnostics Sector, Amoco Technology Company.

COMPUTATIONAL TOOLS TO SUPPORT DATABASE INTEGRATION THROUGH DATA REPRESENTATION

The key to integrating data/information from the various experimental and computational areas described above is dependent upon the ability to: 1) represent the data in a form which minimizes biases; and 2) optimize access to structural, functional and physico-chemical data through the use of common algorithms. This general approach at database integration is indicated in the figure above, where we note the use of data representation and abstraction as a means of integrating information from different databases. We believe that our approach is consistent with object-oriented methodologies rather than using discrete pointers between databases, and we believe will be more adaptable to address complex relationships which will be needed to solve new problems. We provide, here, an introduction to this method for data integration.

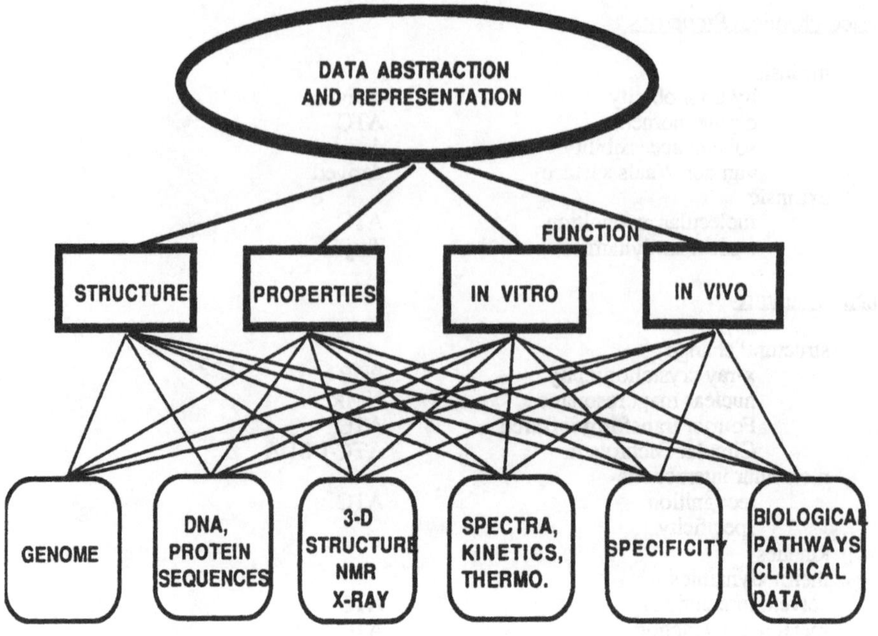

Figure 3. Databases

We describe our methods for representation, analysis and comparison in terms of their dimensionality, i.e. one-dimensional representations relate descriptors to a relative position within the nucleic acid/amino acid sequence, using the sequence to derive the list or ordering; two-dimensional representations also use a sequence-based ordering to generate a matrix containing descriptors which relate pairs of sequence elements, i.e. residues, which may not be contiguous; three-dimensional representations relate descriptors to positions relative to a defined coordinate system, typically Cartesian or polar coordinate space; and, four-dimensional representations involve descriptors which incorporate time with lower dimension descriptors. Structural, functional and physico-chemical descriptors using each of these formats are described below, along with their respective advantages and limitations. An important benefit results from the regions of overlap or redundancy which exist amongst these methods of representation as it enhances the opportunity to make observations and to validate their consistency.

1. <u>One-dimensional representations</u> use the position within a sequence to represent a descriptor based on that specific sequence element, or evaluated within a window of sequence elements which bounds that specific sequence element. With this representation we use descriptors for:

1. structure- Linear distance plot (LDP); bond dihedral angle plot(BDA)
2. physico-chemical properties- as derived from statistical analysis of observed protein/nucleic acid structures, and including hydropathy, flexibility, bulk, surface area, pK, etc.
3. functional properties- relative dipole orientation of sequence elements

Note: the LDP of trypsin is shown in the figure below.

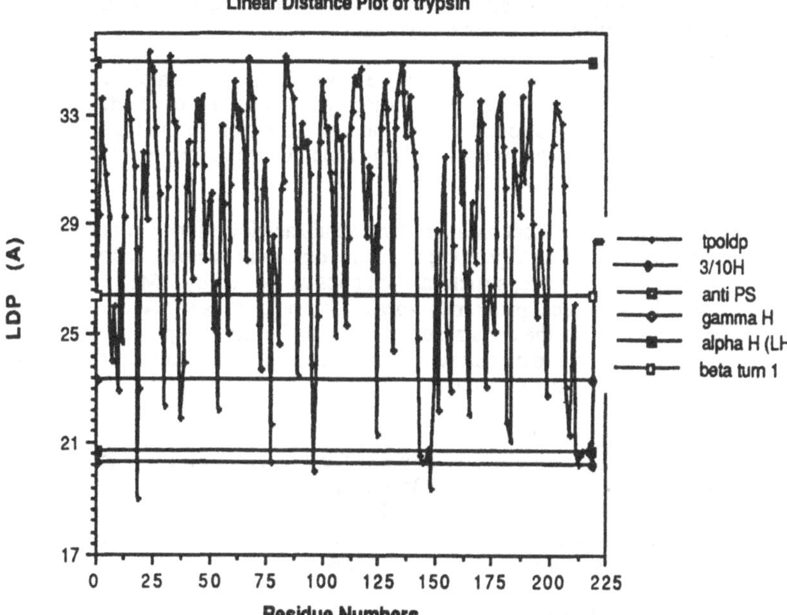

Figure 4. Trypsin Plot

2. Two-dimensional representations are based on the generation of an n-dimensional square matrix, where n is the number of elements within the sequence list. With this representation format we use descriptors for:

1. structure- distance matrix, partitioned distance matrix, both intra- and inter-molecular
2. physico-chemical properties- weighted hydrophobicity, base-pairing potential
3. function- electrostatic potential matrices for backbone, side-chains including dipole-dipole, charge-dipole and charge-charge interactions, dipole-interaction matrix

3. Three-dimensional representations involve the direct referencing of the structure/function/property descriptors to an external (cartesian) or internal coordinate (polar) reference frame. These representations comprise the more typical manner for viewing the three-dimensional structure with interactive, three-dimensional computer graphics displays. The descriptors which we use in this form include:

1. structure- atomic coordinates (orthogonal reference frame), fractional atomic coordinates (crystallographic reference frame), polar plot around designated atom/point centers, molecular graphics representation
2. physico-chemical properties- solvent accessible surface, van der Waals surface, molecular dipole orientation

3. function- hydrodynamic shape, molecular electrostatic potential surface, ion/solvent channels, internal cavities

4. Four-dimensional representations involve those descriptors which contain an element of time in conjunction with other relevant descriptors. Thus any of the descriptors described above could be utilized in a four-dimensional representation due to fluctuation of structure, function or physico-chemical properties with time. Those descriptors which we identify to have specific temporal components are:

1. structure- dynamical nature of a molecule as viewed by computation of a molecular dynamics trajectory; structural disorder observed in crystallographically determined structures; structural dynamics as viewed by NMR
2. physico-chemical properties- measurable properties which change with time, principally due to the occurrence of functionally-derived processes
3. function- enzyme cascades, metabolic pathways, allosteric processes as mapped through sparse matrix methods

5. Petri Net Analysis

The Petri net is presented as a model for a discrete-event system. It is utilized in our representation and analysis of (bio)chemical pathways to enable quantitative pathway comparison, discrete-event simulation and analysis of complex pathway behavior. Petri Net methodology supports application of this approach in systems where incomplete data may exist, e.g. kinetic parameters or even pathway linkages, to allow for semi-quantitative modeling.

A Petri net (PN)' is a graph formed by two kinds of nodes, called places (p_j) and transitions (t_j). Directed edges, called arcs, connect places to transitions, and transitions to places. A non-negative integer number of tokens is assigned to each place, and it can vary based on the state of the Petri net. Each arc has a weight, a positive non-zero integer, assigned to it. Pictorially, places are represented by circles, transitions by bars or boxes, arcs by lines ending in an arrow, and tokens as black dots placed in the circles. Generally if there is no arc-weight specified on the graph we assume it to be equal to one.

APPLICATION OF THE DISTRIBUTED DATABASE

The virtual database, described above, exists as a series of domain-specific databases, residing on a variety of computers, at a series of international sites and linked through a set of tools which focus on data representation and abstraction. The utility of this construct can best be seen by examining an application of its wide-ranging capabilities. The problem we present addresses the analysis of Factor IX, an enzyme in the blood coagulation cascade for which certain chromosomal abnormalities, i.e. natural single site mutations, yield a coagulation deficiency termed hemophilia Bm.

Information which exists, initially, is that Factor IX contains a protease domain which appears to be a serine protease and that the gene that physically maps to this condition, and therefore is responsible for the DNA coding for this enzyme is located on the X chromosome. We describe the linkages, accomplished with the representations and tools described above, to a variety of specific databases and the composite information which has been derived from the virtual database.

Structural information at the level of primary structure enabled the mapping of the protease domain of Factor IX (FIX) to the family of trypsin-like enzymes. Three-dimensional structural information existed for several members of this family, and the identification of a structural core of conserved three-dimensional structure and conserved primary structure could be observed and mapped to FIX. Identification of structurally variable regions of these enzymes, and their mapping to observed in vitro function (e.g. binding of calcium ions), suggested a potential site for modification of FIX which might relate to physiological function. Analysis of trypsin-like enzymes indicate a correlation with physiological function and the ability to modulate inhibitor/substrate interactions through binding to this site. Representation and searching of the CCDB revealed presence of candidates for binding to this site which could modulate in vitro and potentially in vivo function of these enzymes. Conformational response of the enzymes, in contrast to their binding activity, specificity and reactivity, could be evaluated against both in vitro and in vivo inhibitors. Spectroscopic measurements, FT-IR and CD, both revealed common patterns of structural response which differ between in vitro and in vivo enzyme response. Correlation of spectroscopy and x-ray crystallographic analysis was made by evaluation of the solution conditions used for crystallization and evaluation of the environments influence on the enzyme's physical state. Molecular dynamics analysis indicated the occurrence of conformational perturbation outside of the classically defined active site in this enzyme family. Experimental observations revealed that these regions were localized, and correlated specifically with interaction with in vivo and not in vitro inhibitors. Mapping of the chromosomal changes which result in clinical observations of hemophilia Bm indicate that these same non-active site regions of the enzyme are identifiable from the sequence-clinical data as well. Evidence was compiled concerning the mechanism of action of the disease, potential means for modulating genetic defects and rationally designing site-directed mutations or drug/modulator design or selection from large databases through the integrated use of the distributed database, relying on the tools for data abstraction and representation to integrate the data. In addition, representation and analysis of the biological pathway in which FIX participates, blood coagulation, and correlation with physical mapping within the region on the chromosome for the FIX gene, suggest alternative sites for modulating and controlling the impact of genetic disease and the potential for secondary or tertiary effects which result from the genetic event itself as well as the proposed means for controlling it. Experiments to verify these hypotheses and extend the existing knowledge base have been planned and the results will further extend the underlying data and concepts used in this application.

CONCLUSIONS

We have described the ability to integrate a wide-range of data and databases, extending over many computer systems, sites, disciplines and areas of concentration, through the use of data abstraction and representation techniques. The ability to define higher order concepts which can unify the underlying data, e.g. structure, properties, function (in vitro and in vivo) serve as a key to developing this approach. The distributed database approach outlined here has proven to be flexible and extendible in that we can readily incorporate new data into the process without interruption of existing linkages. The complexity of the problems we have applied this approach to, and the ability to identify linkages in information which are not readily apparent even to specialists in a particular application, suggest the potential for generalization of this approach. We are currently extending our research into database mining using neural network methodologies to use this virtual database concept.

REFERENCES

1. Bernstein, F.C., T.F. Koetzle, G.J.B. Williams, E.F. Meyer, Jr., M.D. Brice, J.R.Rodgers, O.Kennard, T. Shimanouchi and M. Tasumi, J. Mol. Biol. 112, 535-542, 1977

2. Wilcox, G.L, Poliac M. and Liebman, M.N., Tetra. Comp. Let., 191-220 1990

3. Liebman, M. N., and Brugge, A. L., Santa Fe Institute Studies in the Sciences of Complexity, Volume VII, eds. G. Bell and T. Marr, Addison-Wesley Longman Publishing Group, 183-202, 1989

4. Liebman, M. N., J. Comp.-Aided Molecular Design 1, 323-341 1987

5. Liebman, M. N., J. Indus. Microbiology 3, 127-137 1988

6. Liebman, M. N., Enzyme 36, 115-140, 1986

7. Prestrelski, S. J., Williams, A. L. and Liebman, M. N. Proteins, 14, 430-439 1992

8. G. C. Klein and M. N. Liebman, unpublished results

9. Prestrelski, S. J., Byler, D. M. and Liebman, M.N., Proteins, 14, 440-450 1992

10. Liebman, M.N., Venanzi, C.A., Weinstein, H., Biopolymers 24, 1721-1758,1985

11. J. Jesson and M. N. Liebman, unpublished results

12. Reddy, V. N., Mavrovouniotis, M. L. and Liebman, M. N., in press

DOMAINS AND SUPERLATTICES IN SELF-ASSEMBLED MONOLAYERS OF LONG-CHAIN MOLECULES

Joseph Hautman and Michael L. Klein

Department of Chemistry, University of Pennsylvania
Philadelphia, Pennsylvania 19104-6323

INTRODUCTION

Self-assembled monolayers (SAMs) are made up of long-chain molecules chemisorbed to a solid substrate.[1,2] Pioneering investigations into fabrication methods and properties of these films [1-10] have generated optimism that they may provide a practical and versatile means of creating chemically tailored surfaces. Such techniques could be used, for example, to facilitate the design of surface properties for applications such as coatings or optical devices, or as substrates for basic research involving wetting [11], adhesion, lubrication or biomolecules [12]. This potential usefulness is, in part, due to the chemisorption of the headgroups which results in films which are more robust and possibly more well characterized than, for example, the physisorbed Langmuir-Blodgett monolayers. Alkylthiol molecules with a variety of headgroups, $S(CH_2)_nX$, have been used to form uniform, close packed and highly oriented self-assembled monolayers on gold substrates.[3-10] Because the molecules are oriented, with the sulfur headgroup bound to the gold surface, the chemical characteristics of the exposed surface are dominated by the properties of the substituted tailgroup. There appear to be few defects detectable to electrochemical and wetting experiments,[13,4] however, as the structures of even the simplest of these layers are probed in more detail, fairly small domains sizes, superlattice structures, and other types of nonuniformities have been reported.[14,15]

The tethering of the tailgroups, *via* a single hydrocarbon chain, to headgroups at fixed chemisorption sites creates possibilities for unique packing structures within the layers. The chain tails have a fixed in plane density and long-range order guaranteed by the connection to the headgroups, however, the *local* packing requirements of the tails may depend dramatically on the substituted tailgroup. The system is likely to be frustrated when the chain tails prefer an in-plane structure that is incommensurate with the symmetry of the headgroup binding sites. This frustration can lead to the formation of complex unit cells, defects and domains at the exposed surface of the film. The nature of these deformations would likely depend on a variety of parameters including the type and strength of the tailgroup interactions, chain length and temperature.

A simple one-dimensional model illustrates the competing interactions present in a tethered system. Consider a row of rigid rods ("chains") with one end ("headgroup") fixed on a linear "surface" with a fixed spacing a. Competing with this imposed length scale is the chain-chain interaction, which involves the whole length of the molecule, and yields a preferred spacing b . The two different length scales do not necessarily lead to frustration in one dimension since the molecules can tilt, as shown in Figure 1(a). However, if we allow a third interaction between the chain "tailgroups", and a corresponding preferred tailgroup spacing c, there *will* be some degree of frustration because uniform tilting does not change the tailgroup spacing. If the tailgroup interaction is strong enough to dominate the chain-chain interaction, the 1-d system could, at low enough temperatures, form domains as shown in figure 1(b). This system is similar to the case of an overlayer of particles at fixed, commensurate, density, however, the tethered system has a limit on the size of the domains since a given tailgroup cannot deviated from the corresponding headgroup position by more than the length of the chain. Without the tethering constraint, one would expect (for the case $c < a$) a condensation of the tailgroups with unlimited cluster size. But. in the tethered case, the chain length, l, limits the domain size to

$$L < 2\,l\,c/|a-c| \quad (1)$$

This is an upper bound which overestimates the actual domain size because chain stiffness and steric hindrance will play some role in limiting the deviations of the chain tail from the headgroup position. In the case $c > a$ one could expect kinking or "2nd layer" promotion with the same upper bound on the domain size.

In the two-dimensional (2-d) case the likelihood of frustration is increased since the chain tilting can accommodate the chain-chain interactions only in the direction of the tilt. One can thus envision the formation of domains, perpendicular to the tilt direction, even in the absence of strong tailgroup interactions (See Figure 2). The tilting of the chains breaks the symmetry of the headgroup lattice, so that any domain formation arising from tailgroup interactions is likely to be anisotropic and correlated to the tilt direction. The above limit on domain size applies to any given direction in the 2-d case, but the relevant length scales can be different in different directions in the plane. The 2-d case is also more complex because, in addition to the spacing, the preferred *symmetry* of the tailgroup lattice can be incommensurate with the headgroup lattice. In the following we explore how these effects are manifested in models of actual self-assembled monolayers with various chain molecules, tailgroups and surface symmetries.

SELF-ASSEMBLED MONOLAYERS: ALKYLTHIOLS ON GOLD

The methyl terminated alkylthiol molecules $S(CH_2)_nCH_3$ self-assemble into a monolayer of thiolate molecules with the sulfur headgroups chemisorbed in a $\sqrt{3} \times \sqrt{3}$ R 30°

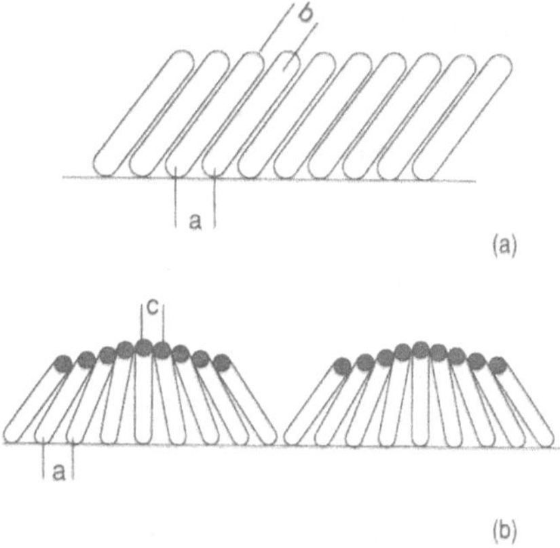

Figure 1. Schematic illustration of the role of competing interactions in tethered systems. The simplified "chain" molecules are bound to a one-dimensional "surface". (a) Two different characteristic lengths "a" for the fixed headgroup distance and "b" for the preferred chain-chain separation are compatible because of the tilt degree of freedom. (b) A third length "c" for the preferred tailgroup-tailgroup distance causes frustration and possible domain formation.

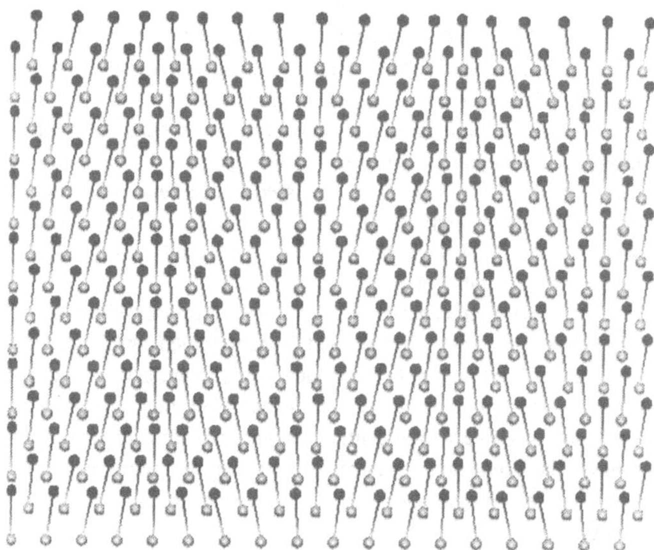

Figure 2. Schematic example of possible domain structure in a two-dimensional layer of tethered chain molecules. The "headgroups" (light circles) are on a perfect triangular lattice while the relatively free "tailgroups" (black circles) may prefer an inhomogeneous density distribution.

triangular lattice on the Au(111) surface.[5] The nearest neighbor (nn) headgroup spacing is 5.0 Å. The chains are close packed, predominantly all-*trans* and, for large enough n, tilted with respect to the surface normal. Models of these monolayers have been developed in earlier simulations which reproduced many of the observed structural and dynamical properties.[16] However, reported correlation lengths (for freshly deposited layers made from C_{14} chains) of ~ 90 Å in x-ray and ~ 40 Å in He beam diffraction experiments suggest a complex domain structure.[14] Furthermore, both probes find evidence of a 4-molecule unit cell modifying the basic $\sqrt{3} \times \sqrt{3}$ symmetry.[14,15]

On the Au(100) surface, LEED experiments indicated that alkylthiol molecules chemisorb at sites which form a square lattice with a 4.54 Å nn spacing.[5] On the other hand, helium beam diffraction, which probes the structure of the exposed chain tails alone,[17] shows a more complex structure indicating that the chain tails adopt an orthorhombic arrangement more similar to the preferred packing in typical alkane crystals. These two results suggested the possiblity of change of symmetry in the inplane structure from the chemisorbed headgroups to the tailgroups at the exposed surface. Several aspects of the structure remain unresolved and a more recent LEED study disputes some of the conclusions of the earlier experiment. [18]

These systems have been studied using Molecular Dynamics (MD) simulations with a simplified model in which the CH_2 or CH_3 group is represented by a single spherical pseudoatom. The C-C bond lengths are constrained in the model, but the chains are otherwise flexible with bond angle bending and torsional degrees of freedom.[19, 20] The details of the SAM model are the same as that described in Ref. 16. The pseudoatoms interact with a Lennard-Jones potential with parameters $\sigma = 3.905$ and $\varepsilon = 59.4$ K, for the CH_2 or $\varepsilon = 88.1$ K, for the CH_3 group. The chain backbone zig-zags with a bending angle minimum at $\theta = 109.5$ degrees. The van der Waals radius of the pseudoatoms and the zig-zag combine to yield an effective chain "thickness" which is close to the typical spacing of the Au chemisorption sites. The symmetry of the binding sites on the different Au surfaces is obtained by suitable choice of boundary conditions, headgroup-headgroup interactions and surface corrugation, which combine to force the headgroups to adopt the prescribed in-plane structure. Our analysis of the simulations has focussed on the behavior of the chains given the imposed geometry for the headgroup binding sites. Subtleties of the surface interaction and, consequently, distortions of the headgroup lattice may influence the packing of the tails[16], but this is not treated reliably in the present model.

The -CN terminated monolayer on the Au(111) surface represents another type of competition between the tailgroup interactions and the requirements of the headgroup lattice. The -CN tailgroup has a dipole moment which leads to strong, and long-ranged, tail-tail interactions. Again we use a pseudoatom model for the CH_2 groups, and the terminal CH_2, C and N are treated as three interaction centers carrying partial charges 0.269, 0.129 and -0.398 e respectively. The interaction parameters were taken from a simulation of acetonitrile which treated the molecule as a rigid body.[21] Here we have allowed a bending motion about the terminal 180 degree bond angle with a force constant derived from the bending frequency of acetonitrile.[22] A modified Ewald method was developed and employed in order to include the long-ranged coulombic interactions in a two-dimensionally periodic system.[23] The potential model for the -CN terminated SAM has been detailed in Ref. 24.

SIMULATION RESULTS

Methyl Teminated Chains on Au(111)

Of the alkylthiol molecules studied here, the CH_3 terminated chains on the Au(111) surface provide the least amount of mismatch between the packing preference of the tails and the headgroup lattice. In this case the headgroup lattice is only a few percent different from the typical lattice constants found in alkane crystals. Part of this mismatch can be accommodated by the tilting of the chains. Although it does not alter the tailgroup-tailgroup distance, uniform chain tilting is a degree of freedom which is always available to reduce a component of the chain-chain distance in the direction of the tilt. Given the small mismatch it is not surprising that domain formation was not observed in the first simulations of this system which consisted of 90 molecules in a simulation cell measuring about 44 Å on a side.[16] Larger simulation cells containing 224 molecules have since been used in MC (Monte

Figure 3. Configuration from a molecular dynamics simulation of 224 $S(CH_2)_{16}CH_3$ chains on a Au(111) surface. The sulfur headgroup is shown as a light sphere and the zig-zag backbone shows the bonds connecting the CH_2 groups.

Carlo) simulations.[25] The results seem to indicate irregular domains containing molecules with a distinct tilt orientation. We have now carried out larger MD simulations, containing 224 molecules in the simulation cell (and periodic boundary conditions). The resulting system, shown in Fig. 3, still does not show evidence of domains of the kind seen in the MC study. At these temperatures the model has been observed in MD to have the freedom to spontaneaously change tilt direction from nn to nnn. In MD this change of state utililizes a collective motion which may be inaccessible to the MC simulation. The apparent domain formation in the MC system may be a manifestation of this shortcoming.

The introductory discussion suggests that for large enough "incommensurability" $|c\text{-}a|/a$ we expect to see domain formation. By artificially decreasing the chain pseudoatom radius, we *were* able to obtain superlattice modulation of the type shown in Fig. 2, but only for unrealistically small values of the Lennard Jones diameter ($q < 3.8$ Å) and very low temperatures. ($T \sim 150$ K).

Methyl Teminated Chains on Au(100)

In the present studies we have simulated the $S(CH_2)_{21}CH_3$ molecules on the Au(100) surface by imposing the square geometry for the chemisorbed S headgroups with nn distance 4.54 Å. Given the uncertainties, mention earlier, regarding the validity of this headgroup lattice we do not represent our results as indications of the actual structure of the monolayer but rather as another example of a possible rearrangement of molecules in response to the different packing requirements of headgroups and chain tails. The molecular dynamics simulations of methyl terminated molecules on the Au(100) surface show the remarkable reorganization of chains from a square symmetry in the headgroup plane to an orthorhombic structure in the tailgroup plane. The structure of the monolayer is illustrated in Figure 4, which shows the time*averaged* positions of the molecular backbones. This simulation was done with 72 chain molecules at T = 300 K. The square headgroup lattice is evident from the open circles representing the position of the sulfur atoms.

The head and tailgroups are isolated in Figure 5 to show the respective square and orthorhombic symmetry. The tailgroup lattice is closer to a triangular symmetry than the structure deduced from fitting to the He-beam diffraction and LEED observations.[17] Note that the molecular configurations shown in Fig. 4 do not represent a unique solution to the required change of symmetry from the headgroup plane to the tailgroups. The rearrangement could, most simply, be accomplished by shifting the tails in every other row by on half of a lattice spacing resulting in a two molecule unit cell. The structure shown in Figure 4 involves at least three different molecular configurations - chains bending to the right and the left, and nearly straight chains. Although there are defects, possibly due to the small size of the system, a repeat distance of three molecules (equivalently, a 3-molecule unit cell) is suggested. This arrangement, perhaps, involves less of an expense in the bending energy of the chain backbones. Further simulation with larger systems is called for to resolve the the structure preferred by this model. It is clear from this structure that it is imperative to include the flexibility of the chain in the modelling of these layers. Even in the absence of *gauche* conformations, the bending of the chain backbone is necessary ingredient, and would likely be required to obtain a good fit to experimental scattering data for a system of the type shown in Figure 4.

Figure 4. Time average atomic positions from a molecular dynamics simulation of $S(CH_2)_{21}CH_3$ chains on a Au(100) surface. The layer is shown from an angle about 10 degrees from the normal to the plane in order to clarify the molecular configurations.

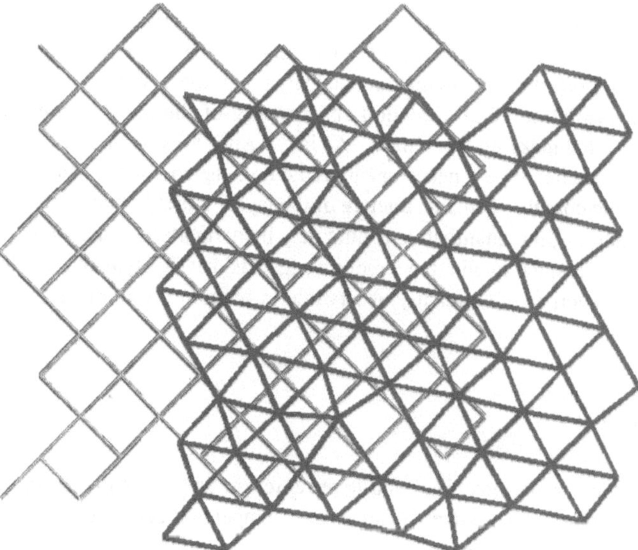

Figure 5. The two lattices formed by the average headgroup (gray) and tailgroup (black) positions for $S(CH_2)_{21}CH_3$ chains on a Au(100) surface. This view is along the surface normal.

-CN Teminated Chains on Au(111)

Earlier simulations of -CN terminated self assembled monolayers indicated that the dipolar tailgroups strongly prefer a ferroelectric packing arrangement in which the dipoles lie in the plane of the surface.[24] The ferroelectric geometry is not by itself incompatible with the triangular headgroup lattice which we have assumed for the Au(111) surface, however, the in-plane orientation of the dipoles requires conformational defects within the molecules, and the preferred tailgroup-tailgroup distance is somewhat smaller than the S-S distance. This mismatch leads to the formation of rather small domains separated by sharp steps. Within the domains, the -CN groups form an orthorhombic lattice which is distorted with respect to the underlying headgroup lattice. Because of the small size (90 chain molecules) of the simulation cell used in the earlier studies, it was not clear how the domain structure might be affected by the imposed periodic boundary conditions. In order to investigate the domain formation more reliably, the present simulations were carried out with a larger system of 224 molecules.

Since the generation of *gauche* defects inhibits the reorganization of the chains, the larger -CN terminated system was first run using a model with a simplified torsional potential to eliminate the formation of *gauche* defects. The modified intramolecular potential for the dihedral angle is a harmonic potential with the same minimum and curvature as the *trans* minimum in the full torsion potential. The molecule retains its torsional flexibility but has no minimum corresponding to the *gauche* configuration. The elimination of *gauche* configurations reduced the entanglement of the chains and permitted a more rapid equilibration. This model was equilibrated with 224 molecules, at 250 K, to the structure shown in Figure 6. Starting from this configuration, the system was then run on with the full torsion potential. While a few (less than 1%) dihedral angles could be found in the gauche configuration, the basic terraced domain structure remained unchanged. The domain boundaries are clearly visible in the larger system. The superlattice consists of striped domains with a four molecule repeat distance. The domain structure is similar to that observed in the smaller system, and the domain size is not commensurate with the size of the simulation cell. This leads to defects in the superlattice but provides evidence that the domain size is not just a product of the simulation boundary conditions. The structure is created by a fairly complex variation of tilt direction and bending of the chain backbone in the four successive rows that make up the domain. As in the Au(100) structure discussed above, the chain flexibility seems to be a necessary ingredient of the structure of the -CN terminated layers. Since the complex domain structure of the layers arises from a competition between different interactions, the quantitative results may be particularly sensitive to details of the model. Nevertheless, the qualitative structure suggests that this type of superlattice in the -CN terminated layers may be readily observable in imaging experiments such as STM or AFM.

SUMMARY

The three simulations discussed here represent only a small sampling of a larger variety of structures which are likely to arise from competition between headgroup geometry, chain packing requirements, and tailgroup interactions in self-assembled monolayres. Given that the headgroup lattice is fixed it is possible that the formation of complex surface

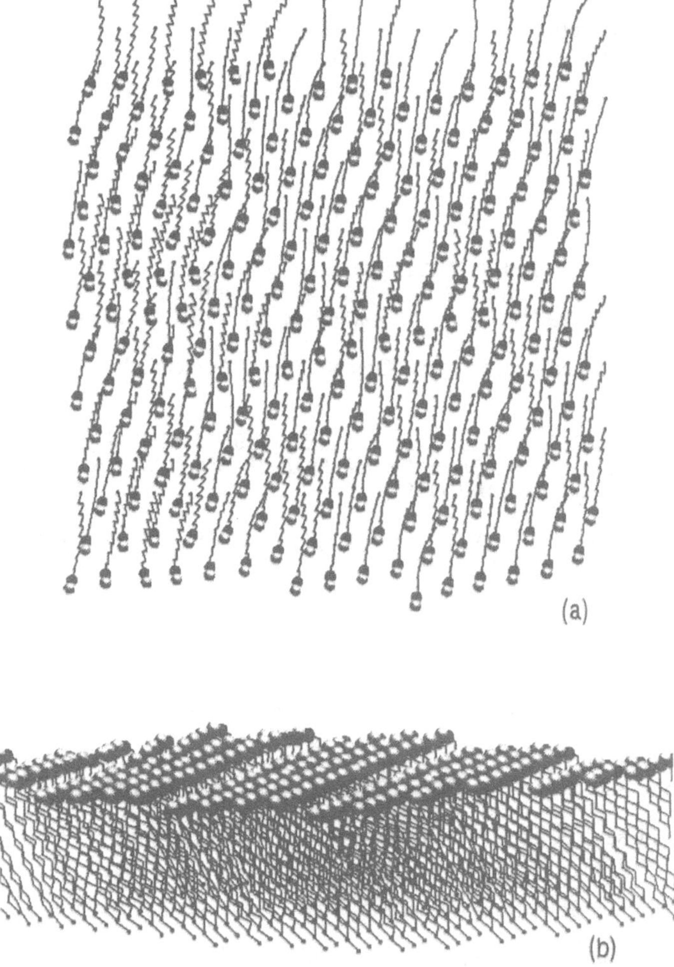

Figure 6. Configuration from simulation of 224 $S(CH_2)_{15}CN$ chains on a Au(111) surface. A "top" view is shown in (a) and the terracing of the -CN surface is elucidated in an oblique view (b) of the monolayer. The C and N atoms of the polar tailgroup are shown (respectively) as black and white spheres. The system shown employs a harmonic torsion potential which does not permit *gauche* defects.

structures may be the rule rather than the exception in these and other chemisorbed systems. The composition of the chains and the substrate, as well as parameters such as chain length and temperature, may all affect the *structure* of the exposed surface. Since the richness of the structural phases and likely temperature dependence offer further possibilities which can be exploited in tailoring surface properties with SAM films, it is important that these factors be considered along with the simple chemistry of the substituted tailgroup. Computer modeling may have an important role to play in sorting through these possible systems. Recent work [26,27] has indicated that the effects of defects and superlattice structures on surface characteristics such as wettability, adhesion, and surface specificity may also be amenable to simulation study.

REFERENCES

1. J. D. Swalen, D. L. Allara, J. D. Andrade, E.A. Chandross, S. Garoff, J. Israelachvili, T. J. McCarthy, R. Murray, R. F. Pease, J. F. Rabolt, K. J. Wynne and H. Yu, *Langmuir*, **3**, 932 (1987); J. Sagiv, *J. Am. Chem. Soc.*, **102**, 92 (1980).

2. A. Ulman "Introduction to Ultrathin Organic Films" Academic Press, SanDiego, CA (1991).

3. R. G. Nuzzo and D. L. Allara, *J. Am. Chem. Soc.*, **105**, 4481 (1983); C. D. Bain and G. M. Whitesides, *Angew. Chem. Int. Ed. Engl.*, **28**, 506 (1989); A. Ulman, *Angew. Chem. Int. Ed. Engl. Adv. Mat.*,(1990).

4. C. D. Bain, E. B.Troughton, Y.-T. Tao, J. Evall, G. M. Whitesides and R. G. Nuzzo, *J. Am. Chem. Soc.*, **111**, 321 (1989).

5. L.S. Strong and G.M. Whitesides, *Langmuir*, **4**, 546 (1988).

6. S.M. Stole and M.D. Porter, *Langmuir*, **6**, 1199 (1990).

7. R.G. Nuzzo, E.M. Korenic and L. H. Dubois, *J. Chem. Phys.*, **93**, 767 (1990).

8. R.G. Nuzzo, L.H. Dubois and D.L. Allara, *J. Am. Chem. Soc.*, **112**, 558 (1990);.L. H. Dubois, B.R. Zegarski and R.G. Nuzzo, *J. Am. Chem. Soc.*, **112**, 570 (1990).

9. C.E.D. Chidsey, G. Liu, P. Rowntree, and G. Scoles, *J. Chem. Phys.*, **91**, 4421 (1989); *Langmuir* (1991).

10. L.H. Dubois and R.G. Nuzzo, Ann. Rev. Phys. Chem., **43**, 437 (1992).

11. P. Silberzan and L. Leger, Phys. Rev. Lett. **66**, 185 (1991).

12. K. Prime, G.M. Whitesides, *Science*, **252**, 1164 (1991); S.M. Amador, J.M. Pachence, R. Fischetti, J. P. McCauley, Jr., A.B. Smith III, and J.K. Blasie, Langmuir, **9**, 812 (1993).

13. M.D. Porter, T.B. Bright, D.L. Allara, C.E.D. Chidsey, J. Am. Chem. Soc. **109**, 3559 (1987).

14. N. Camillone, C.E.D. Chidsey, G. Liu, and G. Scoles, J. Chem. Phys. **98**, 3503 (1993).

15. P. Fenter, P. Eisenberger, and K. Liang, Phys. Rev. Lett., **70**, 2447 (1993).

16. J. Hautman and M.L. Klein, J. Chem. Phys. **91**, 4994 (1989); **93**, 7483 (1990).

17. N. Camillone, C.E.D. Chidsey, G. Liu and G. Scoles, J. Chem. Phys. **98**, 4234 (1993).

18. L.H. Dubois, B.R. Zegarski, and R.G. Nuzzo, J. Chem. Phys. **98**, 678 (1993).

19. J.-P. Ryckaert and A. Bellemans, J.Chem. Soc. Faraday Discuss. **66**, 95 (1978).

20. P. van der Ploeg and H.J.C. Berendsen, J. Chem. Phys. **76**, 3271 (1982).

21. H.J. Bohm, I.R. MacDonald, and P.A. Madden, Mol. Phys.,**49**, 347 (1983); D.M.F.Edwards, P.A. Madden, and I.R. MacDonald, Mol. Phys. **51**, 1141 (1984).

22. W.D. Rothschild J. Chem. Phys. **57**, 991 (1972).

23. J. Hautman and M.L. Klein, Mol. Phys., **75**, 379 (1992).

24. J. Hautman, J. Bareman, W. Mar, and M.L. Klein, J. Chem. Soc., Faraday Trans, **87**, 2031 (1991).

25. J.I. Siepmann and I.R. MacDonald, Langmuir, **9**, 2351, (1993).

26. J. Hautman and M.L. Klein, *Phys. Rev. Lett.*, **67**,1763 (1991) .

27. W. Mar and M.L. Klein, to be published.

MANIPULATING WETTING AND ORDERING AT INTERFACES BY ADSORPTION OF IMPURITIES

Donald J. Olbris[1] and Yitzhak Shnidman[2]

[1]Department of Physics and Astronomy
University of Rochester
Rochester, NY 14627
[2]Research Technical Support Services
Eastman Kodak Company
Rochester, NY 14650

INTRODUCTION

Many-body systems in contact with a heat reservoir and applied ordering fields typically exhibit a variety of phases characterized by a set of order-parameter fields $\varphi_i(\vec{r})$. In the continuous limit of a single-site mean-field approximation, the equilibrium state of the system is achieved by a certain set of order-parameter fields $\varphi_i(\vec{r})$. Assuming a slow spatial variation of the fields, a mean-field free energy of the form

$$G=\int d\vec{r}\Bigg[\sum_i \tfrac{1}{2}L[\{\varphi_i(\vec{r})\};\{\varepsilon_k\}][\nabla\varphi_i(\vec{r})]^2+\operatorname*{Tr}_{\{\varphi_i(\vec{r})\}}\Big(E[\{\varphi_i(\vec{r})\};\{\varepsilon_k\}]P[\{\varphi_i(\vec{r})\};\{\varepsilon_k\}]\Big)$$
$$+k_BT\operatorname*{Tr}_{\{\varphi_i(\vec{r})\}}P[\{\varphi_i(\vec{r})\};\{\varepsilon_k\}]\log P[\{\varphi_i(\vec{r})\};\{\varepsilon_k\}]+\sum_i h_i(\vec{r})\varphi_i(\vec{r})\Bigg] \quad (1)$$

is minimized at fixed values of the reservoir temperature T, interactions ε_k, and ordering fields $h_i(\vec{r})$. The trace Tr sums over all possible configurations of the order-parameter fields.

Equation (1) is composed of three terms. The first two represent the contributions of the self-consistent order-parameter interactions $E[\{\varphi_i(\vec{r}_1)\},\{\varphi_i(\vec{r}_2)\},\cdots,\{\varphi_i(\vec{r}_n)\};\varepsilon_k]$ to the mean configurational internal energy U, averaged over P, the probability density of the order-parameter fields, expanded and diagonalized to the lowest (second) order in their gradients. The next term corresponds to $-TS$, where $S=-k_B\operatorname{Tr}P\log P$ is the entropy. The last term represents coupling to applied ordering fields. The requirement that, at equilibrium, the set of order parameters should minimize (1) for fixed values of the

Theoretical and Computational Approaches to Interface Phenomena
Edited by H.L. Sellers and J.T. Golab, Plenum Press, New York, 1994

temperature, interaction parameters, and ordering fields is a manifestation of the maximum entropy principle of equilibrium statistical mechanics.

If the applied fields are translationally invariant, then in the bulk, far from system boundaries and interfacial regions (which have a vanishing contribution to the free energy in the thermodynamic limit), one has

$$\nabla \varphi_i = 0. \tag{2}$$

Assuming this property for a given parameter set, the variational problem stated above may admit either a unique solution, or multiple degenerate solutions. Each solution corresponds to a bulk phase, so that multiple solutions signify that, for a specified set of parameters, the system may coexist in two or more bulk phases separated by an interfacial region where the φ_i become position-dependent. The spatial distribution of the phases in the system is sensitive to boundary conditions and applied fields.

At equilibrium, an interfacial region (where φ_i are position-dependent) may exist either between two macroscopic regions of degenerate bulk phases, or between a boundary surface and a bulk phase. On the scale of the entire system, these surfaces and interfaces can be, in general, quite convoluted and curved. However, locally they are assumed to have a planar geometry, interpolating between two bulk phases, or a boundary surface and an adjoining bulk phase. In either case, an excess interfacial free energy is associated with this region, given by the difference between the position-dependent free energy (1) in that region and the translationally invariant free-energy of the bulk phase. Such an interfacial free energy will be denoted by $\gamma_{\alpha\beta}$, where the index α denotes either a bulk phase or a boundary surface, and the index β denotes the adjoining bulk phase.

Interfacial ordering plays an important role in such diverse phenomena as static and dynamic wetting, adsorption, layering, anchoring, adhesion, catalysis, and formation of Langmuir, Langmuir-Blodgett, and self-assembled monolayers. Ordering degrees of freedom may include positional, orientational, torsional, conformational, and other degrees of freedom, either of small isotropic or anisotropic molecules, or of oligomers, polymers, and colloids, as well as their collective degrees of freedom (e.g., curvatures). They play an important role in industrial and biological processes such as detergency, coating, and adhesion; manufacturing of plastic, sealants, paints, and liquid-crystalline displays; oil extraction, printing, photography, and xerography. The field is truly multidisciplinary, attracting the attention and efforts of workers in such diverse disciplines as physics, chemistry, chemical and mechanical engineering, metallurgy, materials science, biology, medicine, and pharmacology.

Compared with the study of bulk ordering phenomena, ordering at surfaces and interfaces poses considerable challenges for the investigator. Experimental difficulties include the weakness of interfacial contribution to the measured signal compared with the bulk contribution, which is almost always present, as well as the necessity for a very high spatial resolution. On the theoretical side, the breakdown of the homogeneity assumption and the resulting space dependence of order parameters, even on the mean-field level of approximation, increase the mathematical difficulty of the variational problem. This has been well recognized for some time, and specialized experimental and theoretical methods were developed to cope with these problems that are summarized in a series of recent books[1-3] and review articles.[4-7]

Here we focus on another source of difficulties that is frequently encountered in practice but has been little studied until recently. It has been recognized very early on that adsoprtion at interfaces may strongly affect the interfacial free energies, thus leading to interesting effects due to interplay of wettability and adsorption.[8-10] Hence interfacial properties may be extremely sensitive to the presence of minute amounts of surface-active impurities. This can happen when surface segregation of the impurities lowers the free energy of the system. Such effects are commonly observed at oil/water interfaces when minute quantities of amphiphilic oligomers or polymers are present.[11] Thus manipulation of the impurities distribution in systems of this kind may provide an unprecedented degree of control of interfacial properties, such as wettabilities, adsorption, and ordering. Achieving such control is a scientific challenge of considerable technological importance.

We have recently constructed and studied simple models of different, but closely related systems, where the impurities are simple hydrogen-bonding molecules, such as water or alcohols, at a molecularly smooth solid-fluid interface of self-assembled monolayers (SAMs). This was achieved by generalizing the well-known Cahn[12] model for wetting of pure fluids at homogeneous substrates. The distribution of the impurities in our model can either be annealed (i.e., in thermal equilibrium with the other degrees of freedom in the system), or quenched (i.e., frozen by some process into a configuration independent of the equilibrium between the other degrees of freedom and the external reservoirs), or both. The determination of the model parameters and the method, and the results of the numerical computations have been described by us in great detail elsewhere[13,14] for the simplest case of interplay of wetting and adsorption of impurities. Here we present the main ideas on which our method is based, as well as some of its more instructive results. We set them in a more general context that will allow us to argue that a host of similar phenomena in different, but analogous, systems can be addressed by similar methods.

HUMIDITY-DEPENDENT WETTABILITIES OF MIXED SELF-ASSEMBLED MONOLAYERS

Interfacial Free Energies and Wettability of Smooth Solid Substrates

Homogeneous Surfaces: Young Equation. Wettability of a solid surface by various liquids can be measured by the cosine of the contact angle of a small sessile drop coexisting with its vapor at the solid substrate. Let us denote the interfacial free energies per area associated with the solid-vapor, solid-liquid, and liquid-vapor interfaces by γ_{SV}, γ_{SL}, and γ_{LV}, respectively. Furthermore, let us define dimensionless solid-vapor and solid-liquid interfacial free energies, using the liquid-vapor surface tension to set the energy scale:

$$\tilde{\gamma}_{SV} = \frac{\gamma_{SV}}{\gamma_{LV}}\,, \tag{3}$$

$$\tilde{\gamma}_{SL} = \frac{\gamma_{SL}}{\gamma_{LV}}\,. \tag{4}$$

If the sessile drop is small enough, so that gravity effects are negligible, and if the three-phase system is in equilibrium, the wettability of a smooth, homogeneous, solid surface is given by[15]

$$\cos\theta = \begin{cases} 1 & \text{if } \tilde{\gamma}_{SV} - \tilde{\gamma}_{SL} \geq 1 \quad \text{(complete wetting)} \\ \tilde{\gamma}_{SV} - \tilde{\gamma}_{SL} & \text{if } \left|\tilde{\gamma}_{SV} - \tilde{\gamma}_{SL}\right| < 1 \quad \text{(partial wetting)} \\ -1 & \text{if } \tilde{\gamma}_{SV} - \tilde{\gamma}_{SL} \leq -1 \quad \text{(complete drying)} \end{cases} \tag{5}$$

Most surfaces are neither completely smooth nor homogeneous, causing metastability and hysteresis effects. This gives rise to advancing contact angles that are larger than the receding ones, with the contact angles in the entire range between them corresponding to metastable, rather than stable equilibrium conditions. In practice, if the hysteresis between advancing and receding contact angles is small, either of them can be used as an approximation for the equilibrium wettability (5).

Heterogeneous Surfaces: Cassie Equation. Wettability of smooth, but chemically heterogeneous surfaces is often modeled by the empirical Cassie equation[1,10,16]

$$\cos\theta = p\cos\theta_p + q\cos\theta_q \,. \tag{6}$$

Here θ is the contact angle of a liquid on the heterogeneous surface composed of microscopic patches of area fraction p of one type of chemical groups, and area fraction q of chemical groups of a second type (where $p+q=1$), and θ_p and θ_q are the contact angles of this liquid on the pure homogeneous surfaces corresponding to $p=1$ and $q=1$, respectively.

Equation (6) has been widely used for a phenomenological description of wettabilities of chemically heterogeneous surfaces. However, it neglects many aspects of statistical physics that could be associated with these systems. Moreover, until recently it had no derivation within the framework of modern theoretical approaches to wetting and interfacial phenomena. For example, according to (6), if one looks at two chemically heterogeneous surfaces that differ by the relative concentrations of their components, one would find the same pattern of group distribution at the molecular level. This picture, being purely geometrical, neglects the delicate effects of different, possibly competing interactions between the various molecules at the solid surface and the two coexisting bulk phases. We have studied systems where such competing interactions lead to a very different wettability behavior, due to interplay between wettability and adsorption of a minority species at the surface. This we achieved by adapting and generalizing a variety of methods developed over the last couple of decades, that allow statistical-mechanical modeling of interfacial problems at different levels of approximation.

Experimental Observations of Wettabilities of Mixed Self-Assembled Monolayers

Mixed Self-Assembled Thiolate Monolayers on Gold. Self-assembled monolayers (SAMs) of alkanethiolates are molecular aggregates formed by strong chemisorption from solution to a gold substrate.[17] Experimental probes indicate that the bonding of the thiolate group to the gold surface is very strong (homolytic bond strength is ~44 kcal/mol[18]). This has been supported recently by model *ab initio* calculations.[19] Electron[20,21] and helium diffraction[22] studies and atomic force microscopy (AFM)[23] of such SAMs show that the symmetry of sulfur atoms is a simple hexagonally close-packed $(\sqrt{3}\times\sqrt{3})R30^\circ$ overlayer with an S...S spacing of 4.97 Å and calculated area per molecule of 21.4 $Å^2$.

Systematic wettability data have been obtained[24,25]for mixed monolayers prepared from a series of mixtures of 11-hydroxyundecane-1-thiolate ($HO{-}(CH_2)_{11}{-}SH$, HUT) and dodecanethiolate ($CH_3{-}(CH2)_{11}{-}SH$, DDT) molecules in tetrahydrofuran (THF) under nitrogen onto gold surfaces. The total concentration of alkanethiolates in THF solution was kept at 1 mM, and the relative concentration of OH-terminated chains was systematically varied in the series. All monolayers showed thicknesses of 14 ± 1 Å as estimated by ellipsometry. The surface concentration of the OH groups was determined by ESCA, and found to be in good agreement with the relative concentrations in the adsorbing solution. This rules out preferential chemisorption of either HUT or DDT to the gold substrate. The energy differences due to the different terminal groups are small relative to the sum of pair interactions between groups along the chains in solution and are dominated by the contributions due to the rest of the chains, which are of the same length. Thus, in this case a random distribution of the OH- and CH_3-groups can be assumed at the surface (so that there is no significant phase separation between HUT and DDT within the monolayer). A recent study[26] of mixed $HS-(CH_2)_{16}-OH$:$HS-(CH_2)_{15}-CH_3$ SAMs using both infrared reflection-absorption spectroscopy (IRAS) and X-ray photoelectron spectroscopy (XPS) suggests true molecular-scale mixing in SAMs composed of chains that are of nearly equal lengths. This lends experimental support to the validity of our assumption. For an earlier discussion of the possibility of phase separation and macroscopic domains in mixed SAMs see Bain *et al.*.[27] A schematic representation of the packing and ordering in such a mixed monolayer, based on experimental and modeling studies, is shown in Fig. 1.

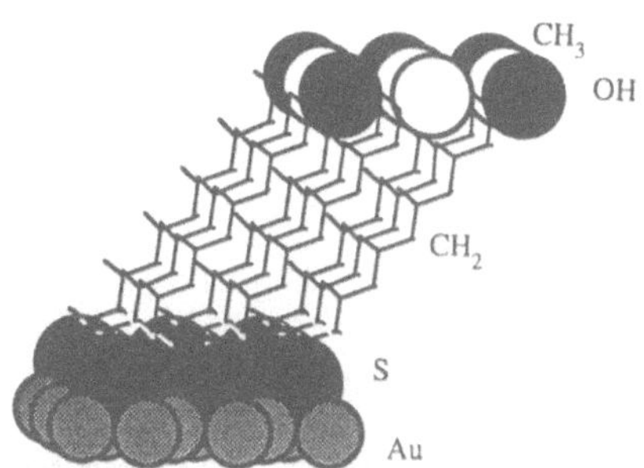

Figure 1. A schematic representation of the packing and ordering in a mixed HUT/DDT SAM, based on FTIR, electron diffraction, and molecular modeling studies.

Wettability Experiments. The wettability by water (H_2O) of freshly prepared—i.e., measured immediately after removal from the solution—monolayers composed of mixtures of HUT and DDT could be easily fitted by the Cassie equation (6), as is shown in Fig. 2.

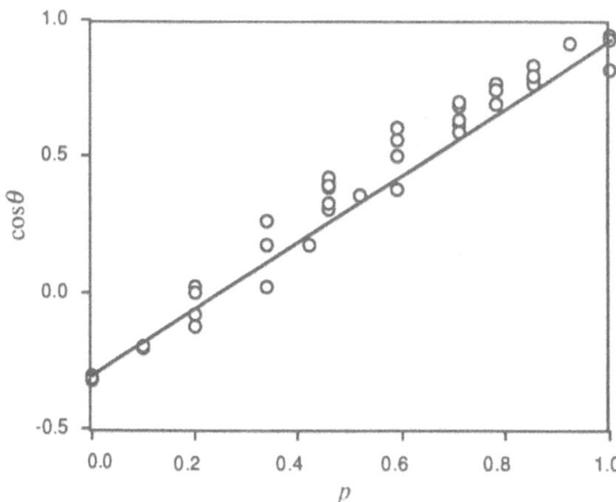

Figure 2. Experimental wettability of mixed HUT/DDT SAMs by water. Solid line is the best fit by Eq. (6).

On the other hand, the wettability of the same series of monolayers by n-hexadecane (HD, $CH_3(CH_2)_{14}CH_3$) exhibited a highly nonlinear behavior about a concentration threshold. HD wettabilities, as observed at room temperature and in ambient atmosphere, are plotted as open circles in Fig. 3 as a function of surface OH-concentration. The sharp variation of the contact angle at ~30-45% surface OH-groups reflects an abrupt change in the solid-vapor (γ_{SL}) and solid-liquid (γ_{SV}) interfacial free energies. Experiments with other organic wetting liquids[24,25] indicate that this interfacial phenomenon is quite general and not limited to wettabilities of such mixed SAMs by HD.

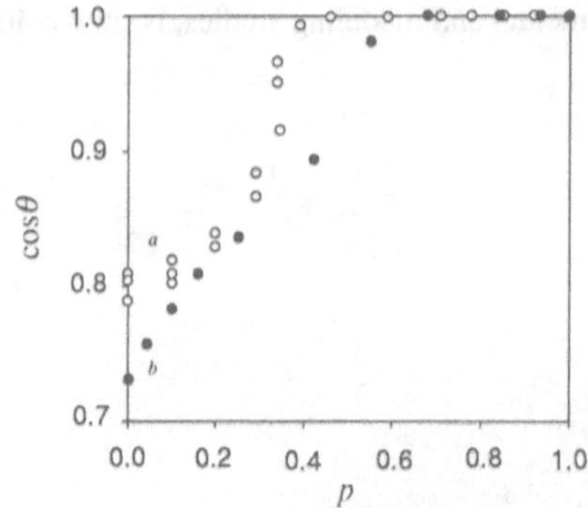

Figure 3. Observed variation of wettabilities of mixed HUT/DDT SAMs by hexadecane with surface OH composition. Advancing contact angles were observed at room temperature ($T \approx 298$ K). (a) In ambient atmosphere, relative humidity $RH \approx 30\%$ (open circles). (b) Controlled relative humidity $RH \leq 2\%$ (filled circles).

INTERPLAY OF WETTING AND ADSORPTION AT HETEROGENEOUS SURFACES

Model of Adsorption at Chemically Heterogeneous Surfaces

Hypothesis and Supporting Evidence. The wettability experiments described above were performed at ambient atmosphere and room temperature, with a typical relative humidity in the laboratory of 30% or higher. Thus, we have advanced the hypothesis[24,25,28] that the abrupt change in the wettabilities is due to formation of adsorbed condensed water layer at highly hydroxylated surfaces by a process of H-bonding. Indirect support for formation of such patches of adsorbed water layers at highly hydroxylated surfaces is provided by x-ray reflectivity[29] and programmed thermal desorption experiments.[30] Inter-

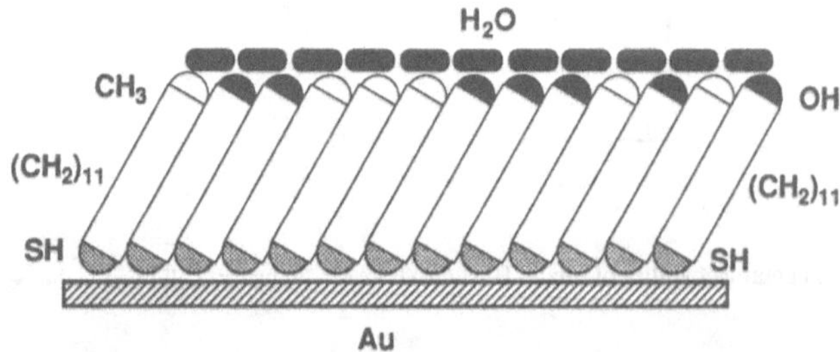

Figure 4. Schematic depiction of a complete hydrogen-bonded water layer forming on top of a mixed HUT/DDT SAM at high enough relative humidities and surface concentrations of OH- versus CH_3-groups.

estingly, the possibility of such a scenario has already been noted by Cassie[15] in his original paper proposing Eq. (6) for wettabilities of heterogeneous surfaces.

Energy minimization at $T = 0$ K using simulated annealing methods[24,25] indicated that, when water molecules adsorb to such hydroxylated surfaces, they prefer sites where they can form multiple H-bonds, bridging between nearest-neighbor surface OH groups. Recently, Klein *et al.*[31] have reported constant-temperature molecular dynamics simulations (at 200 and 300 K) of 90 water molecules placed over homogeneous 90-chain patches of SAM with either methyl or hydroxyl termini. To prevent evaporation, a reflective barrier was placed 25 Å above the SAM surface. This corresponds to a relative humidity above saturation. Under these conditions, methyl-terminated surfaces exhibit bunching of the water molecule into a fluctuating droplet with an average contact angle of 120-135°, while a flat, thin patch of H-bonded water forms on the hydroxylated monolayer.

According to our hypothesis,[19,20,23] at high enough relative humidity and at room temperature the water will evaporate from a methyl-terminated thiolate monolayer, but a thin condensed, adsorbed layer of water will form on the hydroxylated one. Starting with 100% OH terminated SAM, decreasing concentration of OH groups at the SAM surface may destroy the formation of a condensed, adsorbed water layer at some stage. This is because weaker dispersive interactions between water molecules and the methyl groups may not be strong enough to sustain such a condensed surface water phase against evaporation and thermal disorder at room temperature.

To test our hypothesis, experiments were carried out to determine the effect of reduced humidity on the observed transition in the wettability.[19] In these experiments, relative humidity was reduced below 2%, and HD contact angles were then studied. The results, shown in Fig. 3 as filled circles, show that the wettability behavior observed at 30% relative humidity has been considerably altered at ≤2% relative humidity. The wettability of the purely methylated ($p = 0$) surface assumes a lower value and increases in almost a linear fashion with increasing p, attaining complete wetting for $p \geq 0.6$.

Theoretical Motivation. An explanation for the observed phenomenon on the macroscopic level is possible within the framework of the modern theories of wetting. These theories typically represent two-fluid systems either by discrete lattice-gas (pseudospin) models, with appropriate short- and/or long-range intermolecular interactions,[32-34] or by continuum Landau-Ginzburg interfacial free energies. The latter can be derived either phenomenologically using general symmetry arguments, or by fitting to experimental data,[35] or as a continuum limit of a mean-field lattice-gas model.[29,36] The presence of a solid substrate is modeled by modification of effective one- and two-body interactions in the vicinity of the surface. Wetting, prewetting and layering phenomena can then be understood within a single comprehensive model in terms of ordering of the various bulk and surface phases. This results in a phase diagram controlled by the magnitude of the bulk and surface couplings and the temperature.[32-36] Our hypothesis implies that, in the vicinity of a solid surface, abrupt changes in the interfacial free energies and the contact angles arise due to a competitive ordering of both polar (water) and hydrophobic (i.e., HD) molecules into various surface and bulk, liquid or vapor phases.

Semi-Infinite Cubic Lattice-Gas Model. The continuous system of fluid molecules in the vicinity of a smooth, but chemically heterogeneous SAM substrate of Figure 1 can be approximated by a discrete, semi-infinite cubic lattice. The fluid-fluid and fluid-solid intermolecular interactions in real fluids coexisting with a solid substrate are often approximated by a sum over effective pair interactions between lattice-gas occupancy variables t_i, where

$$t_i = \begin{cases} 1 & \text{if lattice site } i \text{ is occupied,} \\ 0 & \text{if lattice site } i \text{ is vacant.} \end{cases} \tag{7}$$

The molecular pair interactions are typically characterized by a strong repulsion at contact, followed by attraction with a deep minimum at a short-range and a slow decay to zero at longer ranges (e.g., the van der Waals or hydrogen-bonding interactions). In the present lattice-gas model, the repulsion is approximated by excluding multiple occupancies at a lattice site. The attractive part of the interaction is further approximated by an effective *short-range* interaction between neighboring occupied sites. The total configurational energy of the system is thus given by the Hamiltonian

$$H = -\varepsilon \sum_{<ij>} t_i t_j - \sum_i \mu_i t_i , \tag{8}$$

where $< ij >$ denotes summation over all pairs of nearest-neighbor sites, ε is the effective short-range pair interaction (for simplicity assumed to be identical within the surface layer, the bulk, and the interpolating layers), and μ_i is the effective local chemical potential at site i. The latter is assumed to be μ_b, the bulk chemical potential of water in air, if i is a bulk site not belonging to the surface layer. The bulk chemical potential enters into the Hamiltonian as a one-body interaction for the lattice-gas occupancy variables. Its physical origin is as a Lagrange multiplier conjugate to the total number of molecules in the mapping from the canonical to the grand-canonical configurational partition function. This, in turn, is mapped on the lattice-gas model. The bulk chemical potential, μ_b, can be calculated within the ideal gas approximation for the bulk water vapor. It depends on the partial pressure of water in the ambient atmosphere and thus on its humidity at a given temperature.

Fluid molecules at the surface of a mixed SAM may interact either with an OH-rich patch, or with a CH_3-rich patch of the substrate. To proceed with the calculations, we need to know the probability distribution of OH-groups on the surface of the organic substrate, given their mean concentration at the surface. Since the chemisorption of thiols on the gold substrate is controlled kinetically, rather than thermodynamically, it is reasonable to assume that the probability distribution of the OH-groups at the surface is quenched, rather than annealed. This means that, for a freshly prepared monolayer, the distribution of OH-groups at the surface can be considered fixed (for the time scale of the wetting experiments), albeit out of thermal equilibrium with the rest of the system (and thus not determined by a thermal equilibrium requirement for the entire system). The two species have roughly equal chain lengths and the different terminal groups contribute a relative small fraction of the total pair interactions. Thus the driving force for phase separation in the organic solution should be fairly small as long as the chain lengths are comparable, though it can become significant when chain lengths are notably different. This lack of phase separation should be reflected in the distribution of terminal groups in the chemisorbed monolayer as well, as is supported by spectroscopic and simulation studies. As we are interested here in the case of equal chain length, we will thus proceed with the simple assumption of independent binary random distribution of OH-groups at each monolayer site. Hence we assign site-independent, bimodal probabilities for effective local chemical potentials at the surface, of the form

$$\mu_i = \begin{cases} \mu_b + \mu_p & \text{with probability } p \\ \mu_b + \mu_q & \text{with probability } q \end{cases} , \tag{9}$$

where p is the fraction of OH groups at the SAM surface, $q = 1 - p$ is the fraction of CH_3-groups, and μ_p and μ_q are enhancements of the effective one-body interaction at a surface site adjacent to a OH- or a CH_3-group, respectively (see Fig. 5). Note that these enhancements represent the total sum of variable-range pair interactions of a fluid molecule with the entire substrate, rather than only the short-range interaction with the closest OH- or a CH_3-group. Hence μ_p and μ_q may, in principle, vary with the length of the alkanethiol chains composing the SAM.

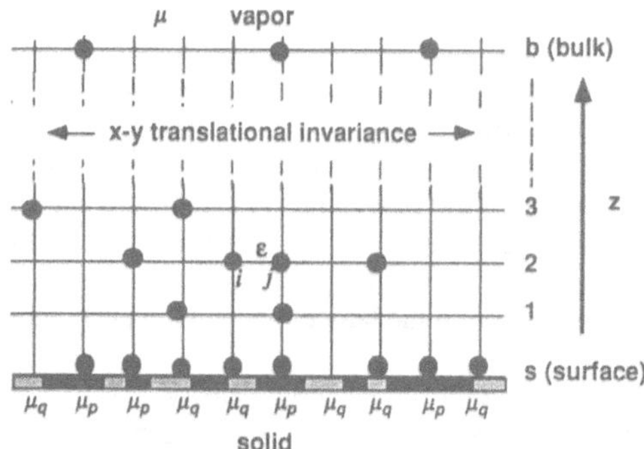

Figure 5. A schematic two-dimensional projection of the semi-infinite cubic lattice-gas model with a quenched bimodal distribution of effective chemical potentials at the surface.

Mean Field Approximations. In the limiting cases $p = 1$ or $q = 1$, which correspond to homogeneous substrates, this semi-infinite lattice-gas model, and its Ising spin equivalent, have been extensively studied. Mean-field approximations demonstrated[31-35] the possibility of layering and prewetting transitions for certain regions in parameter space away from coexistence. Subsequent theoretical work addressed the effect of quenched surface impurities on the wetting transition.[37,38] Further work on the effect of surface heterogeneities on wetting, prewetting and layering transitions has been reviewed by Forgacs *et al*.[7]

We have previously studied the semi-infinite mean-field lattice-gas model for one-component undersaturated vapor in the vicinity of a mixed SAM with a quenched bimodal distribution of OH-groups.[24,25,28] Mean-field calculations, both on the level of the simple Bragg-Williams[39] and the more accurate single-site cluster approximation suggested by Boccara and Benyoussef,[40,41] demonstrated the possibility of a layering transition in the vapor-substrate system and explored its variation with the surface composition. The same approximations provided a discrete version of the system free energy, averaged over the quenched distribution of surface groups, from which a continuum free energy (1), as well as the appropriate pairwise interfacial free energies for the substrate-liquid-gas system can be derived. We have discussed such derivations in great detail elsewhere.[13,14]

Wettability of a Heterogeneous Surface by a One-Component Liquid

Cahn's Form of the Excess Interfacial Free Energy. Consider a one-component molecular fluid phase at liquid-gas coexistence ($\mu_b = \mu_{coex}$), in the vicinity of a heterogeneous substrate with a quenched binary random distribution of molecular groups of type (9). Assuming effective short-range interactions between the fluid molecules and the substrate, the excess interfacial free energy for the interfacial region can then be decomposed as follows:

$$G_{ex}[\psi(z),\psi_s]=G_\psi[\psi(z)]+\bar{\gamma}_\psi(\psi_s), \tag{10}$$

where $\psi(z)$ is the order parameter field for liquid-gas phase separation, and $\psi_s=\psi(z)|_{z=0}$ is its value at the surface. The term $G_\psi[\psi(z)]$ in (10) represents the interactions between the fluid molecules for $T<T_c$ at coexistence. It has the structure

$$G_\psi[\psi(z)]=\int_0^\infty dz\left[f[\psi(z)]-f(\psi_b)+\tfrac{1}{2}L\left(\frac{d\psi}{dz}\right)^2\right], \tag{11}$$

where ψ_b is one of the two bulk phase (liquid or vapor) values in the limit $z\to\infty$. At coexistence, $f(\psi)$ has the general form of a double-well potential exhibiting two degenerate global minima corresponding to the coexisting bulk phases (liquid and vapor). It can be derived from a continuum limit to a complete mean-field approximation for a semi-infinite lattice-gas model. The coefficient L in (11) is the square of the correlation length in the system. In principle, L is a function of ψ as well, the form of which is calculable using the mean-field approximation to the lattice model. However, here we will assume for simplicity that it is a constant.

An approximate Landau-Ginzburg free energy density $f^{(4)}[\psi(z);r,u]$ can be obtained by expanding the complete lattice-gas form $f(\psi)$ to leading order in ψ:

$$f^{(4)}[\psi(z);r,u]=\tfrac{1}{2}r\psi^2(z)+\tfrac{1}{4}u\psi^4(z). \tag{12}$$

It can be shown that $u>0$ and $r\sim(T-T_c)/T_c$ near the critical temperature $T=T_c$. Thus $G_\psi[\psi(z)]$ can be approximated by

$$G_\psi^{(4)}[\psi(z);r,u,L]=\int_0^\infty dz\left[f^{(4)}[\psi(z);r,u]-f^{(4)}[\psi_b;r,u]+\tfrac{1}{2}L\left(\frac{d\psi}{dz}\right)^2\right]. \tag{13}$$

For $r<0$ the free energy density function $f^{(4)}(\psi;r,u)$ has two degenerate minima representing the two coexisting phases at $\psi_b=\pm\psi_0(r,u)$, where

$$\psi_0(r,u)=\sqrt{\frac{-r}{u}}\,. \tag{14}$$

The Landau-Ginzburg form $f^{(4)}(\psi;r,u)$ is clearly an approximation to the complete mean-field lattice-gas form $f(\psi;T,\varepsilon)$. This approximation is more accurate for temperatures close to the mean-field critical temperature. It becomes less accurate an approximation to $f(\psi;T,\varepsilon)$ at lower temperatures, but even at those temperatures it retains the qualitative feature of two degenerate minima at values of ψ close to those of the coexisting bulk vapor and liquid phases in the mean-field lattice-gas approximation. The main qualitative difference is that the function $f(\psi;T,\varepsilon)$ obtained from a lattice-gas mean-field approximation is typically defined over the interval $-1\le\psi\le 1$, whereas the function $f^{(4)}(\psi;r,u)$ is formally defined over the a stretched domain $-\infty<\psi<\infty$. One can either choose to use the full mean-field lattice-gas function to determine the microscopic parameter ε by fitting that function to the available experimental information for the bulk phases, or to fit the parameters r and u of the truncated expansion $f^{(4)}(\psi;r,u)$ directly to the same experimental information. Though less accurate, the second choice makes the calculations much simpler.

The surface term $\bar{\gamma}_\psi(\psi_s)$ in (10) represents the contribution of effective short-range fluid-solid interactions to $G_{ex}[\psi(z),\psi_s]$, as expressed by order parameter interactions at the surface and averaged over the quenched random distribution of ordering fields at the fluid-substrate boundary. Again, the complete form of this function can be obtained from the continuum limit of a discrete mean-field lattice-gas approximation. Expanding to second order in ψ_s, $\bar{\gamma}_\psi(\psi_s)$ has the form

$$\gamma_\psi^{(2)}(\psi_s;a_1,a_2)=a_1\psi_s+a_2\psi_s^{\,2}, \tag{15}$$

up to a constant independent of ψ_s that contributes equally both to the solid-liquid and the solid-vapor interfacial free energies and thus does not play any role in determining the wettability. Here a_1 is a measure of the interactions between the solid molecules and a single fluid molecule, and a_2 is a measure of how the presence of the heterogeneous solid substrate affects the effective two-body interaction between fluid molecules adjacent to the solid surface. Determination of the interfacial profile and the associated free energy amounts to minimization of (10) subject to the boundary conditions

$$\lim_{z\to\infty}\frac{d\psi(z)}{dz}=0; \qquad \lim_{z\to\infty}\psi(z)=\psi_b. \tag{16}$$

Variational Determination of Wettability. Minimization of (10) with respect to $\psi(z)$ and $d\psi/dz$ leads to the Euler-Lagrange equation

$$-L\frac{d^2\psi}{dz^2}+\frac{\partial f}{\partial\psi}=0. \tag{17}$$

Integrating this equation once, we get

$$\tfrac{1}{2}L\left(\frac{d\psi}{dz}\right)^2=f[\psi(z)]-f(\psi_b), \tag{18}$$

where the choice of the integration constant is dictated by the boundary conditions (16). Using

$$\int_0^\infty dz\left(\frac{d\psi}{dz}\right)^2=\int_{\psi_s}^{\psi_b}d\psi\left(\frac{d\psi}{dz}\right), \tag{19}$$

it follows that

$$G_\psi[\psi(z)]=\int_{\psi_s}^{\psi_b}g_\psi(\psi)d\psi\,, \tag{20}$$

where

$$g_\psi(\psi)=\pm\sqrt{2L[f(\psi)-f(\psi_b)]}. \tag{21}$$

Approximating $f(\psi)$ by the Landau-Ginzburg form (12) results in a simpler expression for $g_\psi(\psi)$:

$$g_{\psi}^{(4)}(\psi;r,u,L)=\pm\begin{cases}+\sqrt{\frac{L}{2u}}\left(u\psi^2+r\right) & \text{for } -\infty<\psi<-\sqrt{\frac{-r}{u}}\\ -\sqrt{\frac{L}{2u}}\left(u\psi^2+r\right) & \text{for } -\sqrt{\frac{-r}{u}}\le\psi\le+\sqrt{\frac{-r}{u}}\\ +\sqrt{\frac{L}{2u}}\left(u\psi^2+r\right) & \text{for } +\sqrt{\frac{-r}{u}}<\psi<+\infty\end{cases} \quad (22)$$

The free energy per area of the liquid-vapor interface (surface tension) is given by

$$\gamma_{LV}=\int_{\psi_V}^{\psi_L} g_{\psi}(\psi)d\psi\ , \quad (23)$$

where ψ_V and ψ_L correspond to the values that ψ assumes in the bulk vapor and liquid phases, correspondingly, and the sign of g_{ψ} is selected by the requirement of positive surface tension. The free energies per area of the solid-vapor and the solid-liquid interfaces are given by

$$\gamma_{SV}=\min_{\{\psi_s\}}\left\{\left|\int_{\psi_s}^{\psi_V} g_{\psi}(\psi)d\psi\right|+\bar{\gamma}_{\psi}(\psi_s)\right\}, \quad (24)$$

$$\gamma_{SL}=\min_{\{\psi_s\}}\left\{\left|\int_{\psi_s}^{\psi_L} g_{\psi}(\psi)d\psi\right|+\bar{\gamma}_{\psi}(\psi_s)\right\}. \quad (25)$$

Substitution of (23)-(25) into Eqs. (3)-(5) results in a calculated value of the wettability. Note that we can either choose to use the full mean-field expressions for $\bar{\gamma}_{\psi}(\psi_s)$ and $f(\psi)$, or approximate them by the truncated Landau-Ginzburg forms (15) and (12), respectively, thus simplifying the wettability calculations.

Coupling of Wettability and Adsorption at Heterogeneous Surfaces

Model Construction. Once more, we consider a system of coexisting liquid and vapor phases near a chemically heterogeneous surface. The order parameter ψ describes relative distribution of vacancies in the two coexisting fluid phases, as in the previous section. Let us assume, however, that a minority species (e.g., water) that is potentially capable of adsorption to the surfaces is solubilized in the two coexisting fluids, which are predominantly composed of another, majority species (e.g., hexadecane). Hence we introduce a second field $\rho(z)$ that describes the local mean concentration of such a minority species of adsorbates.

As before, let us assume that the excess interfacial free energy of the solid-fluid interface is separable into two distinct contributions, one composed of effective short-range interactions of the fluid molecules with the solid substrate, and the other caused by the slowly varying z-dependent contributions due to fluid-fluid interactions:

$$G_{ex}[\rho(z),\psi(z),\rho_s,\psi_s]=G_z[\rho(z),\psi(z)]+\bar{\gamma}_s(\rho_s,\psi_s)\ . \quad (26)$$

Each of those two contributions depends now on both $\psi(z)$ and $\rho(z)$. Formally, each one can be further decomposed into a sum of three terms. Thus, for $z>0$:

$$G_z[\rho(z),\psi(z)]=G_{\psi}[\psi(z)]+G_{\rho}[\rho(z)]+G_{\rho\psi}[\rho(z),\psi(z)]\ . \quad (27)$$

The term $G_{\psi}[\psi(z)]$ in (27) depends on $\psi(z)$ alone, corresponding to the limit of vanishing concentration of minority species, reduces to the case discussed in the previous section. These terms may again assume either the lattice-gas form or the truncated Landau-Ginzburg form. To simplify calculations, we choose the latter for describing the majority species (e.g., hexadecane).

The second term in (27) depends on $\rho(z)$ alone and corresponds to the limit where sites occupied by the majority species are replaced by vacancies. We are interested here in the situation when the bulk contaminant (e.g., water) concentration is very low. In this case, the terms that depend on $\rho(z)$ alone correspond to an excess interfacial free energy of an undersaturated vapor phase of contaminants in contact with the heterogeneous surface. Previous calculations[24] lead us to expect a first layering transition in the regime of interest for undersaturated water vapor in contact with the mixed SAMs considered here. Hence, in this case, the minority species profile cannot be slowly varying, and we assume instead that it has a step-function form

$$\rho(z)=\begin{cases}\rho_s & z=0\\ \rho_b & z>0\end{cases}. \tag{28}$$

Thus it has a value ρ_s at the surface and a different constant value ρ_b in the bulk, with a discontinuous jump between the two values. Due to the assumed profile (28), we have

$$G_{\rho}[\rho(z)]=\int_0^{\infty}dz\left[f[2\rho(z)-1]-f(2\rho_b-1)\right]=0, \tag{29}$$

for $z>0$. The function f is the mean-field free energy in its complete form, defined by substituting interactions involving the majority species with the corresponding ones for the minority species at $\mu_b<\mu_{coex}$. The assumption (28) simplifies the ensuing variational problem enormously. It will not be valid in other regions in the parameter space, where the pure adsorption model shows a *sequence* of layering transitions followed by complete wetting. In that case, either an assumption of a more general form of the profile is needed, or the profile has to be calculated numerically from the full coupled Euler-Lagrange equations for the $\psi(z)$ and $\rho(z)$ profiles and their derivatives.

The last term in (27) corresponds to interactions between minority and majority species. Let us consider the dominant contribution to the term that couples the two fields. We are interested in the case where the bulk contaminant concentration is very small, so we may assume that the dominant contribution to this term is of first order in ρ_b. Thus the coupling term contribution to the excess interfacial free energy from fluid-fluid interactions has the form

$$G_{\rho\psi}[\rho(z),\psi(z)]=\int_0^{\infty}dz\left\{f_{\rho\psi}[\rho(z),\psi(z)]-f_{\rho\psi}[\rho_b,\psi_b]\right\}, \tag{30}$$

where, to the first two leading orders in ψ,

$$f_{\rho\psi}[\rho(z),\psi(z)]=\int_0^{\infty}dz[c_{b1}\rho_b\psi(z)+c_{b2}\rho_b\psi^2(z)] \tag{31}$$

expresses the free energy cost of solubilizing impurities in the two coexisting fluid phases composed predominantly of the majority species. Note that the term quadratic in ψ is symmetric about the vapor and liquid phases, while the term linear in ψ is antisymmetric.

The surface contributions $\bar{\gamma}_s(\rho_s, \psi_s)$ to the excess interfacial free energy (26) can similarly be decomposed into three additive terms

$$\bar{\gamma}_s(\rho_s, \psi_s) = \bar{\gamma}_\psi(\psi_s) + \bar{\gamma}_\rho(\rho_s) + \bar{\gamma}_{\rho\psi}(\rho_s, \psi_s). \tag{32}$$

For simplicity, we choose to approximate the term $\bar{\gamma}_\psi(\psi_s)$ by its truncated expansion $\gamma_\psi^{(2)}(\psi_s; a_1, a_2)$ as defined in (15). The second term can be obtained from a mean-field lattice-gas model. A Landau-Ginzburg approximation to this term has to retain powers of at least fourth order in ρ_s to reflect the possibility of an adsorbate layering transition. Finally, the coupling term $\bar{\gamma}_{\rho\psi}(\rho_s, \psi_s)$ can be approximated by averaging the interactions of the majority species over patches of substrate that are either "bare" or "dressed" by adsorbates:

$$(1-\rho_s)(a_1\psi_s + a_2\psi_s^2) + \rho_s(b_1\psi_s + b_2\psi_s^2). \tag{33}$$

Here a_i are surface parameters for bare patches and b_i are surface constants for dressed patches, with $(1-\rho_s)$ and ρ_s corresponding to probabilities for a site to belong to a bare or dressed patch, respectively. Both a_i and b_i are, in general, p-dependent for a mixed surface.

Introducing new constants $c_{si} = b_i - a_i$, the surface layer contribution to the ψ-dependent part of the interfacial free energy is expressed as $\gamma_\psi(\psi_s; a_1, a_2) + \gamma_{\rho\psi}(\rho_s, \psi_s)$, where $\gamma_\psi(\psi_s; a_1, a_2)$ is given by (15) and

$$\gamma_{\rho\psi}(\rho_s, \psi_s) = \rho_s(c_{s1}\psi_s + c_{s2}\psi_s^2), \tag{34}$$

to second order in ψ_s.

Lowest-order coupling terms similar to (31) and (34) were previously used in other Landau-Ginzburg models of both bulk and interfacial systems containing a concentration of minority species (e.g., impurities[42] or surfactants[43,44]) affecting the order parameter field (related, e.g., to a fluid-fluid phase separation,[43,44] or the polymer adsorption profile[44]).

Equation (27) can be rewritten in a form that has the same ψ-dependence as in (13), except for a ρ-dependent renormalization of parameters:

$$G_z[\rho(z), \psi(z)] = G_\psi^{(4)}[\psi(z); \tilde{r}, u, L] , \tag{35}$$

where

$$\tilde{r} = r + 2c_{b2}\rho_b . \tag{36}$$

The effect of the bulk coupling term that is bilinear in ρ and ψ is to shift the location of liquid-vapor coexistence surface, which is determined by the condition of vanishing of the ρ-dependent coefficient in the expansion of the free energy density in ψ.

Similarly, (32) can be rewritten in a more compact form:

$$\bar{\gamma}_s(\rho_s, \psi_s) = \bar{\gamma}_\rho(\rho_s) + \bar{\gamma}_\psi^{(2)}(\psi_s; \tilde{a}_1, \tilde{a}_2), \tag{37}$$

up to an additive constant independent of ρ_s and ψ_s, where

$$\tilde{a}_i = a_i + c_{si}\rho_s , \tag{38}$$

and $i = 1, 2$.

Determination of the interfacial profile and the associated free energy proceeds by solving for ψ_b and ρ_b in the bulk, followed by minimization of (26) with respect to $\psi(z)$, $d\psi/dz$, and ρ_s, subject to the assumption (28) and the boundary conditions (16).

Variational Determination of Adsorbate Surface Coverage and Wettability. Let us first calculate the equilibrium values for the bulk phases far away from the surface. The homogeneous density of the bulk grand potential *at coexistence* of the vapor and liquid phases is given by

$$g(\rho_b,\psi_b)=\tfrac{1}{2}r\psi_b^2+\tfrac{1}{4}u\psi_b^4+f(2\rho_b-1)-\mu_b\rho_b+c_{b2}\rho_b\psi_b^2 . \tag{39}$$

Here we assumed a truncated Landau-Ginzburg form (12) for the part that depends on ψ_b only, so that r and u depend on microscopic interactions involving majority species (e.g., hexadecane) only. On the other hand, we chose to retain the full lattice-gas mean-field form for the contribution involving interactions between minority species (e.g., water) only, with μ_b denoting the bulk chemical potential for the minority species. Finally, the last term is the dominant contribution from interactions coupling between the majority and minority species, as in (31), but dropping the term that is first order in ψ_b, since it is absorbed in the redefined chemical potential at liquid-vapor coexistence when minority species are present.

Minimization of (39) with respect to ψ_b and ρ_b leads to the following system of nonlinear algebraic coupled equations

$$\tilde{r}\psi_b+u\psi_b{}^3=0, \tag{40}$$

$$\frac{\partial f(2\rho_b-1)}{\partial\rho_b}-\mu_b+c_{b2}\psi_b^2=0. \tag{41}$$

Solutions of (40) have a form identical to (14), except that r is replaced by $\tilde{r}$. Thus for $u>0$ and $\tilde{r}<0$ it also has two degenerate minima representing the two coexisting homogeneous bulk phases. The liquid phase minimum is at $\psi_L=\tilde{\psi}_0$ and the vapor one at $\psi_V=-\tilde{\psi}_0$, where

$$\tilde{\psi}_0=\sqrt{\frac{-\tilde{r}}{u}} . \tag{42}$$

Substituting those two values for ψ_b into (41), we get two nonlinear equations determining the values of ρ_b, the bulk concentration of impurities (e.g., water) in the bulk liquid and vapor phases, respectively. The stable solution of that equation corresponding to a global minimum of the bulk grand potential determines ρ_b.

Once the values of ρ_b and ψ_0 are determined, we proceed to minimize (26) with respect to $\psi(z)$ and $d\psi/dz$. Using (35)-(36), this leads to the following Euler-Lagrange equation:

$$-L\frac{d^2\psi}{dz^2}+\tilde{r}\psi+u\psi^3=0. \tag{43}$$

Repeating the steps (18)-(19) with (43), we obtain:

$$G_\psi^{(4)}[\psi(z);\tilde{r},u,L]=\int_{\psi_s}^{\psi_b}g_\psi^{(4)}(\psi;\tilde{r},u,L)d\psi , \tag{44}$$

where the function $g_{\psi}^{(4)}(\psi;r,u,L)$ is defined by (12)-(13).

Thus the excess interfacial free energy (26) can be expressed now in a form that depends only on the values of the fields ρ and ψ at the surface and in the bulk:

$$G_{ex} = \int_{\psi_s}^{\psi_b} g_{\psi}^{(4)}(\psi;\tilde{r},u,L)d\psi + \tilde{a}_1\psi_s + \tilde{a}_2\psi_s^2 + \bar{\gamma}_\rho(\rho_s) . \tag{45}$$

Minimizing (45) with respect to the variables ρ_s and ψ_s, we obtain the system

$$g_{\psi}^{(4)}(\psi_s;\tilde{r},u,L) + (a_1 + c_{s1}\rho_s) + 2(a_2 + c_{s2}\rho_s)\psi_s = 0, \tag{46}$$

$$\frac{\partial\bar{\gamma}_\rho(\rho_s)}{\partial\rho_s} + c_{s1}\psi_s + c_{s2}\psi_s^2 = 0. \tag{47}$$

Equation (47) determines the function $\rho_s(\psi_s)$ with the help of a Maxwell's equal-area construction. Substitution of $\rho_s(\psi_s)$ into (46) determines the solution set $\{\psi_s\}$.

The truncated form of the free energy of the liquid-vapor interface leads to

$$\gamma_{LV} = \frac{2}{3u}\sqrt{-2\tilde{r}^3 L} . \tag{48}$$

The free energies of the solid-vapor and the solid-liquid interfaces are given by

$$\gamma_{SV,L} = \min_{\{\psi_s\}}\left\{\left|\int_{\psi_s}^{-\bar{\psi}_0,\bar{\psi}_0} g_{\psi}^{(4)}(\psi;\tilde{r},u,L)d\psi\right| + \tilde{a}_1\psi_s + \tilde{a}_2\psi_s^2 + \bar{\gamma}_\rho[\rho_s(\psi_s)]\right\} . \tag{49}$$

The wettability of the coupled system is determined by substituting (48)-(49) in (3)-(5).

INTERPLAY OF WETTING AND ADSORPTION: RESULTS AND DISCUSSION

Wettability by Pure Water Coexisting with its Vapor

We present here results obtained with Boccara-Benyoussef level of mean-field lattice-gas approximation. The truncated Landau Ginzburg form (44) was used for the ψ-dependent part of the excess interfacial free energy, while the complete form (30) was used for the ρ-dependent part. The parameters in the model were set to allow comparison of the calculated results to the experimentally measured wettabilities on mixed self-assembled monolayers. Complete details are given in Ref. 13. Figure 6 shows the calculated wettability of pure water on a mixed surface, which is in excellent agreement with the measured values (Fig. 2).

Effect of Relative Humidity on Wettability by Hexadecane

In Fig. 7, the calculated wettability of the heterogeneous SAMs by hexadecane at $T = 298$ K is plotted against p, the percentage of OH groups on the surface, for various values of RH, the relative humidity. The variation of the HD wettability with p is nearly linear until the complete wetting region is reached for curves (a) and (b). Curve (c), corresponding to 30% humidity, shows the nonlinear region seen in the experimental graph of curve (a) of Fig. 3. With increasing p, the slowly varying quasilinear region at low p (which has the same slope as in curves (a) and (b)) is followed beyond a certain p-

threshold by a rapidly varying quasilinear transition region rising to complete wetting. As the humidity increases, this transition region moves to lower values of p.

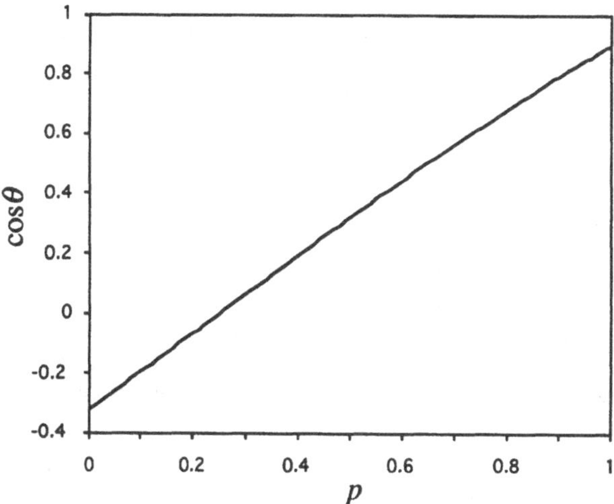

Figure 6. Computed wettability of mixed HUT/DDT SAMs by water, as a function of p, the relative concentration of OH groups at the surface, at $T = 298$ K, using the complete form of the excess interfacial free energy (10) without performing a Landau-Ginzburg expansion.

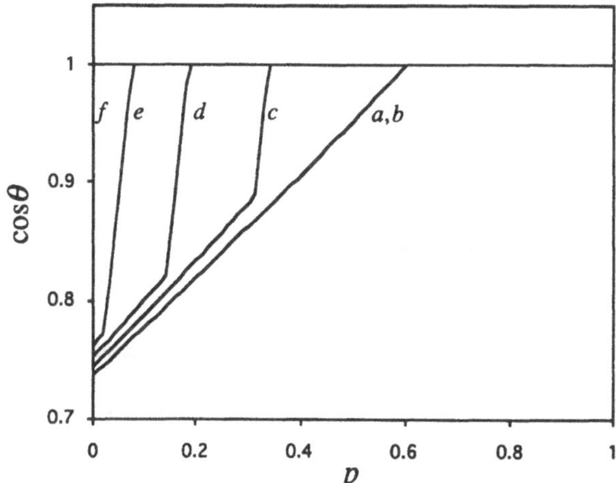

Figure 7. Wettability of mixed HUT/DDT SAMs by hexadecane as a function of p, computed at $T = 298$ K for different relative humidities (RH). (a) $RH = 2\%$, (b) $RH = 15\%$, (c) $RH = 30\%$, (d) $RH = 45\%$; (e) $RH = 60\%$, and (f) $RH = 75\%$.

At low humidities (curves (a) and (b) in Fig. 7) we expect to see little or no effect of water adsorption. The slow, nearly linear variation in p can be assumed to be characteristic of the wettability of the bare surface of the mixed SAM by hexadecane alone. This region still exists in curves (c)-(e), although it gets narrower with increasing humidity. In each of these curves, the bare hexadecane wettability region is followed by a quasilinear region where the contact angle makes a rapid change to zero, corresponding to complete wetting. This region reflects the formation of an adsorbed water layer dressing the surface of the mixed SAM beyond a certain humidity-dependent threshold in p. An example is shown in Fig. 8, which shows a graph of the surface water coverage for 30% humidity. Curve (a) corresponds to the water coverage between the hexadecane liquid phase and the heterogeneous surface; curve (b) corresponds to the water coverage between the hexadecane vapor phase and the mixed SAM. Both show a first-order transition between low density and high density states. The threshold value of p is slightly different for the two. The water layer forms more easily (at lower p) between the liquid and surface than between the vapor and surface. For a narrow range of p, the hexadecane liquid sees a water dressed surface, while the hexadecane vapor sees a bare surface. This accounts for the much more rapid quasilinear rise in the wettability in this region, culminating in the complete wetting region. For comparison with uncoupled pure water model in the undersaturated regime, curve (c) shows the water coverage with no hexadecane present. For this humidity and this range of surface fields, no complete water layer forms in the undersaturated atmosphere.

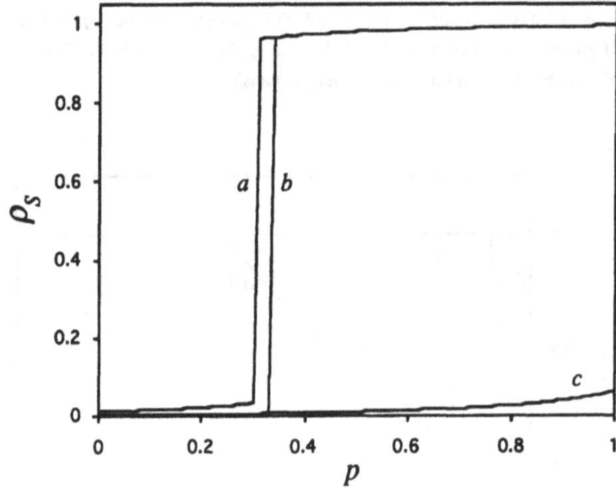

Figure 8. Surface water coverages of mixed HUT/DDT SAMs under hexadecane fluid bulk phases with water impurities, as functions of p, computed at $T = 298$ K and $RH = 30\%$. (a) Coupled model, water coverage under bulk liquid phase. (b) Coupled model, water coverage under bulk vapor. (c) Water coverage under both bulk fluid phases at vanishing surface couplings.

Mapping of Different Wetting Regimes Caused by Adsorbate Layering

Calculations of the temperature dependence of wettabilities and surface water coverages at fixed surface composition and relative humidity, which we do not present here due to space limitations (for details see Ref. 13), show regions of nonmonotonic behavior. This behavior demonstrates the competition between two effects. In a pure wetting fluid, increasing temperature tends to push the system towards large wettabilities, culminating in complete wetting. The water layer, however, which also has the effect of increasing the wettability, compared to the wettability of the bare SAM, tends to disorder and "melt" at higher temperatures. So, at certain surface compositions, when the temperature is low, the water layers have formed both under the liquid and under the vapor bulk phases, and the hexadecane completely wets the mixed surface. At slightly higher temperatures, the water layers melt, one by one. Hence the wettability drops a bit, and the hexadecane will form a sessile drop (partial wetting) instead of completely wetting the substrate. Then, as temperature continues to rise, the system again wets the surface, as it would have at higher temperatures even without the presence of the water layer.

Our numerical calculations of hexadecane wettabilities and surface water layering can be summarized as follows. There are four distinct regions of wetting exhibited by most of the graphs of the calculated wettabilities. At low p there is a region where the water layer does not form, and the hexadecane wets the dry, mixed surface. At high p, there are two complete wetting regions, one where the water layer has formed, and one where it has not. At intermediate values of p, there is a region of rapid variation of the wettability with p. This is the region where a water layer has formed, but the hexadecane does not completely wet the surface. We present calculated planar sections of these regions in the $p-RH$ and $p-T$ planes in Figs. 9, 10, and 11.

Figure 9 shows the boundaries of the four wetting regions in the $p-RH$ plane at $T = 298$ K. The points are the actual data; the boundaries are determined from examination of the values of the wettabilities and the values of the surface water coverages. The lines connecting the points are interpolated between the data points. In regions (c) and (d), the hexadecane completely wets the surface. In region (c), a water layer has formed between both the hexadecane liquid and surface and between the the hexadecane vapor and surface. At low humidities and high p, in region (d), however, it is possible that the hexadecane wets the mixed surface completely even in the absence of water. Region (a) is the hexadecane wetting region. No water layer has formed in this region; the hexadecane sees only the bare mixed SAM. Region (b) is the water-influenced wetting region. In this area, the adsorbed water layer has formed between the hexadecane liquid and surface, but not

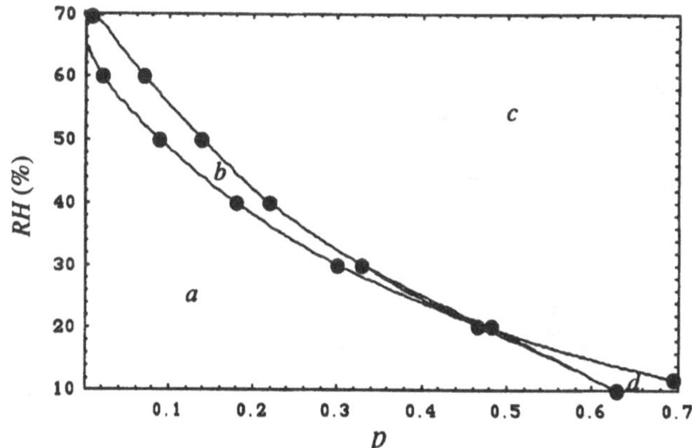

Figure 9. Diagram of wetting regions in the $p-RH$ plane at $T = 298$ K. (a) Slowly varying wettability of the bare heterogeneous surface by hexadecane. (b) Rapidly varying wettability of the heterogeneous water-dressed surface by hexadecane. (c) Complete wetting region, in the presence of a surface water layer. (d) Complete wetting region, in the absence of a surface water layer.

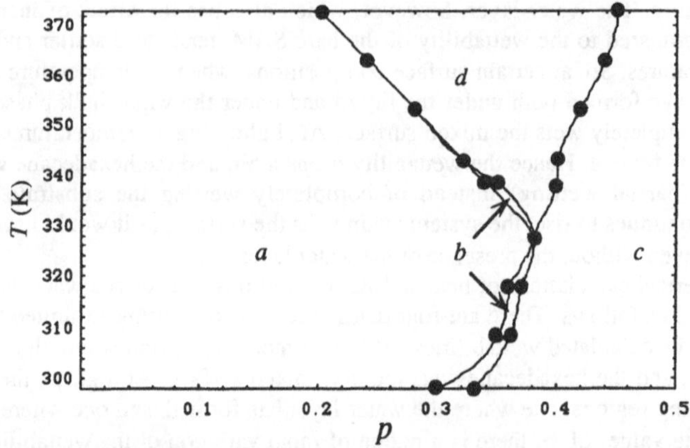

Figure 10. Diagram of wetting regions in the p–T plane at $RH = 30\%$, using the same notations as in Fig. 9.

between the hexadecane vapor and surface. This region exhibits rapid variation of the cosine of the contact angle with p. As one would expect, the region is widest at higher humidity, where the water layer forms easily even at low p, when the surface is not very hydrophilic. At low humidity, the water-influenced region (b) narrows and eventually disappears, while region (d) appears.. For $RH \leq 15\%$, no water layer forms at all, and the hexadecane interacts with the dry heterogeneous surface.

Figure 10 shows the boundaries of the wetting regions in the $p-T$ plane at 30% relative humidity. Again, the points are actual data; the lines interpolate between the data points. It is not clear if the points at $T = 343$ K should be separated in p or not, as they are separated by only $\Delta p = 0.01$, which is the spacing between data points in p. The most interesting results do not depend on that, however. Again we see a large hexadecane wetting region (a) on the left-hand side. As expected, the combined area of the complete wetting regions (c) and (d) increases as temperature increases. It is interesting to note one feature of the graphs. As explained above, there is a region of nonmonotonic variation of wettability with temperature. If we pick an intermediate value for p, such as $p = 0.36$, and follow a line starting from $T = 298$ K to $T = 373$ K, we see this effect. Following such a line, one would start in the complete wetting region (c), pass into the water-influenced

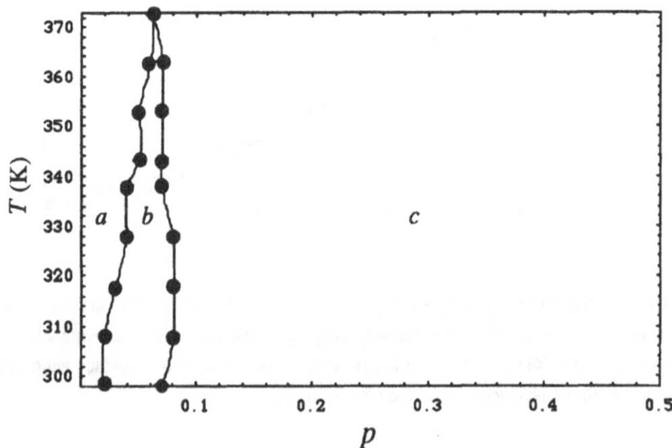

Figure 11. Diagram of wetting regions in the p–T plane at $RH = 60\%$, using the same notations as in Fig. 9.

region (b), then into the hexadecane region (a), before returning to the water-influenced region (b) (possibly) and then returning again to complete wetting, this time in region (d). Figure 11 shows the $p-T$ plane again, but this time at 60% relative humidity. The water-influenced region (b) has widened (as expected with increasing humidity), and the nonmonotonic region has disappeared. The increase in humidity biases the competition between increasing wetting and layer formation toward the latter. By the time the temperature is high enough to disorder the water surface layer in the higher humidity, the hexadecane liquid will completely wet the dry surface anyway.

APPLICABILITY RANGE AND GENERALIZATIONS

Applicability Range

The model above is based on a mean-field approximation that generalizes Cahn's wetting model[12] by introducing a second scalar field representing the solubilized adsorbate concentration. The coupling of the two fields controls the solubility of the minority species. Adsorption of the minority species is allowed by assuming a step-function profile for its concentration. At the heterogeneous surface, the adsorbate coverage is determined by the mean site occupancy in a lattice-gas model with a quenched binary random distribution of solid surface composition. The adsorbate surface concentration undergoes a layering transition that is self-consistently coupled to the magnitude of the primary field, and thus to the wettability.

We were led to develop the model above by experimental observations of anomalous wettabilities of hexadecane on mixed HUT/DDT SAMs. In these simple systems, water molecules solubilized in an organic fluid, such as hexadecane or its vapor, play the role of minority adsorbate species. As demonstrated above, our model captures the salient features of the interplay of wettability and adsorption in observed at mixed monolayers of mixed HUT/DDT SAMs chemisorbed on gold. However, we believe that the utility of the theory presented in this paper is not limited to modeling interplay of wetting and adsorption of mixed CH_3/OH-terminated SAMs. It should apply in a straightforward fashion to any system where two fluid phases with a minority species of solubilized adsorbates coexist near a smooth, but heterogeneous solid substrate. Such systems are, in fact, quite common. There are abundant examples of solid substrates with varying degrees of distribution of OH-groups at their surface. The surface OH composition is typically sensitive to the synthesis process, as well as to post-synthesis chemical and physical treatments, such as chemical etching, plasma treatment, etc. Water and other molecules that are capable of hydrogen-bonding, can readily adsorb to such surfaces. These adsorbates may either originate from contact with ambient atmosphere, or be residual from preparation of the substrates and/or the wetting fluids. Once there, extraordinary precautions are needed to purify the system of these impurities; such precautions are not taken in most wettability and wetting experiments. Hence the usefulness of the model presented here in describing those other systems, though SAMs provide perhaps an unprecedented degree of experimental control of both the chemical heterogeneity and the smoothness of the surface.

The model presented here suffers from some obvious limitations. We have relied on several approximating assumptions in the construction and solution of our model, that naturally limit its domain of applicability. First, we used a semi-infinite cubic lattice-gas model with effective one- and two-body interactions to represent much more complicated molecular interactions between the fluid molecules in continuous space, and between them and the solid substrate. Those interactions have a long-range component in their pair interactions. The projection of such interactions onto an effective short-range model may be an oversimplification that could be removed at the next level of sophistication. This can be done by generalizing the present model to include the long-range part of the pair interactions, as was previously done in other lattice-gas models for wetting,[45] or, alternatively, to account for the long-range part of the interactions on the level of a continuum interfacial free energy functional through a disjoining pressure term.[4,35] The latter approach has been recently adopted for studying wetting of self-assembled monolayers by treating them as stratified solids.[46]

Second, being a static mean-field approximation by design, our model it is not capable of describing or predicting effects stemming from thermal and quenched disorder fluctuations, as well as dynamic effects. These include such physically significant phenomena such as wandering, pinning and instability of the triple line, trapping in multiple metastable minima of the free energy adjacent to the global one, and the associated hysteresis between advancing and receding contact angles.

Lastly, the current model relies on a perturbative treatment of the minority species, coupling the two fields only to the first order in the impurity density. While this approximation is sufficient for the experimental situation addressed here, it certainly breaks down for other experimental situation of interest, such as interplay of wetting and adsorption at such mixed SAMs by a two-liquid system, e.g., where one of the liquids is hexadecane and the other is water.

Possible Extensions and Generalizations

Rough Surfaces. Here we limited our attention to wettability of surfaces that are molecularly smooth, but chemically heterogeneous. However, in many experimental systems the chemical heterogeneity is coupled with a significant degree of molecular roughness, and both influence wettability.[1,4] Here again, SAMs may serve as useful experimental system of reference. One obvious step in this direction is to use mixed SAMs, but with chains of different lengths. If the length difference is small, a quenched random distribution of the type given by (9) can still be used, and the lattice may then be terminated by a jagged, random surface. If the chain lengths are markedly different, they will tend to segregate, both in the bulk solution, and within the mixed SAM.[26,47] Furthermore, in this case the substrate cannot be assumed to be a rigid jagged surface, since thermal disordering of the longer chain will lead to their rearrangement, with a mean order determined by thermal equilibrium. In that case the substrate chains will have to be modeled explicitly, e.g., by lattice mean-field methods.[48,49] This will change the details of the calculation, but not the general approach described here.

Interplay of Wetting and Surfactant Adsorption at an Interface . The interplay of wetting and adsorption in the systems described here bears some resemblance to the wetting of the interface between two fluids (e.g., the water-air interface) by a droplet of an immiscible third fluid (e.g., oil), when concentration of surfactants is varied below the critical micelle concentration. In this case the surfactants play the role of the impurity adsorbates. Hence we should expect a similar interplay between the appropriate wettability and the coverage of the surfactant monolayer at the water-air interface.

Interplay of Polymer Adsorption with Contaminant Adsorption. At homogeneous substrates, de Gennes' mean-field theory of polymer adsorption from semidilute polymer solutions[50,51] has a form very similar to Cahn's mean-field theory of wetting.[12] Thus it is described by an interfacial free energy similar to that given by (10)-(12) and (15), where $\psi^2(z)$ is proportional to $c(z)$, the polymer concentration profile, and $a_1 = 0$. Since the system retains complete symmetry with respect to $\pm\psi$ even in the presence of the surface, it does not make sense to talk about varying wettability. However, the polymer adsorption profile can be determined by very similar methods to the one described here. This has recently been done by Andelman and Joanny[44] for the case of polymer adsorption to a (bare) surface with a quenched distribution of chemical heterogeneities, as well as for the case of polymer adsorption to a Langmuir monolayer. The latter is capable of undergoing a liquid-gas transition that is somewhat analogous to the water layering considered here. A closer analogy to this paper will be polymer adsorption from a semidilute, hydrophobic polymer solution near a mixed SAM of the type considered here, in the presence of hydrophilic impurities at a controlled relative humidity, which we have recently explored in detail on the level of a Bragg-Williams mean-field approximation. In that case no regime analogous to the partial wetting regime is found, due to the higher symmetry of the polymer adsorption problem. However, we could still find transitions between different regimes in the polymer adsorption *profile*, correlated with the *layering* of impurities (e.g., water).

Humidity Control of Anchoring Transitions. Here we dealt with systems where the wetting liquid was *isotropic*, and thus described by a *scalar* field. However, a closely related phenomenon occurs in systems where the wetting liquid is a nematic liquid crystal. The ordering parameter field in these systems is a *tensor* rather than a scalar quantity. The interface between the solid substrate and the nematic can then be modeled by a generalization of the Cahn model due to Sen and Sullivan.[52] The solid substrate may impose a specific *anchoring* direction to the nematic director that minimizes the total interfacial free energy. Recent experiments[53,54] demonstrate that transitions between different anchoring directions can be induced by varying relative humidity of water, or other hydrogen-bonding molecules, in the vapor phase coexisting with the nematic phase. This has been attributed to the adsorption of those H-bonding molecules at the solid substrate. It has also been observed that some of the same systems exhibit a transition from complete to partial wetting as the relative humidity is varied, i.e., the nematic film breaks into droplets. Previous attempts to model these phenomena either assumed a phenomenological expansion of the surface contribution to the free energy in the angular degrees of freedom describing the orientation of the nematic director at the surface,[55] or assumed a slow variation of the adsorbate concentration at the surface.[56] We believe that a generalization of the present approach by incorporating coupling between orientational and positional degrees of freedom could be an adequate model for these systems. This allows for the possibility of a discrete layering transition in the adsorbate coverage at the surface that is essential for modeling the observed phenomenon.

We note that the surfaces used in the humidity-controlled anchoring studies above were all cleaved, inorganic, ionic crystal surfaces. Such surfaces, while smooth and crystallographically well defined, have a fixed composition. On the other hand, SAMs present an easy route for engineering smooth, but chemically heterogeneous surfaces with controlled local chemical potentials. Thus using SAMs as substrates may open new possibilities for studying interplay of wetting, adsorption and anchoring. On the other hand, wettability studies have long been used in SAM research to characterize the surface properties of the monolayers. Using nematic liquid crystals as the wetting liquid has the capability of providing important new information about surface properties of SAMs, such as their symmetry, that are not accessible by isotropic wetting liquids.

ACKNOWLEDGMENTS

We are grateful to Eastman Kodak Company for financial support, and to Prof. Yonathan Shapir of the University of Rochester for his interest and encouragement. Our work on modeling the interplay of wetting and adsorption at heterogeneous surfaces originated from our desire to understand the anomalous wettability behavior observed experimentally in mixed SAMs by Abraham Ulman and Steve Evans. We gratefully acknowledge numerous discussions with them and with Jim Eilers and Ravi Sharma (all of them at Eastman Kodak Company at the time), with whom one of us (Y. S.) collaborated in an earlier effort to understand and model such behavior. Discussions between Y. S. and D. Andelman, and between Y. S. and R. M. Hornreich, E. I. Katz and V. V. Lebedev, were instrumental for establishing possible generalizations of our model to humidity controlled polymer adsorption, and anchoring transitions, respectively. Lastly, it is our pleasure to thank Harrell Sellers and Joe Golab for their hard work in organizing this conference and editing the current volume, thus providing us with an excellent avenue for disseminating our research.

REFERENCES

1. A.W. Adamson, "Physical Chemistry of Interfaces," 5th edition, John Wiley, New York (1990).
2. J.S. Rowlinson and B. Widom, "Molecular Theory of Capillarity," 1st paperback edition, Clarendon Press, Oxford (1989).
3. "Wettability," J.C. Berg, ed., Surfactant Science Series, Vol. 49, Marcel Dekker, New York (1993).
4. P.G. de Gennes, *Rev. Mod. Phys.* 57:827 (1985).
5. S. Dietrich, Wetting phenomena, *in:* "Phase Transitions and Critical Phenomena," Vol. 10, eds. C. Domb and J. Lebowitz, Academic Press, London (1988).
6. M. Schick, Introduction to wetting phenomena, *in:* "Liquids and Interfaces," J. Charvolin, J.F. Joanny, and J. Zinn-Justin, eds., North Holland, Amsterdam (1990).
7. G. Forgacs, R. Lipowsky, and Th.M. Niewenhuizen, The behavior of interfaces in ordered and disordered systems, *in:* "Phase Transitions and Critical Phenomena," Vol. 14, C. Domb and J. Lebowitz, eds., Academic Press, London (1991).
8. J.W. Gibbs, "The Collected Works of J. Willard Gibbs," Vol. 1, Thermodynamics, Yale University Press, New Haven (1928).
9. D.H. Bangham and R.I. Razouk, *Trans. Faraday Soc.* 33:1459 (1947).
10. R E. Johnson, Jr., and R.H. Dettre, Wetting of low-energy surfaces, *in:* Ref. 3.
11. A. Bose, Wetting by solutions, *in:* Ref. 3.
12. J.W. Cahn, *J. Chem. Phys.* 66:3667 (1977).
13. D J. Olbris, A. Ulman, and Y. Shnidman, *to be published.*
14. D.J. Olbris and Y. Shnidman, *to be published.*
15. For derivation, see e.g., Ch. X *in:* Ref. 1.
16. A.B.D. Cassie, *Discuss. Faraday Soc.* 3:11 (1948).
17. A. Ulman, "An Introduction to Ultrathin Organic Films From Langmuir–Blodgett to Self–Assembly," Academic Press, Boston (1991).
18. L.H. Dubois and R.G. Nuzzo, *Annu. Rev. Phys. Chem.* 43:437 (1992).
19. H. Sellers, A. Ulman, Y. Shnidman, and J.E. Eilers, *J. Am. Chem. Soc.* 115:9389 (1993).
20. L. Strong and G.M. Whitesides, *Langmuir* 4:546 (1988); C.E.D. Chidsey and D.N. Loiacono, *Langmuir* 6:709 (1990).
21. L.H. Dubois, B.R. Zegarski, and R.G. Nuzzo, *J. Chem. Phys.* 98:678 (1993).
22. C.E.D. Chidsey, G.-Y. Liu, P. Rowntree, and G. Scoles, *J. Chem. Phys.* 91:4421 (1989).
23. C.A. Alves, E.L. Smith, and M.D. Porter, *J. Am. Chem. Soc.* 114:1222 (1992).
24. A. Ulman, S.D. Evans, Y. Shnidman, R. Sharma, J.E. Eilers, and J.C. Chang, *J. Am. Chem. Soc.* 113:1499 (1991).
25. A. Ulman, S.D. Evans, Y. Shnidman, R. Sharma, and J.E. Eilers. *Adv. Colloid Interface Sci.* 39:175 (1992).
26. L. Bertilsson and B. Liedberg, *Langmuir* 9:141 (1993).
27. C.D. Bain, E.B. Troughton, Y.-T. Tau, J. Evall, G.M. Whitesides, and R.G. Nuzzo, *J. Am. Chem. Soc.* 111:321 (1989).
28. A. Ulman, S.D. Evans, and Y. Shnidman, *in:* "Dynamical Phenomena at Interfaces, Surfaces and Membranes, Les Houches 1991", D. Beysens, N. Boccara, and G. Forgacs, eds., (Nova Science Publishers, New York, 1993).
29. S. Garoff, E.B. Sirota, S.K. Sinha, and H.B. Stanley, *J. Chem. Phys.* 90:7505 (1989).
30. L.H. Dubois, B.H. Zegarski, and R.G. Nuzzo, *J. Am. Chem. Soc.* 112:570 (1990).
31. J. Hautman and M.L. Klein, *Phys. Rev. Lett.* 67:1763 (1991).
32. M.J. de Oliviera and R.B. Griffiths, *Surf. Sci.* 71:687 (1978).

33. R. Pandit and M. Wortis, *Phys. Rev. B* 25:3226 (1982).
34. R. Pandit, M. Schick, and M. Wortis, *Phys. Rev. B* 26:5112 (1982).
35. G.F. Teletzke, L.E. Scriven, and H.T. Davis, *J. Coll. Int. Science* 87:550 (1982).
36. H. Nakanishi, M.E. Fisher, *Phys. Rev. Lett.* 49:1565 (1982).
37. G. Forgacs, H. Orland, and M. Schick, *Phys. Rev. B* 25:3192 (1985).
38. N.M. Švrakic, *J. Phys. A* 18:L891 (1985).
39. W.L. Bragg and E.J. Williams, *Proc. Royal Soc. A* 150:552 (1935).
40. A. Benyoussef and N. Boccara, *J. Phys. France* 44:1143 (1983).
41. For a previous application to a wetting problem, see A. Maritan, G. Langie, and J.O. Indekeu, *Physica A* 170:326 (1991).
42. B.I. Halperin, P.C. Hohenberg, and S.K Ma, *Phys. Rev. B* 10:139 (1974); P.C. Hohenberg and B.I. Halperin, *Rev. Mod. Phys.* 49:435 (1977).
43. M. Laradji, H. Guo, M. Grant, and M.J. Zuckerman, *J. Phys. A* 24:L629 (1991).
44. D. Andelman and J.F. Joanny, J. *Phys. II France* 3:121 (1993).
45. C. Ebner, W.F. Saam, and A.K. Sen, *Phys. Rev. B* 32:2776 (1985).
46. F. Brochard-Wyart, P.G. de Gennes, and H. Hervet, *Adv. Colloid Interface Sci.* 34:561 (1991).
47. J.I. Siepmann and I.R. McDonald, *Molec. Phys.* 75:255 (1992).
48. J.M.H.M. Scheutjens and G.J. Fleer, *J. Phys. Chem.* 83:1619 (1979); ibid, 84:178 (1980).
49. F. Aguilera-Granja and R. Kikuchi, *Physica A* 189:81 (1992).
50. P.G. de Gennes, "Scaling Concepts in Polymer Physics,"Cornell University Press, Ithaca, (1979).
51. P.G. de Gennes, *Adv. Colloid Interface Sci.* 27:189 (1987).
52. A.K. Sen and D. E. Sullivan, *Phys. Rev. A* 35:1391 (1987).
53. P. Pieranski and B. Jerome, *Phys. Rev. A* 40:317 (1989).
54. J. Benchofler, B. Jerome, and P. Pieranski, *Phys. Rev. A* 41:3187 (1990).
55. J. Benchofler, J.-L. Duvail, L. Masson, B. Jerome, R.M. Hornreich, and P. Pieranski, *Phys. Rev. Lett.* 64:1911 (1990).
56. P.I.C. Texeira and T.J. Sluckin, *J. Chem. Phys.* 97:1510 (1992).

DENSITY FUNCTIONAL DESCRIPTION OF METAL-METAL AND METAL-LIGAND BONDS

D. R. Salahub, M. Castro[1], R. Fournier[2], P. Calaminici[3], N. Godbout, A. Goursot[4], C. Jamorski, H. Kobayashi[5], A. Martínez[6], I. Pápai[4,7], E. Proynov, N. Russo[3], S. Sirois, J. Ushio[8], and A. Vela[6]

Département de chimie, Université de Montréal, C.P. 6128, succursale A, Montréal, Québec H3C 3J7, Canada

INTRODUCTION

Problems related to surfaces (such as the catalytic processes which often involve metal-ligand interactions) present considerable challenge to computational scientists. From the point of view of solid state physics, difficulties arise because of the low symmetry, compared to bulk materials, of the relevant model systems. From the point of view of chemistry, difficulties arise from the sheer number of atoms needed for clusters or supermolecules to model realistically systems of interest. These difficulties are compounded when transition metal atoms are involved because of the large number of valence electrons, the importance of electron correlation, spin polarization effects, etc. Taken together these impose severe computational restrictions for the treatment of these many-electron systems. As we will show in this contribution, density functional theory (DFT) is a practical first-principles approach for the study of the electronic structure of large and complicated systems and also a very useful tool for the study of interactions between ligands and metals in various states of aggregation (atoms, clusters, and infinite surfaces). DFT has been widely used in solid state physics since the introduction of the Xα method by Slater in the early 50's[1]. About twenty years later,

[1]On leave from Facultad de Química, UNAM, México D. F., 04510 México.
[2]Steacie Institute, NRC, 100 Sussex Drive, Ottawa, Ontario K1A 0R6, Canada.
[3]Dipartimento di Chimica, Universitá della Calabria, I-87030 Arcavacata di Rende (CS), Italy.
[4]École Nationale de Chimie, 8 rue de l'École Normale, F-34075 Montpellier Cedex, France.
[5]Department of Chemistry, Kyoto Prefectural University, Shimogamo, Kyoto 606, Japan.
[6]On leave from UAM-Iztapalapa, A. P. 55-534 México D. F., 09340 México.
[7]Institute of Isotopes of the Hungarian Academy of Sciences, H-1525 Budapest, P.O.B. 77, Hungary.
[8]Hitachi Ltd., Central Research, Kokubunji-shi, Tokyo 185, Japan.

Theoretical and Computational Approaches to Interface Phenomena
Edited by H.L. Sellers and J.T. Golab, Plenum Press, New York, 1994

in the early 70's, the first DFT method generally applicable to finite systems of interest for chemistry was devised by Slater and Johnson[2]. Over the following twenty years, DFT methods in chemistry have gradually evolved to include many of the standard features of *ab initio* methods: basis sets, algorithms for geometry optimization, etc .... In parallel, more sophisticated treatments of exchange and correlation were developed. Indeed, the use of non-local functionals allows, in many cases, quantitative predictions of total energy differences (binding energies, ionization potentials, etc.). With these advances, DFT has gained much popularity recently and there has been an explosive growth in the number of DFT applications to chemistry in the last two or three years[3]. In electronic structure theory, the study of surfaces is, in a sense, at the interface of physics and chemistry: insight can be gained both from few-atom systems and from infinite ideal surfaces. We think that it is profitable to take both points of view and to study systems of any size within a single conceptual framework and with consistent numerical methods. In our work, we use the conceptual framework of DFT and the numerical methods implemented in the program deMon. We approach the subject from a "chemical" point of view and the applications will follow the reverse order of historical development of DFT: from systems involving very few atoms, to clusters, and finally to infinite surfaces and the bulk.

KOHN-SHAM THEORY AND THE GAUSSIAN-DFT CODE deMon

The Kohn-Sham equations[4] are the cornerstone for the development of modern DFT. Furthermore, their practical solution has given rise to a new branch of computational chemistry with its own specialized tools, the DFT-based methods, which can be applied to a wide range of systems (condensed phases, surfaces, gas-surface interactions, molecules and atomic clusters) and are accurate and efficient enough to provide insight into the properties of more and more complex systems [5].

DFT methodology was introduced into chemistry about twenty years ago with the implementation of the $X\alpha$-Scattered-Wave technique by Johnson [2], as suggested by Slater [1]. Here, a complex arrangement of numerical solutions to sphericalized atomic problems along with a partial wave expansion was used as a basis set, while the exchange-correlation energy was approximated by the scaled exchange energy of a homogeneous electron gas. Much basic insight into the electronic structure and spectroscopy of transition-metal (TM) complexes, clusters, chemisorption models and the like was obtained with this simple technique. However, apart from the lack of electronic correlation, the use of the *muffin-tin* potential (spherical near the atomic sites and constant between those sites) precluded the possibity of geometry optimization.

Inspired by the *ab initio* quantum chemistry techniques, Sambe and Felton [6] proposed the use of Gaussian-type orbitals (GTO) for the construction of the $X\alpha$ molecular orbitals (MO) ψ_i in the framework of local density functional theory. The resulting LCAO-$X\alpha$ technique, and its Kohn-Sham extension and improvements made by Dunlap *et al.* [7], opened a plethora of possibilities in molecular-type calculations. In fact, the use of Gaussians in DFT offers these advantages [8]: i) there is a wealth of experience from the *ab initio* Hartree-Fock methodology in the use of GTO basis sets; ii) the Gaussian approach can be implemented in DFT as a highly efficient computational method, see below; iii) accurate analytic calculations of total energies, energy gradients (for geometry optimizations), and density gradients (for nonlocal corrections) are possible; and iv) the model core potential (MCP) algorithms can be readily incorporated.

Additionally, there are two other sets of Gaussian expansions used in the present approach, one for the electron density, n, and one for the exchange-correlation potential, v_{xc}, which are convenient for what follows.

The use of GTOs for the expansion of the MO's ψ_i,

$$\psi_i(\mathbf{r}) = \sum_k C_{qi}\chi_q(\mathbf{r}) \tag{1}$$

leads to matrix Kohn-Sham (KS) secular equations, which determine the C_{qi}'s,

$$\sum_q (H_{pq} - \epsilon_i S_{pq})C_{qi} = 0 \tag{2}$$

with

$$H_{pq} = \int \chi_p^*(\mathbf{r})\hat{h}_{KS}\chi_q(\mathbf{r})d\mathbf{r} \tag{3}$$

and

$$S_{pq} = \int \chi_p^*(\mathbf{r})\chi_q(\mathbf{r})d\mathbf{r}, \tag{4}$$

that can be treated analytically. Here, the effective one-electron hamiltonian $\hat{h}_{KS}$ is the sum of a kinetic energy operator, the external potential (usually just the potential due to nuclei, but it could be any scalar potential), the classical potential due to the electronic charge density, and the exchange-correlation potential:

$$\hat{h}_{KS} = -\frac{1}{2}\nabla^2 + v(\mathbf{r}) + \int \frac{\rho(\mathbf{r}')d\mathbf{r}'}{|\mathbf{r}-\mathbf{r}'|} + v_{xc}(\mathbf{r}) \tag{5}$$

The first three terms are analogous to a Hartree hamiltonian: the electrons are described as essentially independent particles which interact only indirectly, through their repulsion to the overall electron charge density. The genuine many-body effects are described via the exchange-correlation potential $v_{xc}(\mathbf{r})$, defined as

$$v_{xc}(\mathbf{r}) = \frac{\delta E_{xc}}{\delta\rho(\mathbf{r})} \tag{6}$$

A remarkable feature of the Kohn-Sham theory is that, for non-degenerate ground states, these many-body effects can, in principle, be treated exactly with a local operator $v_{xc}(\mathbf{r})$. In standard *ab initio* methods, improving the description of correlation is usually done by improving the form of the trial wavefunction while always working with an "exact" hamiltonian. In DFT, better descriptions of correlation are achieved by improving the form of the operator $v_{xc}(\mathbf{r})$ while always keeping simple Hartree-like procedure, equations and orbitals. Of course, a change in the form of $v_{xc}(\mathbf{r})$ implies a change in the orbitals. Moreover, the accuracy of calculations in both traditional *ab initio* and DFT methods also depends on computational details, in particular on basis sets, but not exactly in the same way.

A simple scheme is obtained if the exchange-correlation energy per particle, ϵ_{xc}, is evaluated for an (interacting) homogenous electron gas of density ρ. This defines the local density approximation (LDA) for exchange and correlation energy

$$E_{xc}^{LDA} = \int \rho(\mathbf{r})\epsilon_{xc}(\rho)\, d\mathbf{r} \ , \tag{7}$$

the corresponding exchange-correlation potential is given by

$$v_{xc}^{LDA}(\mathbf{r}) = \frac{\delta E_{xc}^{LDA}}{\delta\rho(\mathbf{r})} = \epsilon_{xc}(\rho(\mathbf{r})) + \frac{\delta\epsilon_{xc}(\rho)}{\delta\rho}\rho(\mathbf{r}) \ . \tag{8}$$

The LDA provides a useful and practical level of approximation for the study of a variety of systems and properties, which ranges from solid state physics to atoms and molecules. The spin-polarized generalization of the LDA, the local-spin density approximation (LSDA) is appropriate for systems with unpaired spins. At this level of approach the calculations are computationally less demanding than those of Hartree-Fock theory and, importantly, some account of correlation is included through v_{xc}. In many instances, in highly correlated systems (such as transition metal systems), it is necessary to go beyond the local approximation. In this contribution, we have used non-local functionals which involve a gradient expansion of the exchange and correlation energies to account for inhomogenieties in the electron distribution. It will be shown below that the introduction of non-local corrections improves dramatically the estimation of binding energies for TM systems.

Following Sambe and Felton [6], the electron density is expanded in an auxiliary basis set of Gaussian-type functions,

$$\rho(\mathbf{r}) = \sum_i a_i f_i(\mathbf{r}) \tag{9}$$

As proposed by Dunlap *et al.* [7], the density fitting coefficients a_i are defined such that the Coulomb energy

$$\int\int \frac{\delta\rho(\mathbf{r})}{|\mathbf{r}-\mathbf{r}'|}\delta\rho(\mathbf{r}')d\mathbf{r}d\mathbf{r}' \tag{10}$$

arising from the difference between the fitted and original density

$$\delta\rho(\mathbf{r}) = \rho(\mathbf{r}) - \rho'(\mathbf{r}) \tag{11}$$

is minimized while maintaining charge conservation. In other words, the Gaussian-based implementation of the Kohn-Sham equations relies on a variational, analytical approximation to the density. Numerical stability of the total energy was attained with this procedure for the Coulomb potential. The remaining exchange-correlation potential, v_{xc}, is expanded in another set of Gaussian-type functions,

$$v_{xc}(\mathbf{r}) = \sum_i b_i g_i(\mathbf{r}) \tag{12}$$

This potential, being a smooth function of the density, is fitted on a set of grid points centered about each atom [9, 10]. The coefficients b_i are determined by a least squares fitting procedure, which minimizes the error in the fitted potentials over the grid points.

An important consequence of this fitting procedure is that the four-index two-electron integrals are translated into three-index two-electron integrals, for both the coulomb

$$\int\int \chi_p^*(\mathbf{r})\chi_q(\mathbf{r})\frac{1}{|\mathbf{r}-\mathbf{r}'|}f_i(\mathbf{r}')d\mathbf{r}d\mathbf{r}' \tag{13}$$

and exchange-correlation

$$\int \chi_p^*(\mathbf{r})\chi_q(\mathbf{r})g_i(\mathbf{r})d\mathbf{r} \tag{14}$$

contributions. The one-electron integrals remain exactly the same as in Hartree-Fock theory, namely the overlap (Eq. 4), and the kinetic energy and nuclear attraction terms

$$\int \chi_p^*(\mathbf{r})(-\frac{1}{2}\nabla^2 - \sum_i \frac{Z_i}{|\mathbf{r}-\mathbf{r}_i|})\chi_q(\mathbf{r})d\mathbf{r} \tag{15}$$

The net result is that the LCGTO-DFT method includes electronic correlation, through v_{xc}, with a computational effort which scales formally as N^2M, where N is the number of functions in the orbital basis and M is the number of functions in an auxiliary basis. This is opposed to the Hartree-Fock method, where there is no electronic correlation and the scaling is formally N^4 due to four-index two-electron integrals; the inclusion of electronic correlation increases the scaling to at least N^5 (for MP2). Therefore, DF-based methods represent a clear first principles alternative for the study of *many-electron* systems.

The computation of the corresponding matrix elements is well-suited for modern vector machines. Very efficient algorithms have been developed for this, as well as new accurate integration schemes, so that good accuracy can be obtained with reasonably sized grids. In the present method they are evaluated analytically, once the coulomb and exchange-correlation potentials or energies have been fitted to auxiliary gaussian basis sets, as discussed above. The fitting for these exchange-correlation terms involves the same type of grid as in numerical integration techniques. So, they can also be integrated numerically to high accuracy. In deMon [8] the two-electron integrals are evaluated with the efficient method of Obara and Saika [11]. This recursive computation of four-index cartesian Gaussian integrals, for the Hartree-Fock method, has been reformulated for the three-index integrals of the present procedure (this part of the code has yet to be vectorized). As a consequence of this, an effective scaling approaching N^2 for large systems is observed, as measured by the CPU time, instead of N^3.

Actually, the numerical errors (due to finite basis sets and grids) could be decreased to reasonably converged limits so that one could be confident that remaining errors are primarily those of the density functional itself and not of the particular numerical machinery used [8]. Indeed, any desired representation for the expansion of the Kohn-Sham orbitals ψ_i might be chosen. The classic quantum chemistry manifold is again on the scene, with its pros and cons. For example, ADF [12], a HFS-LCAO technique, can use Slater-type functions which provide a more compact expansion. DMOL [20] uses flexible numerically defined basis sets, but here the final analysis can sometimes be less familiar. The use of a plane wave representation, as in the Car-Parinello [21] and CORNING [22] codes, has the merit that it is independent of the nuclear positions, which facilitates the calculation of forces in the geometry optimizations and/or molecular dynamics simulations. However, one needs to use a large number of plane waves to describe localized systems, such as first row atoms, not to mention TM atoms. NUMOL [23] is a promising basis set free code. Many other quantum chemistry programs (CADPAC, Gaussian, CRYSTAL, CRS4) implement the Kohn-Sham equations, each with its own combination of grid, basis sets, pseudopotentials, energy functionals ...

A Gaussian representation, as in DGAUSS and deMon, has the advantage of being more familiar to quantum chemists. The same Hartree-Fock optimized basis sets such as 6-31G**, developed by Pople *et al.* [24], can be used in DF calculations to produce good geometries. But accurate predictions of binding energies and/or reaction energies require the use of DF optimized Gaussian-type basis sets. The LSD basis sets are constructed with an optimization procedure [25] based on the Tatewaki and Huzinaga algorithm [26]. Starting with contracted Hartree-Fock optimized basis sets [27], contractions are optimized one at a time (all contractions but one are "frozen") by varying the exponents and coefficients so as to minimize the total atomic energy [28]. When all contractions have been optimized once, the process is repeated in a cyclic fashion until the total energy does not decrease appreciably between two cycles. This yields optimized minimal basis sets which are decontracted for molecular calculations. For example, from B to Ne, the bases are constructed from nine s-type, and five p-type primitive Cartesian Gaussians; the (63/5) set is then split into (621/41). To further improve the molecular results, a d polarization function is added and the basis set is turned into a (621/41/1*) set. The polarization functions were chosen from the previous experience in LCGTO-LSD calculations and from Pople's 6-31G* basis sets. This brings the LSD basis sets to the quality of DZVP (double-zeta plus polarization for the valence electrons). Indeed, these DZVP basis sets are characterized by a high quality of the valence orbitals, as inferred from comparison with the energies and (shapes) of orbitals obtained by the numerical solution of the Kohn-Sham equation, and exhibit only small Basis Set Superposition Errors (BSSE) since they give a good representation of the core orbitals. The accuracy of the DZVP sets may be increased as needed. For example, from B to Ne, this is done by adding one s- and one p-type primitive to (621/41/1*) which results in (721/51/1*) basis sets, the so called DZVP2. The DZVP and DZVP2 basis sets are the current defaults in **deMon** for all atoms up to Xe [28].

The auxiliary basis sets for the electron density, ρ, and the exchange-correlation potential, v_{xc}, and energy are derived from the orbital basis sets, DZVP or DZVP2, according to a procedure similar to that developed by Dunlap. The smallest exponent of the orbital set is used to start the expansion for the electron density set. Since the charge density is the sum of the squares of the molecular orbitals, this small exponent is multiplied by two. The auxiliary set is built using s-type functions and blocks of s-, p-, and d-type functions with shared exponents. This facilitates the calculation of molecular integrals. A similar procedure is used for the exchange-correlation set. It is built from the electron density set by dividing the exponents by 3 since the exchange correlation goes essentially as $\rho^{1/3}$ [7]. In this way, auxiliary basis sets were generated to supplement the orbital bases DZVP (basis A1) and DZVP2 (basis A2). The notation we use for auxiliary bases is (k_1,k_2;l_1,l_2), where the semi-colon separates the charge density (CD) from the exchange correlation (XC), $k_1(l_1)$ is the number of s-type gaussians in the CD(XC) basis and $k_2(l_2)$ is the number of s-, p-, and d-type gaussians constrained to have the same exponent in the CD(XC) basis.

A fundamental point in this computational methodology is the advent, and the systematic implementation in the codes, of gradient corrections to the total energy, using the proposed *nonlocal* functionals. In the LCGTO code **deMon**, the LSDA is introduced through the Vosko, Wilk, and Nusair parametrization of the homogenous gas correlation energy [29]. This approach represents one of the best analytic functional forms for the LSD potentials. In this scheme there is no empirically adjustable parameters. Several options exist for including nonlocal spin-density corrections. The two schemes first implemented in **deMon** use the gradient correction to correlation that was proposed by Perdew[30], but they differ in their treatment of exchange. The gradient correction to exchange due to Perdew and Wang[31] is used in the "PP" option and those of Becke[32] in the "BP" option. The nonlocal potential, obtained through the functional derivatives of the energy, is included in the Kohn-Sham potential during the Self-Consistent-Field (SCF) procedure. Our nonlocal calculations are fully self-consistent by default. In some cases it can be sufficient to evaluate the nonlocal corrections in a perturbative way, by using the LSD potential throughout the SCF iterations and introducing gradient corrections only in the energy evaluation. The nonlocal corrections for the exchange and correlation energies are essential if results of chemical accuracy are desired in the computation of binding energies and in the study of chemical reactions (activation energies). This will be stressed in the next section for some TM-Ligand, and hydrogen-bond systems.

Efficient algorithms for locating extrema of potential energy surfaces (PES) using analytical gradients have been implemented into *ab initio* methods over the last twenty years. There has been very many successful applications to s- and sp-valence electron systems. Applications to systems with TM atoms, however, are much more difficult since Hartree-Fock theory is usually a bad starting point and the cost of multideterminantal methods becomes rapidly prohibitive as the number of TM atoms increases. The LCGTO-DF method offers an attractive alternative in such cases. As discussed above, it includes electronic correlation in a form that does not lead to the scaling problem of traditional *ab initio* techniques. Furthermore, the use of GTO functions allows analytical evaluation of the first derivatives of the total energy with respect to nuclear displacements. The working equations and/or algorithms have been obtained [33, 34] in a way similar to those in the Hartree-Fock [35] and MCSCF [36] methods. The LSD energy gradient can be written as

$$-\frac{\partial E_{LSD}}{\partial x} = F_{HFB} + F_D \tag{16}$$

where F_{HFB} is the Hellman-Feyman force with a correction to account for the incompleteness of the orbital basis:

$$F_{HFB} = \sum_{pq} P_{pq}\{\frac{\partial h_{pq}}{\partial x} + \sum_i a_i[\frac{\partial(pq)}{\partial x}||i] + [v_{xc}\frac{\partial(pq)}{\partial x}]\}$$

$$+ \frac{\partial U_n}{\partial x} - \sum_{pq} W_{pq}\frac{\partial S_{pq}}{\partial x}\ , \tag{17}$$

whereas F_D accounts for incompleteness of the density fit

$$F_D = \sum_i a_i[\frac{\partial f_i}{\partial x}||(\rho - \rho')] \tag{18}$$

with ρ being the exact density and ρ' the fitted density. Clearly, in the case of a perfect fit, F_D vanishes. In Eq. (14), p and q denote GTO functions, P_{pq} is the associated density matrix element, and h_{pq} is the matrix element for the sum of the kinetic energy and nuclear attraction operators plus, possibly, an external electric or magnetic field. Density fitting basis functions are denoted by i and the a_i are the associated density fit coefficients. U_n denotes the nuclear-nuclear repulsion term. The symbol || denotes the Coulomb operator $1/|\mathbf{r}-\mathbf{r}'|$. Finally, W_{pq} is an energy weighted density matrix element as in the Hartree-Fock gradient formula [37]. In essence, the expressions for F_{HFB} and F_D contain derivatives of three-index integrals, which can be calculated analytically in the same way as those of the SCF procedure. The exchange-correlation contributions to the energy gradients are calculated by numerical integration on an augmented set of grid points, using the FINE grid defined in **deMon**. This is done for each new geometry, at the end of the SCF step. Spurious one-center contributions to the exchange-correlation forces are eliminated in a similar way as has been done by Versluis and Ziegler [38].

A full derivation of the LCGTO-LSD second and third derivatives has been performed by Fournier [33]. The following remarks are in order: i) the LCGTO-LSD energy derivatives are readily generalized to XC energy functionals depending on derivatives of the density to any order; ii) these derivatives, as implemented in the present methodology, include an exact correction for the incompleteness of the density fit basis; iii) the second derivatives (the Hessian matrix) are currently updated numerically. It appears that LCGTO-LSD analytic second derivatives will offer a moderate gain in efficiency compared to numerical differentiation but that the savings will be considerable for third derivatives. Work in this direction is currently in progress in our laboratory and elsewhere [39, 40]. In addition to potential energy surface walkers, target properties are IR and Raman intensities, and polarizabilities [41].

The functionality of the DF techniques is rapidly approaching the standards of traditional *ab initio* techniques. Results from academic and commercial versions of Gaussian-based DF methods are more and more frequently compared in the literature, along with their *ab initio* counterparts. These DF computational techniques have entered the mainstream of molecular electronic structure calculations.

TRANSITION METAL-LIGAND BONDING

Our qualitative understanding of the bonding at surfaces and in complexes and clusters derives largely from studies of relatively few and simple model systems. Complexes consisting of only one metal atom and one ligand molecule are particularly useful in this respect (see, for example, Ref. [42]). We will outline in this section results of DFT calculations for various monoligand complexes of TM atoms. We first discuss metal carbonyls as they occupy a central position in inorganic chemistry and surface science. They are among the most studied systems and display most clearly the σ-donation and π-backdonation bonding mechanism on which much chemical intuition is based.

TM Monocarbonyls, M-CO (M=Sc–Cu)

Figure 1 illustrates the trends in some properties of high-spin and low-spin monocarbonyl complexes of first-row TM atoms calculated with deMon. A few remarks are in order before discussing Fig. 1. The *bond strength* is the energy of the complex at its minimum calculated relative to the M + CO fragments in electronic states which correlate with that of the complex. The *binding energies* reported in Fig. 1 are calculated relative to ground state metal atom and CO molecule; they include an empirical correction for the $3d^{n-2}4s^2$—$3d^{n-1}4s$ splitting which is incorrectly predicted by the calculations. Indeed, calculations on the isolated atoms are plagued with rather large errors due to limitations of the functionals used and to the difficulties of a proper treatment of symmetry and spin within DFT (see, for example, Ref. [43, 44, 45, 46]). Because of this, there is some uncertainty on the precise values of the DFT M-CO binding energies. The energies were calculated with perturbative BP gradient corrections. The extent of backdonation was calculated from Mulliken populations of σ and π symmetries on the M and CO fragments.

The *increase* in CO bond length and the metal-CO bond strength both correlate very well with backdonation. The binding energy, on the other hand, shows little correlation with backdonation because of the large scatter in TM atoms' promotion energies. For the metal atom the $3d^{n-1}4s$ configuration is optimal for bonding. It has minimum occupation of the $3d\sigma$ orbital and maximum occupation of the $3d\pi$ orbital. Such a configuration gives a much smaller repulsion with the carbon lone pair 5σ orbital than the $3d^{n-2}4s^2$ configuration[49], and it favours CO-to-metal σ donation and metal-to-CO π backdonation[50, 51]. The metal atoms' excitation energies to states that lead to strong bonding with CO vary greatly across the row and this variation prevents a simple correlation of binding energy with backdonation. The calculated binding energies (in kcal/mol) for CrCO, FeCO, NiCO and CuCO are, in that order (experimental values are in parentheses): 13 (≈0 [52]), 14 (10.5±3.5 [53], 8.1±3.5 [54]), 54 (40.5±5.8 [54]) and 19 (6.0±1.2 [55]). The calculated values for ScCO, TiCO and VCO are 9, 16 and 26 kcal/mol respectively, which compare rather well with estimates based on *ab initio* calculation by Barnes *et al.* [56] (6, 14 and 21 kcal/mol) and by Jeung [57] (4, 7 and 15 kcal/mol). Although the BP gradient-corrected functional presents a great improvement over the LSD functional in all cases [58, 59], it still seriously overestimates some binding energies (CrCO, CuCO, NiCO).

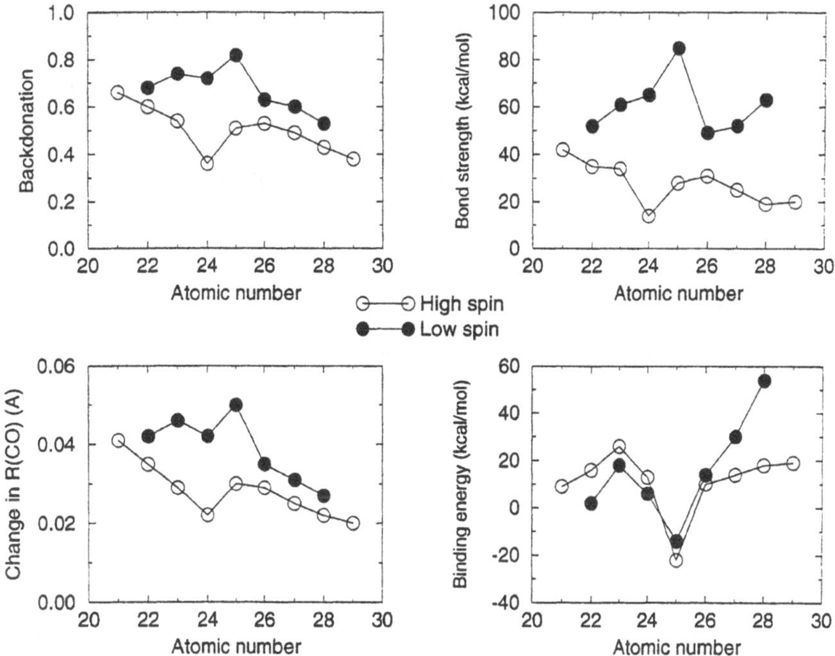

Figure 1. Trends in calculated properties of first-row transition-metal monocarbonyls in their high-spin and low-spin states.

An interesting result of our DF calculations is that the high-spin state complexes for *all* elements to the right of vanadium (CrCO, MnCO, FeCO, CoCO, NiCO and CuCO) have bent equilibrium geometries; they have MCO angles of 142, 156, 158, 153, 148 and 143 degrees respectively[51, 58, 60]. We attribute this to the fact that the σ repulsion, which is relatively large in these high-spin complexes, is decreased upon bending. Recent coupled cluster calculations confirm the prediction of a bent geometry for CuCO, albeit with a larger MCO angle (153 degrees) and a much softer bending mode[61]. On the other hand, ESR experiments show that CuCO is linear when formed in an inert gas matrix[62, 63]. Moreover, photoelectron spectra are consistent with linear geometries for both the low-spin and high-spin states of FeCO[53] and CoCO[64]. Further work, both theoretical and experimental, seems in order to resolve the question of bent *vs* linear geometry of TM monocarbonyls.

Recently, we investigated the neutral, positive, and negative species of FeCO [59]. The binding energy of Fe-CO was performed through the use of three different basis sets for the Fe atom: a compact (43321/4211*/311+) set [25], the default DZVP2 (63321/5211*/41+) set [28], and a large uncontracted (20s/12p/9d) basis set supplemented with one s, one p and one d diffuse function [65] (basis "C"). In the three cases, the non-spherical non-local binding energies [59] were equal to 30, 33, and 34 kcal/mol, respectively. The quality of our basis sets (optimized explicitly for DFT calculations) is apparent from the relative insensitivity of the binding energy (and other properties) calculated with these various basis sets. This gives us further assurance that the results are converged with respect to addition of basis functions. The changes in the FeC stretch and the harmonic bending frequencies that we calculate in going from neutral FeCO to the anion ($FeCO^-$) agree well with a photoelectron spectroscopic study[53]. For the FeC stretch we calculate 658 and 566 cm^{-1} (experimental: 530 and 465), and for the bending mode we obtain 368 and 272 (experimental: 330 and 230). Our calculations reproduce the observed trend in dissociation energies ($D(FeCO^+) > D(FeCO^-) > D(FeCO)$). The calculated electron affinity and ionization potentials of FeCO are 1.18 eV and 7.48 eV: the experimental values are 1.157±0.005 eV [53] and 6.7±0.2 eV [54].

NiC_2H_4 and NiC_2H_2

The NiC_2H_4 and NiC_2H_2 complexes are prototypes of a TM π-bonded to double and triple CC bonds. The infrared spectra obtained in matrix isolation studies of NiC_2H_4[66, 67] and NiC_2H_2[68] provide a testing ground for theory. We calculated the harmonic frequencies and infrared intensities in the ground state for these two complexes at their DFT-optimized geometries using the BP potential and basis sets of roughly triple-zeta-plus-polarization quality [69, 70].

First, it should be said that the force fields of the two ligand molecules are correctly described by DFT. The average percent deviation of BP harmonic frequencies from *harmonic* experimental values is 3.1% for C_2H_4 and 3.8% for C_2H_2. The calculated CC stretch frequencies are particularly accurate, with errors of only 1640−1655=−15 cm^{-1} for ethylene and 2016−2011=+5 cm^{-1} for acetylene. The calculated shifts in the frequencies of C_2H_2 upon isotopic substitution are generally within 10% of experiment[70]. The results for the complexes are summarized in Table 1 which lists the experimentally observed *fundamentals* together with the BP calculated *harmonic* frequencies. The excellent agreement between theory and experiment is partly fortuitous — the effect of anharmonicity would somewhat increase the discrepancy. The calculated intensities are qualitatively correct: in general, the calculated intensities corresponding to observed bands are large and the others are small. Overall, the BP-calculated frequencies and intensities are accurate enough for assigning many observed bands and, if this accuracy is reproducible, for predicting the main features in IR spectra of other complexes.

The agreement between the calculated binding energies and their experimental values (in parentheses) is better than for the monocarbonyls: 40 kcal/mol[69] (35.5±5[71]) for NiC_2H_4 and 46 kcal/mol (46±6) for NiC_2H_2 [70]. In addition to the singlet ground state, calculations predict that there are bound triplet states with a well depth of about 13 kcal/mol in both cases. This prediction is entirely consistent with data on the kinetics of association of nickel atoms with C_2H_4 and C_2H_2[70].

Table 1. Calculated harmonic, and observed fundamental, frequencies of NiC_2H_4 and NiC_2H_2 (cm^{-1}).

NiC_2H_4

Mode	Calc.[(a)]	Exptl[(b)]	Exptl[(c)]
CH stretch	2995	—	2960
CC stretch	1461	1465	1496
CH_2 scissor	1158	1156	1158
CH_2 wag	910	901	900
NiC stretch	539	498	376

NiC_2H_2

Mode	Calc.[(d)]	Exptl[(e)]
CC stretch	1655	1640
CH in-plane bend	825	844
CH out-of-plane bend	661	657
NiC stretch	541	545

(a) see Ref. [69]; (b) see Ref. [67]; (c) see Ref. [66]; (d) see Ref. [70]; (e) see Ref. [68].

Monoligand Complexes of Copper Atom, Dimer and Trimer

Studies of the bonding of ligands to TM atoms are useful but many aspects of cluster and surface chemistry have no counterpart in the single TM atom models. In that respect, small ligands bonded to TM dimers and trimers represent a natural first step toward cluster chemistry. To keep things as simple as possible, we chose one of the easiest TM atoms to investigate (copper) and studied trends among the Cu–L, Cu_2–L and Cu_3–L series. We fully optimized the geometries starting from a number of initial structures (end-on, side-on, etc ...).

In the first case we considered, L=C_2H_2, two main conclusions were reached[72]. First, the Cu_n–C_2H_2 binding energy increases with n: 10 kcal/mol for n=1, 15 kcal/mol for n=2 and 32 kcal/mol for n=3. This comes about because, as n increases, 4s4p hybridization is more facile and there is a greater polarization of the electron density on the copper atoms away from acetylene thus reducing the σ repulsion. Moreover, the empty 4p and 4s4p-hybrid orbitals in Cu_2 and Cu_3, which are low-lying compared to the 4p of the copper atom, can mix to some extent with 3d orbitals and get involved in σ-acceptor and π-donor bonding orbitals. Second, we found that the *cis*- and *trans*-1,2-dicuproethylene isomers of $Cu_2C_2H_2$ are *almost* as stable as the π complex. We also find that the Cu_3 ring relaxes significantly from its isosceles geometry upon bonding to C_2H_2 and adopts a nearly equilateral geometry. Therefore, Cu_2 and Cu_3 give some indication of the possible importance of structural relaxation or rearrangement of clusters upon association reactions. Preliminary results of calculations on copper atom and dimer bonded to CO, NH_3 and C_2H_4 show that the end-bonded geometry of Cu_2L

is always favoured. We also find that D(Cu_2-L) is larger than D(Cu-L) by about 10 kcal/mol in all cases, consistent with recent experiments showing that Cu_2, Ag_2 and Au_2 are more reactive than the corresponding atoms[73, 74].

The Chemisorption of CO_2 by Pd(110)

As we mentioned above, there have been many studies of the bonding of CO to transition metal centers. However, comparable studies involving CO_2 are more scarce. Brosseau *et al.* [75, 76] have studied the adsorption of CO_2 on a Pd(110) surface by means of HREELS spectroscopy. There has been a theoretical study of the bonding of CO_2 on Ni surfaces [78] but no calculations were reported for CO_2 bonded to palladium atoms or surfaces.

Sirois and Salahub [79] have used **deMon** to study the interaction of CO_2 with Pd_n ($n \leq 4$) clusters. Both local and nonlocal functionals were used. The core electrons of the palladium atom were represented by a MCP which includes relativistic effects. The $4p^6 4d^9 5s^1$ electrons were included explicitly in the calculation, for which a palladium (+16)(2211/2111/21) orbital basis set was constructed. Basis sets of 6-31G** quality were used for C and O atoms. Several coordination modes of the CO_2 moiety were studied. In all cases, the geometry of Pd_n-CO_2 ($n \leq 4$) was fully optimized. A vibrational analysis was done for the lowest energy structures.

The main results for Pd-CO_2 are given in Table 2. The dihapto CO (η^2-CO) coordination mode, of C_s symmetry, was found to have the lowest energy. There is a strong charge transfer from Pd to CO_2 (about 0.7 electrons) which produces a large structural change in CO_2: it goes from a linear to a bent geometry. In this η^2-CO mode, or mixed Pd-C/Pd-O bond, there is an elongation (with respect to free CO_2) of the C=O bond involving the oxygen atom nearest to the metal center. The two C=O bond lengths, in Pd-CO_2, are no longer equal, they are 1.20 and 1.24 Å. The Mulliken population analysis shows that the metal atom is positive (+0.7), the carbon atom is positive (+0.1), the oxygen atom nearest to the Pd center is negative (-0.5), and the other oxygen atom is also negative, but to a lesser extent (-0.3 electrons). There are two types of interactions in the η^2-CO structure. There is a repulsive electrostatic interaction between the metal and the carbon centers, and an attractive electrostatic interaction between the oxygen and the palladium atoms. The latter contributes to the total energy lowering of the η^2-CO structure. The vibrational analysis yields no imaginary frequency and shows that η^2-CO is a minimum. The calculated frequencies associated with metal-carbon and metal-oxygen modes are 203, 373, and 561 cm^{-1}. In particular, 373 cm^{-1} corresponds to the symmetric elongation of the Pd-C and Pd-O bonds. This structure favors the occurrence of a *mixed* carbon-oxygen bond in Pd-CO_2. The bending, symmetric, and asymmetric vibrational modes of CO_2 in Pd-CO_2 have calculated frequencies of 689, 1212, and 2018 cm^{-1} respectively. The experimental frequencies of CO_2 on a H_2O/Pd(110) surface are [75, 76] 786, 1199, and 1631 cm^{-1} respectively. This suggests a C_s symmetry because the asymmetric mode at 1631 cm^{-1} is dipole allowed in C_s symmetry and dipole forbidden in C_{2v}. Two of the observed bands are in agreement with our calculations. However, there is a large discrepancy for the asymmetric mode (calculated: 2018 cm^{-1}, experimental: 1631 cm^{-1}). This is directly related to the size of the Pd_n cluster. The asymmetric mode is shifted to lower values as the size of the cluster is increased. For example, in Pd_4-CO_2 that frequency is equal to 1843 cm^{-1}, which is closer to the experimental finding. Calculations for an infinite CO_2/Pd(110) surface using **Bloch-deMon** (see below) are in progress in our laboratory; this will complement our studies using small Pd_n clusters.

Table 2. Bond lengths, bond angles, relative energies, and vibrational frequencies for various coordination modes of $Pd\text{-}CO_2$.

	η^1 (Pd-C)	η^2 (Pd-CO)	η^1 (Pd-O)
Bond Lengths (Å)			
Pd-C	1.99	1.96	
$C\text{-}O_1$	1.24	1.22	1.18
$C\text{-}O_2$	1.20	1.22	1.17
$Pd\text{-}O_1$			2.07
Bond Angles			
$O_1\text{-}C\text{-}Pd$	79.7	104.6	
$O_2\text{-}C\text{-}Pd$	129.9	104.6	
$O_1\text{-}C\text{-}O_2$	150.4	150.8	179.0
$Pd\text{-}O_1\text{-}C$			146.0
	Relative Energy (Kcal/mol)		
	0.0	+2.2	+13.8
	$(0.0)^a$	$(+8.4)^a$	
	Frequencies $(cm^{-1})^e$		
	203	-112	97
	373	311	333
	561	586	538
	$689\delta^b$	$707\delta^b$	543
	$1212\nu^c$	$1245\nu^c$	1321
	$2012\nu^d$	$1974\nu^d$	2398

a) Nonlocal results.
b) Bending, c) symmetric, and d) asymmetric stretch of CO_2 in $Pd\text{-}CO_2$.
e) Experimental values from Ref. [76] are 786, 1199 and 1631cm^{-1} respectively.

The monohapto C (η^1-C) coordination mode was found to be a transition state, 2.2 kcal/mol, for the LSDA, and 8.4 kcal/mol, at the nonlocal level, above the (η^2-CO) minimum. (The nonlocal approach, a higher level of theory, enhances the stability of the η^2-CO structure). Similarly, the η^1-O coordination mode is 13.8 kcal/mol, for the LSDA, less stable than the η^2-CO ground state. Finally, the structure which corresponds to the η^2-OO coordination mode collapses into the η^2-CO minimum during the optimization procedure. In other words, there is no minimum on the PES of $PdCO_2$ with the η^2-OO mode of bonding. In summary, the $Pd\text{-}CO_2$ system shows quite a different picture from that found for $Ni\text{-}CO_2$ where CO_2 is chemisorbed in the dihapto OO (η^2-OO) coordination mode.

Formate Chemisorption on Rh_4 and Rh_2Sn_2

The effect of the presence of Sn atoms on the interaction of formate (HCOO) with Rh clusters has been investigated [80] using the model metal clusters shown in Figure 2. Rh_4 represents a model of the pure transition metal catalyst, while the two Rh_2Sn_2 clusters are models of bimetallic catalysts. The geometry of the formate chemisorbed in a bidentate bridge form (see Figure 3.) on each cluster has been fully optimized and the harmonic vibrational frequencies of the chemisorbed formate have been calculated. The results are listed in Table 3.

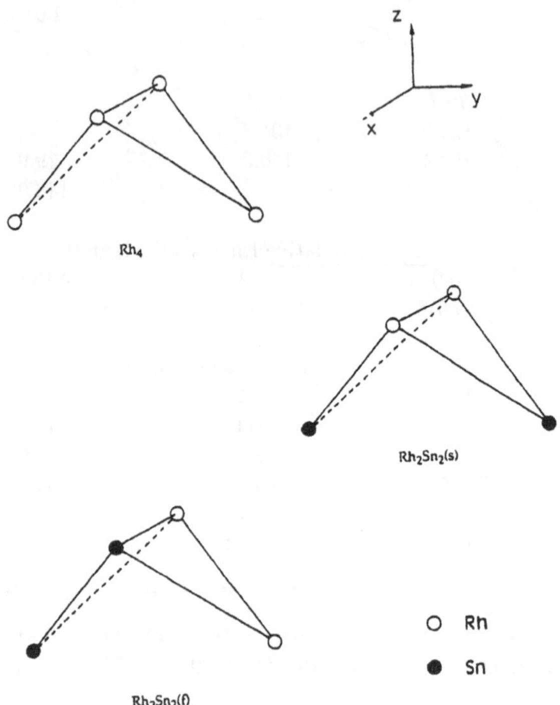

Figure 2. Metal clusters for modeling the adsorption of formate on Rh and $Rh_{0.5}Sn_{0.5}$ surfaces.

It is apparent from these results that the presence of the Sn atoms in the metal cluster does not alter the internal equilibrium properties of the adsorbed formate, however, it weakens the metal-formate bond: in the $Rh_2Sn_2(OOCH)$ clusters, the metal-oxygen bond is lengthened, the corresponding bond order is decreased, and the metal-formate frequencies shift to lower values, as compared to those in the pure rhodium cluster. Based on this finding, we can give a possible explanation for the improved selectivity of bimetallic RhSn catalysts for reactions where the decomposition of formate surface intermediates is an "undesirable" reaction channel. As the influence of the Sn atoms increases, the formation and further decomposition of the formate species is suppressed, shifting the composition of products in favour of other reactions, which are not associated with the formate intermediates.

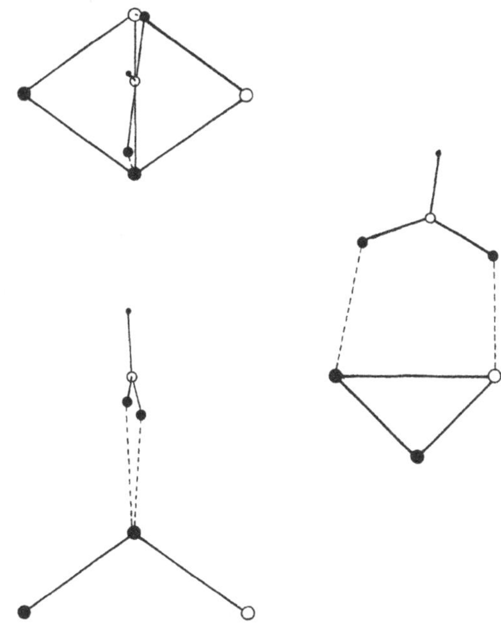

Figure 3. Geometry of formate chemisorbed on Rh_4 and Rh_2Sn_2

Table 3. Bond lengths (in Å), bond orders, and harmonic frequencies for Rh_4F, $Rh_2Sn_2(s)F$, and $Rh_2Sn_2(f)F$ (F = OOCH).

System	Rh_4F	$Rh_2Sn_2(s)F$	$Rh_2Sn_2(f)F$
Bond distances			
Rh-O	2.009	2.094	2.030
Sn-O	–	–	2.254
C-O	1.267	1.263	1.267, (1.260)[a]
C-H	1.118	1.119	1.118
Bond orders			
Rh-O	0.73	0.49	0.59
Sn-O	–	–	0.30
C-O	1.50	1.55	1.52, (1.60)[a]
C-H	0.93	0.92	0.92
Vibrational Mode[b]			
ω_1 CH str	2989 (2910)	2967	2973
ω_2 CO str	1302 (1339)	1320	1323
ω_3 OCO bend	736 (790)	738	739
ω_4 MO str (tz)	369 (350)	323	381[c]
ω_{11} MO str (tauy)	382	239	258[d]

[a] Values corresponding to CO(Rh) and CO(Sn) respectively.
[b] M = metal atom; in parentheses are the experimental values. [c] RhO str [d] SnO str.

CO on Rh and Pd.

Goursot *et al.* have studied recently how to model the chemisorption of CO on Rh and Pd surfaces [81] using a *cluster approach*. First, the method was tested on the Pd_2 and Rh_2 dimers. Theoretical effort devoted to the study of TM dimers has shown that the level of correlation needed to describe these units is very high [100]. For instance, for Rh_2 and Pd_2, early *ab initio* results yielded unrealistic equilibrium bond lengths (more than 0.15 Å longer than the bulk values), associated with very low dissociation energies [82, 83]. In Table 4 are shown the calculated spectroscopic constants for the ground states (GS) of Pd_2 and Rh_2. Here, MCPs (Rh^{15+}, Pd^{16+}) have been used for Rh and Pd atoms. The scalar relativistic effects are incorporated into the MCP's [84]. The 4p, 5s, 5p, and 4d orbitals were treated explicitly. The contraction pattern of the valence electron orbital basis sets is (2211/2111/121) for Rh and Pd. All-electron (5211/411/1) basis sets were used for carbon and oxygen atoms. The PP nonlocal functional was used [31, 30] Although *ab initio* CI of appropriately high level and DF calculations are two different approaches to the problem of electronic correlation, they produced similar GS properties for Pd_2 and Rh_2. The MRSDCI [87, 88] equilibrium bond lengths (2.48 and 2.28 Å), stretching frequencies (160 and 267 cm^{-1}), and D_es (0.85 and 2.10 eV) for Pd_2 and Rh_2, respectively, compare reasonably well with the DF values. The two approaches reproduce the relative bond strength of Rh_2 versus Pd_2, which is a favorable factor for a comparative study of the chemisorption of CO on Rh and Pd [81]. For this purpose, tetrahedral Rh_4 and Pd_4 clusters were chosen as models for (111) surfaces. Although these clusters are very small, we have found that they are adequate to delineate the salient features of the bonding for the present case (further discussion of cluster size dependence may be found, for example, in Refs. [89, 90, 91]. The metal-metal distances have been fixed to the bulk values, 2.69 Å for Rh_4 and 2.75 Å for Pd_4. The calculated adsorption energies for Rh_4CO and Pd_4CO clusters are compared in Table 5. For Pd_4CO the 3-fold site is the most favored, whereas the binding energies for the various sites on Rh_4CO are very close to each other, with the 3-fold site being slightly less favored. The calculated values compare surprisingly well with the experimental results, since adsorption energies at low coverage should be somewhat larger than at half- or high-coverages, as has been measured.

Table 4. Calculated ground-state properties of Rh_2 and Pd_2.

System	State	r_e(Å)	ω_e(cm^{-1})	D_e^a(eV)
Pd_2	$^3\Sigma_u^+$	2.46	209	1.35 (0.74/1.13[b])
Rh_2	$^5\Pi_g$	2.23	260	3.13 (2.97[c])

[a] With respect to GS atoms ^1S Pd and ^4F Rh (nonspherical).
[b] Experimental D_e from Ref. [85]
[c] Experimental D_e from Ref. [86]

Table 5. Calculated M-CO bond strengths (in kcal/mol) for Rh and Pd.

	Pd_4CO			Rh_4CO		
	top	bridge	3-fold	top	bridge	3-fold
D_e(calcd)	30.4	41.2	54.2	53.3	54.4	49.5
D_e(Exp.)	(20-25)[a]	(35-40)[b]	(35-40)[b]			

[a] From Ref. [92] [b] From Ref. [92, 93]

In essence, these DF results show a strong site preference for Pd_4 and a very weak one for Rh_4. The electronic configurations of the metal atoms (bonded to CO) vary strongly with the adsorption site [81], but its variation with the nature of the metal is negligible, except for the one d-electron difference. The adsorbate-induced changes in the configuration of the metal atom are found to be specific to the adsorption *site* and to the position of the metal atom relative to the adsorbed *molecule.* This peculiarity can explain the site preference, different for Rh and Pd surfaces, if we realize that the metal atom at the adsorption site retains some characteristics of an isolated Rh or Pd atom. Indeed, Rh has a $4d^8 5s^1$ GS configuration, and its lowest excited configuration, $4d^9$, is only 0.35 eV higher, which means that the energetic cost for changing to an intermediate configuration (acquired at a given site) will be relatively low. A quite different picture is shown by Pd: its lowest excited state is 1 eV higher in energy [81] and CO adsorption at the bridge or, especially, the top site requires admixture from these higher-lying states, corresponding to a larger promotion energy and a reduced binding energy.

Transition Metal-Ligand Bonding: Summary

It is fortunate that experiments on complexes in the gas phase are becoming more common and more accurate. These experiments yield data on properties (bond energies, vibrational frequencies, ...) which are directly comparable to results of calculations. In the near future it should become possible to make many detailed comparisons and assess the reliability of computational methods. At the present time, the following tentative conclusions about the accuracy of properties have emerged from DF calculations.

1. Metal-ligand binding energies calculated with gradient corrections are sometimes within 5 kcal/mol of experimental values (e.g., NiC_2H_4), sometimes too large by as much as 15 kcal/mol (e.g., CuCO), but never too small in the cases we have examined so far.

2. Calculated harmonic frequencies of C_2H_4 and C_2H_2 bonded to a nickel atom are very close to the observed fundamentals, typically within 5%. Experimental data is not as firm for metal monocarbonyls but the discrepancy with our calculations seems somewhat larger.

3. Conceptual and practical difficulties in DFT calculations for isolated TM atoms present major obstacles for accurate predictions of binding energies involving a TM atom in the dissociation limit.

4. When calculated frequencies of ligands bonded to very small clusters are compared to experiments on adsorbate/surface systems, it seems that the main source of discrepancy is the cluster model, *not* the DFT calculation. If one takes into account trends with increasing cluster size, quantitative comparisons with experiment become possible for the geometrical parameters and vibrational modes associated with the adsorbate.

The most serious limitation to the accuracy of DFT calculations still comes from the exchange-correlation functional. There is a clear need for continuing the search for better functionals. We also think that it will be important to test these functionals not only against small organic molecules (e.g., the G2 database [94]), but also against "really difficult" species — those involving TM atoms among others — that are well characterized experimentally. In that respect, spectroscopic investigations of TM dimers, trimers and complexes are extremely important to guide developments in theoretical methods. We would like to stress again two consequences of the low computational cost of the LCGTO-DF method.

- The modest cost of calculations on complexes with only one or two TM atoms allows many cases to be treated and trends to be isolated.
- Fairly elaborate cluster models can be tackled: they can mimic flat surfaces or steps, pure metals or alloys, the presence of a support (e.g., Al_2O_3), promoters (e.g., a potassium atom) or poisons (e.g., a sulphur atom) etc ... This, coupled with the conceptual simplicity of DFT, provides a general method to obtain insight into some important aspects of catalytic processes.

METAL CLUSTERS

Recent development in beam techniques have allowed the synthesis and characterization of metallic clusters of well defined size. In this way, the ionization potentials (IP) and the bond dissociation energies (BDE) of several metal clusters have been determined [95]. In the case of TM clusters, a complex magnetic behaviour has been revealed [96] that was not anticipated [97].. These clusters often exhibit superparamagnetism and generally bear magnetic moments that are larger than those in the bulk. In general, the size dependence of the properties imply non-trivial changes in electronic structure and an evolution toward metallic behaviour (with itinerant electrons) in which the spin-polarization effects may play an important role [99].

Cluster properties may well depend sensitively on the geometry. But experimental data on the geometry of metal clusters is scarce. Even for the dimers the experimental data are not yet complete and some R_e values still show large uncertainties [100]. Fortunately, significant progress has been made in the last few years in highly accurate spectroscopic characterization of metal dimers [101, 102]. The prospect of extending such studies to trimers or even larger clusters is very exciting but, at present, experimental information about cluster structure is very limited and mostly based on indirect evidence [103]. Further complication arises from the possible co-existence of two or more isomers, with possibly very different properties, for some cluster sizes (see, for example, Ref. [104]). The assumption that a state of quasi-equilibrium has been reached and that the species probed are clusters in their ground state with a well-defined structure can then be misleading.

In the study of clusters, one of the most interesting questions is that of how many stable isomers exist and how large is their energy separation. This is crucial for a correct assignment of the *true* ground states: geometrical and electronic structures. But it is only recently that first-principles calculations have shown a real potential for determining the lowest energy structures of relatively large clusters [105, 106] We will address this question, first, for aluminum clusters which, given the relatively modest number of electrons and basis set requirements, are amenable to an extensive search of geometries and electronic states and then turn to TM clusters.

Aluminum Clusters

Using deMon, we determined the preferred geometries for the lowest energy states of small aluminum clusters Al_n and Al_n^+, (n=2-4). The calculations were of two types: all-electron (AE) [107] and calculations using a MCP for the $1s^2 2s^2 2p^6$ *core* electrons, while the valence $3s^2 3p^1$ electrons were included explicitly in the calculation [108]. For each Al atom, we used the DZVP (6321/521/1*) and the (311/31/1*) orbital basis sets, for the AE and for the MCP calculations, respectively. We used the PP functional. A full optimization procedure (geometry and electronic structure) was performed for Al_n and Al_n^+ (n≤4). In the two approaches, AE and MCP, the geometry optimization was started from the same initial geometries. No symmetry constraints were used and we tried several candidates in order to determine the lowest energy structure. Symmetries, bond distances and multiplicity (M=2S+1) are shown in Table 6.

Table 6. Symmetries, bond lengths (R_e), multiplicities, and relative energies, for the lowest energy states of Al_n and Al_n^+ (n=2-4).

System	Symmetry	R_e (Å)	Multiplicity	Δ E (eV)
Al_2	$D_{\infty h}$	2.53 (2.48)[a]	3	0.0 (0.0)[a]
	$D_{\infty h}$	2.78 (2.71)	1	0.45 (0.52)
Al_3	D_{3h}	2.56 (2.52)	2	0.0 (0.0)
	C_{2v}	2.63 (2.57), 3.11 (2.90)	4	0.15 (0.19)
Al_4	D_{2h}	2.56 (2.55)	3	0.0 (0.0)
	D_{4h}	2.69 (2.62)	3	0.01 (0.11)
	D_{2h}	2.55 (2.56)	1	0.27 (0.34)
Al_2^+	$D_{\infty h}$	3.39 (3.27)	2	0.0 (0.0)
	$D_{\infty h}$	2.57 (2.56)	4	2.90 (2.51)
Al_3^+	D_{3h}	2.75 (2.64)	3	0.0 (0.0)
	$D_{\infty h}$	2.91 (2.78)	3	0.13 (0.38)
	D_{3h}	2.56 (2.58)	1	0.37 (0.43)
Al_4^+	D_{2h}	2.71 (2.67)	4	0.0 (0.0)
	D_{4h}	2.78 (2.71)	4	0.03 (0.02)
	T_d	2.60 (2.54), 3.50 (3.30)	2	0.06 (0.06)

[a] MCP results are in parenthesis.

The total energy differences between the most stable and the first excited state for each considered structure is also given. The AE and MCP calculations yield the same ground state geometry for Al_n and Al_n^+. The local minima are practically the same. For Al_4 and Al_{4+}, there is degeneracy between the rhombus and the square structures; see Table 6.

The near degeneracy between the rhombus and the square geometries has some physical consequences for the properties of Al_4. Experimentally, a range of values 6.5-7.87 eV, has been determined for the IP of Al_4, rather than a single value [110]. The AE and MCP calculations [107, 108] produce ranges of 5.94-8.05 eV and 6.16-7.90 eV, respectively, for the adiabatic IP of Al_4 (these ranges were determined taking into account all stable structures that there were found for the tetramer; see Refs. [107, 108] for more details). It is important to underline that all the experimental and theoretical values reported in literature until now for Al_4 are within this range.

Finally, an AE potential energy surface corresponding to a symmetrical distortion that transforms the square (triplet) into the rhombus, was calculated for Al_4. The change of the binding energy as a function of the Al-Al bond distance, for the bond angles shown in the inserted box is depicted in Figure 4; a FINE grid was used. As we can see, the surface is very flat and we can not say that the rhombus is more stable than the square. For such a flat surface, numerical noise in the calculations becomes troublesome and we have found difficulties when the bond angle was equal to 75 degrees. It has been observed [69] that structures which correspond to stationary states on a flat potential energy surface can be very sensitive to the grid used for the fitting and numerical integration. All these features complicate enormously the vibrational analysis for Al_4. Indeed, the question even arises of whether any one structure is pertinent. Calculations performed with an extended grid are necessary in order to eliminate these errors and such calculations are in progress.

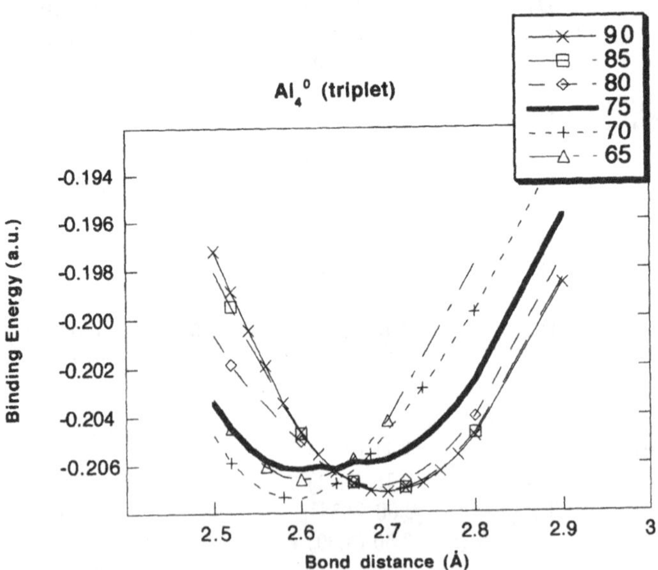

Figure 4. Potential energy surface for a square to rhombus transition of Al_4, see text.

Transition Metal Clusters

As mentioned above, modern quantum theoretical tools have proven their ability to determine the lowest energy structures of relatively large clusters. Particularly s- and sp-valence electron systems have been well studied [105, 111, 112]. Due to the complexity of the TM atomic forces, which arise from the complex TM-TM exchange-correlation interactions, calculations on TM clusters are often done with the constraints of *frozen* bond lengths and/or bond angles, usually equal to the bulk values [113].

In the last examples reviewed here, the ability of deMon to handle structural relaxation in TM clusters is shown for: a) Iron clusters [114], Fe_n, up to n=5, by means of all electron calculations, using a (43321/4211/311) orbital basis set [25], and b) Niobium clusters [115], Nb_n, up to n=7, by means of MCP calculations. We will also discuss briefly some results that have been obtained recently [116] for Co_n and Ni_n ($n \leq 5$). The PP functional was used in all cases. Several candidate structures were tried in each case so as to identify different minima on the PES.

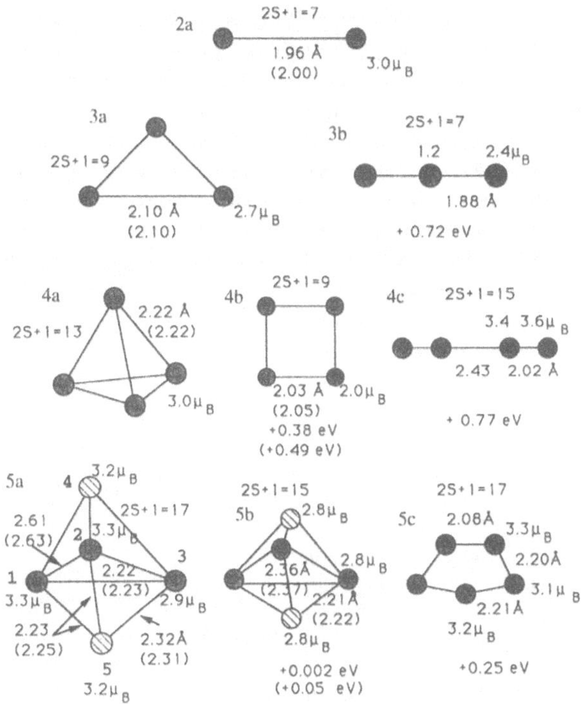

Figure 5. The lowest energy structures of Fe_n ($n \leq 5$). Ground states are 2a, 3a, 4a, and 5a for n=2,3,4 and 5 respectively. Also indicated are the LSD and NLSD (in parentheses) relative energies (eV/atom), the multiplicity, the spin per atom and the equilibrium bond lengths (NLSD in parenthesis).

The ground states (structures 2a, 3a, 4a, and 5a in Fig. 5) of Fe_n ($n \leq 5$) are ferromagnetic, with a high number of nearest neighbor bonds (NNB) and with average magnetic moments, μ, enhanced over the bulk value, of $2.2\mu_B$, by 22-45%. Additionally, for a given cluster (for example, structure 5a) there is an uneven distribution of atomic moments, which depends on the chemical environment. In 2a-4a the R_es are much shorter than the shortest distance in the bulk, 2.48 Å, whereas in 5a there are various short values, not much different than that of 4a, along with R_es close to, but bigger than, 2.48 Å. The two forces, chemical and magnetic, drive the iron clusters towards ferromagnetic ground states with a maximization in the chemical bond formation.

The calculated nonlocal adiabatic IPs 7.0, 6.20, 6.20, and 6.52 eV, for Fe_2, Fe_3, Fe_4, and Fe_5 respectively, are close to theirs experimental counterparts [99] 6.3, 6.4-6.5, 6.3-6.5, and 5.9-6.0 eV. The introduction of NL gradient-type corrections shows a dramatic improvement in the estimation of the binding energies. But even at this level of theory there remains an overestimation in the computed values, with respect to the experimental results. For example, the binding energy of Fe_2 is 4.05, 3.24, and 2.08 eV, at the LSDA, NL (PP), and non-spherical-NL levels of theory; all these values show a huge discrepancy with respect to the experimental result [117] 1.30 eV. The calculations of the binding energies involves the energy of the open-shell multiplets of the isolated iron atom, and this has proven to be difficult to calculate with very high accuracy within Kohn-Sham DFT in its present state.

Similarly, all-electron deMon calculations, at the LSDA and NL (PP) levels, yield structures like 2a, 3a, 4a, and 5a (in Fig. 5) for the ground state geometries of the corresponding clusters of Co_n and Ni_n, n≤5 [116]. In going from Fe_n to Co_n to Ni_n the atomic magnetic moments are substantially reduced, as expected. For example, for the tetrahedral ground states of Fe_4, Co_4 and Ni_4, the magnetic moments per atom are 3.0, 1.5, and 0.5 μ_B, respectively. Overall, these results stress the importance of the d electrons on the properties of clusters of ferromagnetic elements: they participate in the binding and they should be included explicitly in the calculations, otherwise a wrong description could be obtained. For example, using Generalized-Valence-Bond techniques [118] and without the explicit inclusion of the d electrons, a planar structure like 5c (in Fig. 5) has been predicted for the ground state geometry of Ni_5. In our calculations, the planar structure (like 5c) of Ni_5 is located $\simeq$ 1 eV above the ground state geometry, a trigonal bipyramid. We have performed a vibrational analysis [116] for the structures 2a, 3a, 4a, and 5a of Fe_n, n≤5 (and also for Co_n and Ni_n); in all cases the results indicate that such structures are indeed minima on the potential energy surfaces, since there were no imaginary frequencies.

In contrast to Fe, Co, and Ni, for copper clusters the binding is dominated by the 4s electrons; the Cu atom has a closed 3d shell. Indeed we have found [119] that the two-dimensional geometries of Cu_4 and Cu_5, a rhombus and a trapezoidal form, respectively, are lower in energy than the three-dimensional structures.

In Fig. 6 are shown the most stable geometries located for Nb_n, up to n=7. These are the most extensive *state-of-the-art* calculations to date for clusters of TM atoms. These results could not have been obtained without the advent of energy gradient techniques in DF. The trends in the experimental binding energies, bond dissociation energies and ionisation potentials of Nb_n are well reproduced by these DF calculations [115].

Metallic Clusters: Summary

DFT techniques, as exemplified by the Gaussian code deMon, are able to handle the structural relaxation in transition metal clusters. These calculations show the importance of geometry optimization on metal cluster properties. The results provide insight as to how the structural, electronic, and magnetic aspects are involved in the stability of these clusters. The maximization of the number of nearest neighbors and the gain in magnetic energy drives the Fe_n, Co_n, and Ni_n ($n \leq 5$) clusters towards compact geometries (the triangle, tetrahedron, and trigonal bipyramid are the preferred geometries for n=3, 4, and 5, respectively). A similar behaviour is displayed by the Nb_n ($n \leq 7$) clusters: the planar structures are approximately one eV less stable than the more compact ones. On the other hand, more open geometries were found for Cu_n ($n \leq 5$) clusters.

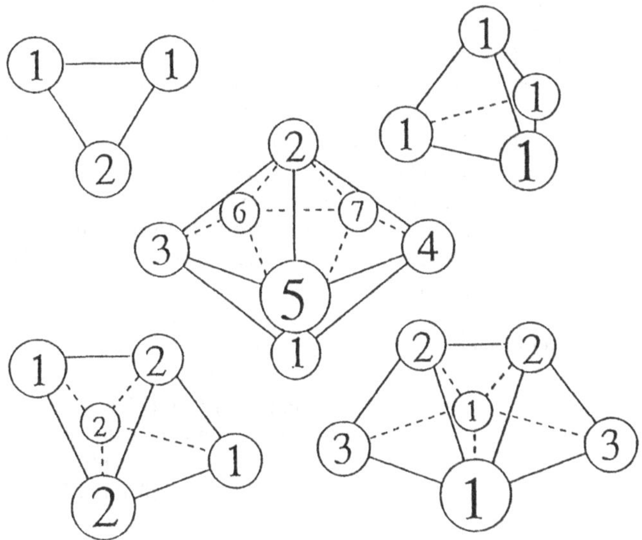

Figure 6. The most stable geometries located for niobium clusters. The labels indicate atoms related under symmetry group operations.

Cluster science is a relatively young field of research. Experimentally, there have been determinations of ionization potentials, bond dissociation energies, and magnetic moments. But there is no direct measurement of geometrical parameters for clusters. So, most of our results for bond lengths and bond angles constitute predictions. The biggest clusters we investigated (n=5, for Fe–Cu, n=7 for Nb) are still too small for a generalization of the trends. The theoretical study of bigger clusters will require the introduction of some simplifications, such as the use of model-core potentials, more compact basis sets, locally concentrated basis sets (for optimizations involving a subset of atoms), etc. as well as more powerful computers, algorithms, and computer codes.

PERIODIC SYSTEMS: Bloch-deMon

We conclude with a brief progress report on some recent developments in the deMon family aimed at treating the "infinite limit" for cluster studies, namely, bulk solids or infinite surfaces. Periodic boundary conditions have been implemented recently. The extension is relatively straightforward since the inclusion of Bloch functions does not entail the evaluation of a new set of integrals. In the new code, called Bloch-deMon, all the machinery and experience already developed in deMon can be used: orbital and auxiliary basis sets, grids, functionals, etc ... with appropriate modifications. This will allow the study of systems such as: polymers, crystals, surfaces, and, particularly important for us, the analysis of surface-ligand interactions.

As a test of Bloch-deMon, we calculated the electronic structure of the face centered cubic (fcc) Pd crystal. For bulk fcc Pd the unit cell includes one Pd atom, and we have used O_h symmetry for the **r**- and **k**-spaces. We have increased, gradually, the interaction range until convergence of the calculated properties was achieved. The cohesive energy and the Fermi energy become constant for an interaction range of 7.3–8.3 Å. The cohesive energy is calculated to be 5.3 and 3.5 eV for the local and nonlocal potentials, respectively. Compared to the experimental value of 3.89 eV [120], we see that the nonlocal potential decreases the overestimated LSDA binding and improves the agreement with experiment. The effects of the number of k-points were also examined. We have used only 19, and then 86, points for the irreducible part of the Brillouin zone. The calculated cohesive energy is the same for the two sets of points. However, the Fermi energy is shifted to lower energy by 0.2 eV upon increasing the number of k-points. The calculated Fermi energies are -5.5 and -4.7 eV for the local and nonlocal potentials, while the experimental value of the work function is 5.12 eV [121]. The cohesive energy has been calculated for lattice parameters varying around the experimental value, 3.89 Å. The energy minimum is found at 4.0 Å, for the nonlocal potential, whereas the local potential calculation gives a value that is too small (3.7 Å). The difference in the energy between 3.89 and 4.0 Å, is only 0.3 kcal/mol (for the nonlocal potential) : the potential energy surface is very flat. Although more elaborate calculations (with more k-points and a longer interaction range) seem to be necessary, the inclusion of the nonlocal corrections works to lengthen the bond, in agreement with our experience in molecular calculations, and the minimum comes close to the experimental value (3.89 Å). The band dispersion relation and the d-band width were calculated and compare well with previously published values.

Magnesium oxide is an interesting material, which dissociatively adsorbs H_2, CH_4 and other alkanes. Common understanding on the mechanisms is that the H-H and C-H bonds are cleaved ionically and the surface atoms at the low coordination number sites work as the active center. A few theoretical works using small cluster models and based on Hartree-Fock and 2nd order Moller-Plesset calculations showed that adsorption to the three-coordinated Mg-O pair, designated as Mg(3)O(3), leads to dissociation of molecules without a barrier, whereas the dissociated products are not stable on Mg(4)O(4) [122].

We have carried out a more extensive study using more cluster models than before and also slab models. The slab models are free from artificial boundaries, and suitable for modeling the high coordination-number sites. The products are unstable on the flat surface (Mg(5)O(5)) and the edge site (Mg(4)O(4)), but stable on the corner site (Mg(3)O(3)). The mixed sites Mg(3)O(4) and Mg(4)O(3) stabilize them weakly. These results are similar to those of previous works and are consistent with the experimental findings. Furthermore the reasons for the site dependence have been elucidated. The band gap between the valence and conduction bands becomes narrower with a decrease of coordination number; the values calculated for models including only the five and three coordination number sites are 3.3 and 1.2 eV, respectively. Measurements by UV diffuse reflectance and photoluminescence spectroscopies show the same trend in band gap, 6.6 and 4.6 eV, respectively [123]

With decreasing number of nearest neighbour counter ions, the electrostatic potential which stabilizes the ionic lattice structure becomes weaker. Compared to their values in the bulk crystal, the energy of the valence band is destabilized and the energy of the conduction band is stabilized. Thus the band gap is reduced, and this is favorable to the dissociative adsorption process including charge transfer and bond alternation between the molecule and surface. We can understand the enhanced activity for the low coordination number sites in terms of the "local enhancement of covalency".

The above studies represent our first attempts to adapt **deMon** to calculate solids, surfaces, and periodic overlayers. Such studies provide a valuable complement to our cluster studies and we expect the use of **Bloch-deMon** to grow substantially in the coming months.

CONCLUSION

We have shown in this chapter how the DFT Gaussian code **deMon** can handle complicated aggregates of atoms. In particular, we have addressed systems containing transition metal atoms with open d shells where the bonding involves both localized nd- and delocalized (n+1)s- electrons. This gives rise to complicated electronic structures which require the use of good basis sets and accurate exchange-correlation functionals. In the TM-L and metallic clusters reviewed here, the inclusion of electron correlation at a high level of theory is a key factor for the accurate estimation of the binding properties. Whereas the LSDA is sufficient for the calculation of bond lengths and bond angles, the introduction of nonlocal corrections (as in the Becke-Perdew and Perdew-Perdew schemes) results in a dramatic improvement of the binding energies. At the nonlocal level the **deMon** results for TM systems are comparable to those performed with traditional *ab initio* quantum chemistry methods including correlation (when available). In fact, DFT methodology has matured and reached a stage of development that has brought it into the "*ab initio*" comunity.

DFT has provided methods of great generality that can be applied to a wide variety of complex systems. The examples we presented here involving the Gaussian DFT code **deMon** illustrate this point:

- small, unsaturated, TM complexes.
- clusters modelling adsorbate/surface systems.
- clusters at their optimized geometries, as they probably exist in supersonic jet experiments.
- bulk and infinite surfaces.

In addition, the literature contains a myriad of other succesful applications of DFT to organic molecules, inorganic complexes, polymers, biomolecules, etc ... The list is long and growing continuously.

Most of our theoretical estimates (equilibrium bond lengths, vibrational frequencies, ionization potentials, etc ...) are close to the corresponding experimental results.

However there is still, often, an overestimation of the calculated binding energies even at the gradient-corrected non-local level. Further improvement would require the construction of even more sophisticated non-local functionals than the current ones [124]. Other sources of errors are related to the details of the calculation, the basis sets, the grid, convergence criteria, etc. The use of extended grids and tight convergence is crucial for a proper description of systems which have flat potential energy surfaces.

The low computational cost of the LCGTO-DFT method, coupled with the conceptual simplicity of DFT, provides a valuable tool for the calculation of the electronic structure of bigger and more complicated systems. With DFT, as implemented in deMon, we are getting closer to the "dream" of quantum chemistry: treating, within a single framework and with appropriate accuracy systems of any size with atoms from anywhere in the periodic table. The future looks very bright, indeed.

ACKNOWLEGEMENTS

Valuable discussions with our current colleagues at the Université de Montréal, Mark Casida, Andreas Koster, Vladimir Malkin, Olga Malkina, Jingang Guan, and Eliseo Ruiz are gratefully acknowledged. Support from NSERC and the Canadian Network of Centres of Excellence in Molecular and Interfacial Dynamics is gratefully acknowledged, as is the provision of computing resources by the Services Informatiques de l'Université de Montréal. M. C. acknowledges a sabbatical leave from Facultad de Química, Universidad Nacional Autónoma de México.

References

[1] J. C. Slater, *The Self-Consistent Field For Molecules and Solids*, vol. 4, McGraw-Hill, New-York, 1974.

[2] K. H. Johnson, Adv. Quantum Chem., **7**, 143 (1973).

[3] There are too many recent applications of DFT to chemistry to cite them all, however, the following references provide a good sample of properties and molecular systems: (a) F. Sim, D. Salahub, S. Chin, and M. Dupuis, J. Chem. Phys. **95**, 4317 (1991); (b) G. Fitzgerald and J. Andzelm, J. Phys. Chem. **95**, 10531 (1991); (c) J. Andzelm and E. Wimmer, J. Chem. Phys. **96**, 1280 (1992); (d) R. Fournier and A. E. DePristo, J. Chem. Phys. **96**, 1183 (1992); (e) F. Sim, A. St-Amant, I. Pápai, and D. R. Salahub, J. Am. Chem. Soc. **114**, 4391 (1992); (f) L. Fan and T. Ziegler, J. Am. Chem. Soc. **114**, 10890 (1992); (g) A. M. Rappe, J. D. Joannopoulos, and P. A. Bash, J. Am. Chem. Soc. **114**, 6466 (1992); (h) D. A. Dixon and J. L. Gole, Chem. Phys. Lett. **189**, 390 (1992); (i) J. Andzelm, C. Sosa, and R. A. Eades, J. Phys. Chem. **97**, 4664 (1993); (j) N. C. Handy, C. W. Murray, and R. D. Amos, J. Phys. Chem. **97**, 4392 (1993); (k) R. D. Amos, C. W. Murray, and N. C. Handy, Chem. Phys. Lett. **202**, 489 (1993); (l) V. Barone, C. Adamo, and N. Russo, Chem. Phys. Lett. **212**, 5 (1993); (m) T. A. Holme and T. N. Truong, Chem. Phys. Lett. **215**, 53 (1993); (n) C. Sosa, C. Lee, G. Fitzgerald, and R. A. Eades, Chem. Phys. Lett. **211**, 265 (1993); (o) L. A. Eriksson, S. Lunell, and R. J. Boyd, J. Am. Chem. Soc. **115**, 6896 (1993); (p) D. P. Chong and A. V. Bree, Chem. Phys. Lett. **210**, 443 (1993); (q) J. Guan, P. Duffy, J. T. Carter, D. P. Chong, K. C. Casida, M. E. Casida, and M. Wrinn, J. Chem. Phys. **98**, 4753 (1993); (r) H. Burghgraef, A. P. Jansen, and R. A. van Santen, J. Chem. Phys. **98**, 8810 (1993); (s) C. Sosa and C. Lee, J. Chem. Phys. **98**, 8004 (1993); (t) G. L. Gutsev, J. Chem. Phys. **98**, 7072 (1993); (u) C. W. Murray, N. C. Handy, and R. D. Amos, J. Chem. Phys. **98**, 7145 (1993); (v) D. P. Chong and C. Y. Ng, J. Chem. Phys. **98**, 759 (1993); (w) C. Lee, G. Fitzgerald, and W. Yang, J. Chem. Phys. **98**, 2971 (1993); (x) L. Deng, T. Ziegler, and L. Fan, J. Chem. Phys. **99**, 3823 (1993). (y) V. G. Malkin, O. L. Malkina, and D. R. Salahub, Chem. Phys. Lett. **204**, 80, (1992). (z) L. A. Eriksson, V. G. Malkin, O. L. Malkina, and D. R. Salahub, J. Chem. Phys. **99**, 9756 (1993).

[4] W. Kohn and L. J. Sham, Phys. Rev. **140**, A1133 (1965).

[5] D. R. Salahub in: *Metal-Ligand Interactions: From Atoms, to Clusters, to Surfaces*, Eds. D. R. Salahub and N. Russo, NATO ASI, Kluwer Academic Publ., Dordrecht, 1992.

[6] H. Sambe, and R. H. Felton, J. Chem. Phys. **62**, 1122 (1975).

[7] B. I. Dunlap, J. W. D. Connolly, and J. R. Sabin, J. Chem. Phys. **71**, 3396, 4993 (1979).

[8] (a) A. St-Amant and D. R. Salahub, Chem. Phys. Lett. 169, 387 (1990).
(b) D. R. Salahub, R. Fournier, P. Mlynarski, I. Papai, A. St.-Amant, and J. Ushio, in *Density Functional Methods in Chemistry*, edited by J. Labanowski and J. Andzelm (Springer, New York, 1991).

[9] A. D. Becke, J. Chem. Phys. **88**, 2547, (1988).

[10] C. Daul, A. Goursot, and D. R. Salahub in: NATO ARW Proceedings on "Grid Quantum Calculations", C. Leforestier, ed., 1993, in press.

[11] S. Obara and A. Saika, J. Chem. Phys. **84** 3963 (1986).

[12] P. M. Boerrigter, G. te Velde, G., and E. J. Baerends, Int. J. Quantum Chem., **33**, 87 (1988); L. Versluis and T. Ziegler, J. Chem. Phys., **88**, 322 (1988).

[13] T. Ziegler and A. Rauk Inorg. Chem. **18**, 1755 (1979).

[14] T. Ziegler Chem. Rev. **91**, 651 (1991).

[15] O. Gunnarsson, J. Harris, and R. O. Jones Phys. Rev. **15**, 3027 (1977).

[16] O. Gunnarsson and B. I. Lundqvist Phys. Rev. B 13, 4274 (1976).

[17] J. Andzelm and E. Wimmer J. Chem. Phys. **96**, 1280 (1992).

[18] D. E. Ellis and G. S. Painter Phys. Rev. **B2**, 2887 (1970).

[19] L. Hedin and B. I. Lundqvist, J. Phys. **C4**:2064 (1971).

[20] B. J. Delley, J. Chem. Phys., **92**, 508 (1990).

[21] R. Car and M. Parinello, Phys. Rev. Lett. **55**, 2471 (1985).

[22] M. P. Teter, M. C., Payne, D. C. and Allan, Phys. Rev., **B40**, 12255 (1989).

[23] A. D. Becke and R. M. Dickson, J. Chem. Phys., **92**, 3610 (1990).

[24] W. J. Hehre, L. Radom, P. v. R. Schleyer, and J. A. Pople, 1986. *Ab initio molecular orbital theory.* John Wiley and Sons, New York. P. C. Hariharan and J. A. Pople 1972. Chem. Phys. Lett. 66, 217.

[25] J. Andzelm, E. Radzio and D. R. Salahub, J. Comp. Chem. 6, 520 (1985).

[26] H. Tatewaki and S. Huzinaga, 1979, J. Chem. Phys. 71:4339.

[27] S. Huzinaga, J. Andzelm, H. Klobukowski, E Radzio, E Sakai, and H. Tatewaki. Gaussian basis sets for molecular calculations. Elsevier, Amsterdam. 1984.

[28] N. Godbout, D. R. Salahub, J. Andzelm, and E. Wimmer, Can. J. Chem. 70, 560 (1992).

[29] S. H. Vosko, L. Wilk, and M. Nusair, Can. J. Phys. 58, 1200 (1980).

[30] J. P. Perdew Phys. Rev. B33, 8822 (1986); Phys. Rev. B34, 7406E (1986).

[31] J. P. Perdew and Y. Wang, Phys. Rev. B33, 8800 (1986).

[32] A. D. Becke Phys. Rev. A 38, 3098 (1988).

[33] R. Fournier, J. Chem. Phys. **92**, 5422 (1990).

[34] R. Fournier, J. Andzelm, and D. R. Salahub J. Chem. Phys. **90**, 6371 (1989).

[35] J. A. Pople, R. Krishnan, H. B. Schlegel, and J. S. Binkley, Int. J. Qu, J. Chem. Phys. **88**, 322 (1988).antum Chem. Symp. **13**, 225 (1979).

[36] P. Pulay, J. Chem. Phys. **78**, 5043 (1983).

[37] P. Pulay, Mol. Phys. **17**, 197 (1969).

[38] L. Versluis and T. Ziegler, J. Chem. Phys. **88**, 322 (1988).

[39] A. Komornicki and G. Fitzgerald, J. Chem. Phys. **98**, 1398 (1993).

[40] B. G. Johnson and M. J. Frisch, Chem. Phys. Lett. **216**, 133 (1993).

[41] J. Guan *et al.*, to be published.

[42] P. E. M. Siegbahn and M. R. A. Blomberg, in *Theoretical Aspects of Homogeneous Catalysis, Applications of Ab Initio Molecular Orbital Theory*, P. W. N. M. van Leeuwen, J. H. van Lenthe, and K. Morokuma Eds., Kluwer Academic Pub. (1993).

[43] T. Ziegler, A. Rauk, and E. J. Baerends, Theoret. Chim. Acta **43**, 261 (1977).

[44] F. Kutzler and G. S. Painter, Phys. Rev. B **43**, 6865 (1991).

[45] A. Görling, Phys. Rev. A **47**, 2783 (1993).

[46] T. V. Russo, R. L. Martin, and P. J. Hay, preprint.

[47] A. D. Becke, Phys. Rev. A **38**, 3098 (1988).

[48] J. P. Perdew, Phys. Rev. B **33**, 8822 (1986).

[49] C. W. Bauschlicher, Jr., P. S. Bagus, C. J. Nelin, and B. O. Roos, J. Chem. Phys. **85**, 354 (1986).

[50] A. Daoudi, M. Suard, and G. Berthier, J. Mol. Struct. (Theochem) **210**, 139 (1990).

[51] R. Fournier, J. Chem. Phys. **99**, 1801 (1993).

[52] J. M. Parnis, S. A. Mitchell, and P. A. Hackett, J. Phys. Chem. **94**, 8152 (1990).

[53] P. W. Villalta and D. G. Leopold, J. Chem. Phys. **98**, 7730 (1993).

[54] L. S. Sunderlin, D. Wang, and R. R. Squires, J. Am. Chem. Soc. **114**, 2788 (1992).

[55] M. A. Blitz, S. A. Mitchell, and P. A. Hackett, J. Phys. Chem. **95**, 8719 (1991).

[56] L. A. Barnes, M. Rosi, and C. W. Bauschlicher Jr., J. Chem. Phys. **94**, 2031 (1991).

[57] G.-H. Jeung, J. Am. Chem. Soc. **114**, 3211 (1992).

[58] R. Fournier, J. Chem. Phys. **98**, 8041 (1993).

[59] M. Castro, D. R. Salahub, and R. Fournier, J. Chem. Phys. (1994).

[60] A bent geometry for the triplet state of NiCO had previously been suggested by Clark *et al.* on the basis of Hartree-Fock calculations, D. T. Clark, B. J. Cromarty, and A. Sgamellotti, Chem. Phys. Lett. **55**, 482 (1978).

[61] C. W. Bauschlicher, Jr., J. Chem. Phys. **100**, 1215 (1994).

[62] J. H. B. Chenier, C. A. Hampson, J. A. Howard and B. Mile, J. Phys. Chem. **93**, 114 (1989).

[63] P. H. Kasai and P. M. Jones, J. Am. Chem. Soc. **107**, 813 (1985).

[64] D. G. Leopold, private communication.

[65] H. Partridge, J. Chem. Phys. **90**, 1043 (1989).

[66] H. Huber, G. A. Ozin, and W. J. Power, J. Am. Chem. Soc. **98**, 6508 (1976).

[67] T. Merle-Mejean, C. Cosse-Mertens, S. Bouchareb, F. Galan, J. Mascetti, and M. Tranquille, J. Phys. Chem. **96**, 9148 (1992).

[68] E. S. Kline, Z. H. Kafafi, R. H. Hauge, and J. L. Margrave, J. Am. Chem. Soc. **109**, 2402 (1987).

[69] I. Pápai, J. Mink, R. Fournier, and D. R. Salahub, J. Phys. Chem., **97**, 9986 (1993).

[70] S. A. Mitchell, M. A. Blitz, and R. Fournier, Can. J. Chem., in press.

[71] C. E. Brown, S. A. Mitchell, and P. A. Hackett, Chem. Phys. Lett. **191**, 175 (1992).

[72] R. Fournier, Int. J. Quantum Chem., in press.

[73] L. Lian, F. Akhtar, P. A. Hackett, and D. M. Rayner, Int. J. Chem. Kin., in press.

[74] L. Lian, P. A. Hackett, and D. M. Rayner, J. Chem. Phys. **99**, 2583 (1993).

[75] R. Brosseau, T. H. Ellis, and Wang, Chem. Phys. Lett, **117**, 118 (1991).

[76] R. Brosseau, PhD thesis, Université de Montréal (1993)

[77] See, H. B. Schlegel, in *Ab Initio Methods in Quantum Chemistry–I*, edited by K. P. Lawley (wiley, New York, 1987).

[78] H. J. Freund and R. P. Messmer, Surf. Science, **172**, 1 (1968).

[79] S. Sirois and D. R. Salahub, to be published

[80] I. Pápai, J. Ushio and D. R. Salahub, Surf. Sci. **282**, 262 (1993).

[81] A. Goursot, I. Papai, and D. R. Salahub, J. Am. Chem. Soc. **114**, 7452 (1992).

[82] I. Shim and K. A. Gingerich, J. Chem. Phys **80**, 5107 (1984).

[83] H. Basch and D. Cohen, Isr. J. Chem. **19**, 233 (1980).

[84] J. Andzelm, E. Radzio and D. R. Salahub, J. Chem. Phys. **83**, 4573 (1985).

[85] S. S. Lin, B. Strauss and A. Kant, J. Chem. Phys. **51**, 2282 (1969).

[86] D. L. Cocke and K. A. Gingerich, J. Chem. Phys. **60**, 1958 (1974).

[87] K. Balasubramanian, J. Chem. Phys. **89**, 6310 (1988).

[88] K. Balasubramanian and D. W. Liao, J. Phys. Chem. **93**, 3989 (1989).

[89] J. Andzelm and D.R. Salahub in: *Physics and Chemistry of Small Clusters*, Eds. P. Jena, B. K. Rao and S. N. Khanna, Nato Advanced Study Institute, Physics, (Plenum, New York, vol. 158, p. 867, 1987).

[90] R. Fournier and D.R. Salahub, Surf. Sci. **245**, 263 (1991).

[91] P. Mlynarski and D. R. Salahub, J. Chem. Phys. **95**, 6050 (1991).

[92] S. Ladas, H. Poppa and M. Boudart Surf. Sci. **102**, 151 (1981).

[93] E. Gillet, S. Channakhone and V. Matolin, J. Catal. **97**, 437 (1986).

[94] B. G. Johnson, P. M. W. Gill, and J. A. Pople, J. Chem. Phys. **98**, 5612 (1993).

[95] S. K. Loh, D. A. Hales, L. Lian, and P. B. Armentrout, J. Chem. Phys. **90**, 5466 (1989); L. Lian, C.-X. Su, and P. B. Armentrout, J. Chem. Phys. **97**, 4072 (1992); *ibid.*, J. Chem. Phys. **96**, 7542 (1992); *ibid.*, J. Chem. Phys. **97**, 4084 (1992); C.-X. Su and P. B. Armentrout, J. Chem. Phys. **99**, 6506 (1993); D. A. Hales, C.-X. Su, L. Lian, and P. B. Armentrout, J. Chem. Phys. **100**, 1049 (1994).

[96] W. A. de Heer, P. Milani, and A. Châtelain, Phys. Rev. Lett. **65**, 488 (1990); J. P. Bucher, D. C. Douglas, and L. A. Bloomfield, Phys. Rev. Lett. **66**, 3052 (1991); I. M. L. Billas, J. A. Becker, A. Châtelain, and W. A. Deheer, Phys. Rev. Lett. **71**, 4067 (1993).

[97] D. M. Cox, D. J. Trevor, R. L. Whetten, E. A. Rohlfing, and A. Kaldor, Phys. Rev. B **32**, 7290 (1985).

[98] S. N. Khanna and S. Linderoth, Phys. Rev. Lett. **67**, 742 (1991).

[99] E. A. Rohlfing, D. M. Cox, A. Kaldor, and K. H. Johnson, J. Chem. Phys. **81**, 3846 (1984).

[100] D. R. Salahub, Adv. Chem. Phys. **69**, 447 (1987), and references therein.

[101] S. Taylor, E. M. Spain, and M. D. Morse, J. Chem. Phys. **92**, 2698 (1990); E. M. Spain and M. D. Morse, J. Phys. Chem. **96**, 2479 (1992); J. M. Behm, C. A. Arrington, J. D. Langenberg, and M. D. Morse, J. Chem. Phys. **99**, 6394 (1993); J. M. Behm, C. A. Arrington, and M. D. Morse, J. Chem. Phys. **99**, 6409 (1993).

[102] B. Simard, P. A. Hackett, A. M. James, P. R. R. Langridge-Smith, Chem. Phys. Lett. **186**, 415 (1991); A. M. James, P. Kowalczyk, R. Fournier, and B. Simard, J. Chem. Phys. **99**, 8504 (1993).

[103] B. J. Winter, E. K. Parks, and S. J. Riley, J. Chem. Phys. **94**, 8618 (1991); E. K. Parks, B. J. Winter, T. D. Klots, and S. J. Riley, J. Chem. Phys. **94**, 1882 (1991); E. K. Parks, T. D. Klots, B. J. Winter, and S. J. Riley, J. Chem. Phys. **99**, 5831 (1993).

[104] P. J. Brucat, C. L. Pettiette, S. Yang, L.-S. Zheng, M. J. Craycraft, and R. E. Smalley, J. Chem. Phys. **85**, 4747 (1986); Y. Hamrich, S. Taylor, G. W. Lemire, Z.-W. Fu, J.-C. Shiu, and M. D. Morse, J. Chem. Phys. **88**, 4095 (1988); J. L. Elkind, F. D. Weiss, J. M. Alford, R. T. Laaksonen, and R. E. Smalley, J. Chem. Phys. **88**, 5215 (1988).

[105] J. L. Martins, J. Buttet, and R. Car, Phys. Rev. B **31**, 1804 (1985);

[106] R. O. Jones. Phys. Rev. Lett. **67**, 224 (1991); J. Chem. Phys. **99**, 1194 (1993).

[107] A. Martínez, A. Vela, D. R. Salahub, P. Calaminici and N. Russo, to be published.

[108] P. Calaminici and N. Russo, Z Phys. D, submitted.

[109] T. H. Upton, Phys. Rev. Lett. **56**, 2168 (1986); J. Chem. Phys. **86**, 7054 (1987).

[110] L. Hanley, S. A. Ruatta and S. L. Anderson J. Chem. Phys. **87**, 260 (1987).

[111] K. Raghavachari and C. M. Rohlfing, J. Chem. Phys. **89**, 2219 (1988);

[112] R. Fournier, S. B. Sinnott, and A. E. DePristo, J. Chem. Phys. **97**, 4149 (1992); erratum, ibid. **98**, 9222 (1993);

[113] H. Tatewaki, M. Tomonari and T. Nakamura, J. Chem. Phys. **88**, 6419 (1989).

[114] M. Castro and D. R. Salahub, Phys. Rev. B **47**, 10955 (1993).

[115] L. Goodwin and D. R. Salahub, Phys. Rev. A **47**, R774 (1993).

[116] C. Jamorski, M. Castro, and D. R. Salahub, to be published.

[117] M. Moskovits, D. P. DiLella, and W. Limm, J. Chem. Phys. 80, 626 (1984).

[118] E. Carter, private communication.

[119] P. Calaminici, M. Castro, D. R. Salahub, and N. Russo, to be published.

[120] C.Kittel, *Introduction to Solid State Physics*, 6th Ed. (Wiley, New York, 1986).

[121] R. C. Weast Ed., *CRC Handbook of Chemistry and Physics* 66th ed. (CRC Press, Boca Raton, 1985).

[122] H. Kobayashi, M. Yamaguchi, and T. Ito, J. Phys. Chem. **94**, 7206 (1990); T.Ito, T. Tashiro, M. Kawasaki, T. Watanabe, K. Toi, and H. Kobayashi, J. Phys. Chem. **95**, 4477 (1991); K. Sawabe, N. Koga, and K. Morokuma, J. Chem. Phys. **97**, 6871 (1992).

[123] A. Zecchina, M. G. Lofthouse, F. S. Stone, J. Chem. Soc., Faraday Trans. 1 **87**, 4411 (1983); E. Garrone, A. Zecchina, F. S. Stone, Phil. Mag., **B42**, 683 (1980); M. Che, A. J. Tench, Adv. Catal. **31**, 77 (1982).

[124] E. Proynov and D. R. Salahub, Phys. Rev. B **49** (1994), in press, and references therein.

DENSITY FUNCTIONAL STUDIES OF BORON SUBSTITUTED ZEOLITE ZSM-5

Mark S. Stave and John B. Nicholas

Molecular Science Research Center
Pacific Northwest Laboratory
Richland, Washington 99352

INTRODUCTION

Zeolites are porous crystalline aluminosilicates.[1-7] Their crystal structures are composed of SiO_4 and AlO_4^- tetrahedra. These basic building blocks interconnect to form networks of cavities that permit the transport of molecules and ions relevant to many environmental and catalytic processes.[8-10] The sizes and shapes of the channels markedly affect the rates at which molecules and ions diffuse. As a result, zeolites can perform highly selective separations. In addition, the channels contain bridging hydroxyl (Si-OH-Al) groups that catalyze a wide variety of chemical reactions. These Brønsted acid sites form when a proton compensates the negative charge that evolves with the isomorphous substitution of Si by Al. The catalytic activity of zeolites is therefore directly related to the amount of Al incorporated into the siliceous framework. The increased presence of Al also structurally weakens zeolites. Therefore, the ratio of Si:Al measures both zeolite acidity and stability. This ratio ranges from 1-25 in commercial zeolite catalysts. Theoretical models of zeolite acid sites increase our understanding of their structure and acidity and facilitate the design of new catalysts and molecular sieves.

Zeolite ZSM-5 is a particularly important industrial solid acid catalyst. It plays a key role in the production of *para*-xylene and in the hydrocarbon cracking process used to increase the octane rating of motor fuels.[4] The siliceous analog of ZSM-5 is silicalite. It has the same structural topology as ZSM-5, but it lacks the catalytic activity that results from the presence of Al in the lattice. As illustrated by Figure 1, these materials have a three-dimensional pore system characterized by straight, parallel channels in the *a*-direction intersected by the so-called zig-zag channels that run in the *b*-direction. Diffusion of adsorbate molecules in the *c*-direction is also possible, but it requires a rather complicated motion involving both the straight and zig-zag channels. The pore openings are defined by 10-membered rings that are wide enough (≈5.5 Å) to allow passage of molecules as large as benzene. Figure 2 depicts a view looking down a straight channel of the zeolite ZSM-5.

In addition to Al, trivalent atoms such as B, Ga, and Fe can be isomorphously substituted for framework Si in a variety of zeolites.[11-24] The resulting changes in adsorption

Theoretical and Computational Approaches to Interface Phenomena
Edited by H.L. Sellers and J.T. Golab, Plenum Press, New York, 1994

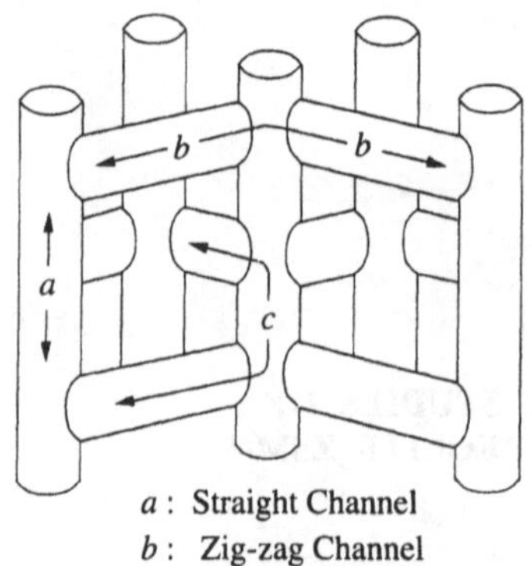

Figure 1. Schematic representation of the intracrystalline channels in silicalite and the zeolite ZSM-5.

and catalytic properties offer potentially new applications. For instance, B-substituted zeolites are critical components in the Assoreni process of methyl butyl ether conversion to methanol and isobutene.[15] In addition, the Amoco processes of xylene isomerization and ethylbenzene conversion use the B-substituted analog of ZSM-5.[15] The extent of structural deformation and the relative acidity of hydroxyl groups depend on the substituent. For instance, based on infrared (IR) spectroscopy of the ν(OH) region and temperature programmed NH_3 desorption, the Brønsted acidity of [B]-ZSM-5 is remarkably lower than that of [Fe]-, [Ga]-, or [Al]-ZSM-5.[17] In fact, the acidity of [B]-ZSM-5 is similar to the very low acidity of terminal hydroxyl groups on the external zeolite surface. This is surprising since the greater electronegativity of B compared to Al might be expected to lead to a greater acidity. It is also interesting to note that the IR spectra of B-substituted zeolites display a doublet at 1380 and 1405 cm^{-1} that is associated with the B-O asymmetric stretch in trigonal BO_3 units.[15, 16] These bands disappear with the adsorption of bases like NH_3 and return after evacuation of the sorbed molecules. The unique properties of [B]-ZSM-5 have led others to postulate that B, in comparison to other trivalent substituents, is small enough to relax into the threefold face of the tetrahedral substitution site.[15, 16] This relaxation would decrease the interaction between B and the bridging hydroxyl and would result in an acidity similar to that of terminal hydroxyls.

Theoretical studies often use cluster models to mimic various features of zeolites.[25] The bonds in the clusters that would normally extend into the framework are terminated, most often by hydrogens or hydroxyl groups. Using quantum mechanical calculations of these zeolite models, we can study the effects of isomorphous substitution on their structure, acidity, and interaction with various adsorbates. Obviously, clusters are only approximate models of zeolites. One must be concerned that the finite boundaries and lack of long range electrostatics in the cluster calculations may adversely affect the results. As an alternative, we have recently used periodic Hartree-Fock methods to perform calculations of the entire zeolite lattice that include effects due to long range electrostatics.[26]

The relative acidity of substituted zeolites can be predicted from the calculated proton affinities (PA) of cluster models.[27-29] Clusters that exhibit a high PA are poor proton donors and, therefore, have low Brønsted acidities. Clusters with a low PA are better proton donors and are more acidic. The inclusion of zero-point vibrational energy differences and temperature

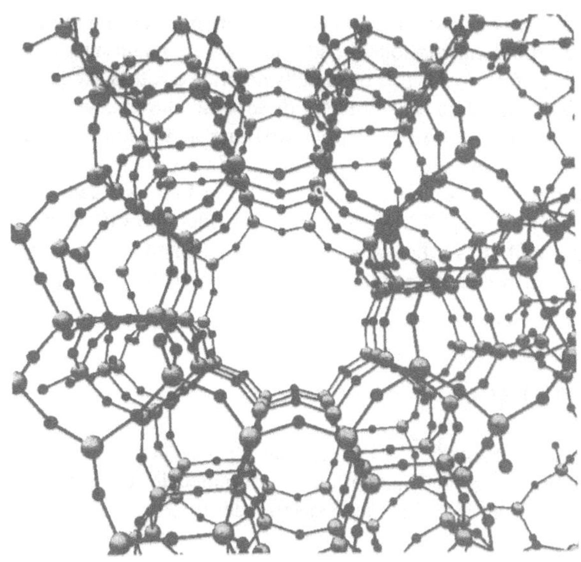

Figure 2. A view down the straight channel in silicalite and the zeolite ZSM-5.

effects is generally not needed to determine trends in acidity. Rather, the PA of a substituted zeolite cluster, ZO^-, can be reasonably approximated by the deprotonation energy (PA_e) of the reaction

$$ZOH \rightarrow ZO^- + H^+ . \qquad (1)$$

Most previous studies of zeolite structure and reactivity have been based on *ab initio* molecular orbital theory.[30] The accuracy of molecular orbital methods strongly depends on the number of basis functions and the level of electron correlation included in the calculation. Unfortunately, the computational demands of these methods, especially those that treat electron correlation, rapidly increase with the size of the system. Since zeolite unit cells typically contain several hundred atoms, molecular orbital cluster calculations large enough to capture even gross features of the true zeolite are very computationally demanding. Density functional theory[31-37] (DFT) provides an expedient, yet relatively accurate treatment of electron correlation and is a viable alternative to molecular orbital theory. In DFT the electron density fundamentally determines the physical state of any N-electron system. Consequently, DFT simplifies electronic structure calculations in comparison to molecular orbital theory which explicitly focuses on an N-electron wavefunction. Although the computational efficiency of DFT strongly depends on implementation, its relative simplicity promises to allow calculations of much larger systems.

Our research involves the study of zeolite chemistry with DFT. To understand the atomic-level details of zeolite catalysis, we must first examine the effect of isomorphous substitution in zeolites. Our initial work[38] focused on disiloxane, $H_3Si\text{-}O\text{-}SiH_3$, and a series of analogs. This molecule is the simplest model of the Si-O-Si bridge found in zeolites. For this series of clusters, we assessed (1) the sensitivity of our method to the quality of basis set expansions and quadrature evaluations and (2) the accuracy of our method in comparison to correlated molecular orbital calculations. In addition, we are examining the effects of using recently proposed energy functionals that potentially describe electron exchange and correlation with greater accuracy. Building on the results for these small clusters, we are now studying the effects of isomorphous substitution on zeolite structure and acidity. In particular, we are examining B-, Fe-, Ga-, and Al-substitution in ZSM-5 cluster models. In this paper, we limit our focus to B-substitution in zeolite cluster models. We first elucidate the reasons for the remarkably low acidity of [B]-ZSM-5. We then consider the effect on the structure of B-substituted zeolite acid sites when a Brønsted base, trimethylamine, interacts with the hydroxyl group. Before proceeding with our discussion of these results, we review our theoretical methods.

THEORETICAL METHODS

Density functional theory[31-34] focuses on the electron density, ρ, as the basic variable that uniquely determines the ground state properties of molecules and solids. In particular, the total energy, E_t, is given as a functional of the electron density,

$$E_t[\rho] = T[\rho] + U[\rho] + E_{xc}[\rho], \tag{2}$$

in which T is the kinetic energy of a system of non-interacting particles of density ρ, U is the classical Coulomb electrostatic energy, and E_{xc} is a many-body term that includes all exchange and correlation effects. The electron density is generally expressed as a sum over occupied orbitals. The orbitals can be expanded either in terms of atom-centered Gaussian,[39, 40] Slater,[41, 42] or non-analytic numerical basis functions,[43-45] or in plane waves.[46-49] In addition, fully numerical density functional methods have been developed that eliminate the need for basis set expansions.[50] Variation of E_t with respect to ρ leads to a set of effective single-particle Schrödinger equations, typically referred to as the Kohn-Sham equations.[32] These coupled differential equations are solved iteratively until the electron density and the energy converge. While Eq. 2 is formally exact, only approximate forms of E_{xc} are currently known. For a given inhomogeneous electron density, ρ, the local density approximation (LDA) integrates the exchange-correlation energy per particle, ε_{xc}, over all space,

$$E_{xc}[\rho] \approx \int \rho(\vec{r})\, \varepsilon_{xc}[\rho(\vec{r})]\, d\vec{r}\,, \tag{3}$$

using a form of ε_{xc} that is essentially exact for a homogeneous electron gas. This approximation assumes the electron density varies slowly in comparison to the exchange and correlation effects. Functionals involving the electron density gradient are often added to the LDA to more accurately model these variations.[37]

For most of the density functional calculations discussed below, we used the DMol program.[45] DMol uses a form of the LDA that is based on the work of von Barth and Hedin.[51] DMol employs numerically tabulated basis functions to solve the Kohn-Sham equations. These functions are defined for each atom on a spherical polar mesh. The radial components are solutions of the LDA equations for neutral, ionic, or hydrogenic atoms, while the angular portion is expressed in terms of spherical harmonics. The occupied orbitals obtained from neutral atom calculations comprise a minimal basis set for molecular systems. Greater variational flexibility beyond this minimal basis set is achieved by including orbitals from calculations for the atomic ions with a +2 charge. For H, however, a nuclear charge of +1.3 is used. This basis set is termed double numerical (DN). A double numerical with polarization (DNP) basis set adds a function on each atom with one angular momentum unit higher than that of its highest occupied orbital. For H, this polarization function is obtained from the calculation with a nuclear charge of +1.3. For the 1st row atoms, the polarization functions are hydrogenic orbitals with a nuclear charge of +5.0. For the 2nd and 3rd row atoms, the polarization functions are orbitals from the atomic ion calculations. The DN and the DNP basis sets are equivalent in size to the widely-used 6-31G and 6-31G** basis sets.[30] Additional functions beyond the DNP basis set are available in DMol for the lighter atoms. Extra 1*s* and 2*p* orbitals for H are obtained from a calculation using a nuclear charge of +4.0. For B and O, hydrogenic calculations with nuclear charges of +5.0 and +7.0 provide additional 1*s*, 2*p*, and 3*d* orbitals. Adding these extra functions to the DNP set forms the basis set we term DNP+.

DMol evaluates the electrostatic potential by decomposing the charge density in terms of atom-centered, multipolar expansions. Our calculations terminate these expansions at angular momenta of $l = 3$ for H and $l = 4$ for all other atoms. We set the self-consistent field parameters so that the total energy converges to better than 1×10^{-6} Hartree. We consider the geometry to be fully optimized when the largest gradient component is less than $\approx 5 \times 10^{-4}$ a.u. The root mean square (RMS) of the gradient is typically twice the magnitude of the largest gradient component with this tolerance.

DMol did not support gradient-corrected functionals at the time of this work. Therefore, we used the deMon program[40] to analyze the effect of adding gradient-corrections to the LDA. Instead of the von Barth and Hedin functional used in DMol, deMon uses the electron correlation functional of Vosko, Wilk, and Nusair[52] for LDA calculations. To potentially achieve better accuracy than possible with the LDA, deMon applies the gradient-corrected correlation functional of Perdew[53] and either the exchange functional of Perdew[54] or that of Becke.[55] We incorporate these corrections in the SCF potential as well as in the calculation of the energy and gradients. In addition to its atomic orbital basis set, deMon employs separate basis sets to fit the exchange-correlation potential and the charge density. The atomic orbitals are expanded in terms of contracted Gaussian functions, while the auxiliary basis sets utilize uncontracted Gaussian functions. We use a double ζ atomic orbital basis set (DZVP) that includes polarization functions on all atoms. For the auxiliary fitting functions, we use the A1 basis set. We set the convergence parameters for the SCF and geometric optimizations to be comparable with our DMol calculations.

The clusters are often difficult to optimize due to the floppy nature of the T-O-T bond angle. For example, less than 1 kcal/mol is needed to expand the Si-O-Si bond angle in disiloxane by 10°.[38] Successful optimization of such floppy internal coordinates requires very fine precision in the energy and gradients. This precision depends on the adaptive numerical quadrature schemes DMol and deMon use to evaluate various matrix elements, the total energy, and the analytical gradients. To estimate the numerical uncertainty due to these quadrature schemes, we rotated the $H_3Si\text{-}O\text{-}AlH_3^-$ analog of disiloxane about an axis bisecting the Si-O-Al bond angle multiple times and calculated the total energy for each orientation. Ideally, the total energy should be invariant to rotation of the molecule in space. The observed variation in energy is a measure of the quality of the grid. Large variations can lead to convergence problems during geometry optimization and will clearly contribute to uncertainties in energetic properties, such as the PA_e. In DMol, the MEDIUM (≈1700 points/atom), FINE (≈3000 points/atom), and XFINE (≈6000 points/atom) grids lead to RMS errors in the total energy of ≈0.05, 0.03, and 0.01 kcal/mol, respectively. deMon's quadrature scheme gives much poorer results; the FINE (≈3000 points/atom) grid leads to an RMS error of ≈0.3 kcal/mol. One of the reasons we chose to use DMol for most of our calculations was its apparently superior quadrature accuracy.

As part of our assessment of the accuracy of DFT, we compare our DFT results to molecular orbital calculations. We previously reported a systematic study of the basis set dependence of restricted Hartree-Fock (RHF) calculations for disiloxane.[56] That work indicated that the *s*- and *p*-space of the wave function is saturated at the double ζ level, while the *d*-space is saturated with the use of two *d*-functions on heavy atoms. Accordingly, we constructed the basis set for the RHF calculations presented in this paper starting from the double ζ split valence basis set of Dunning.[57, 58] H is treated in a 4s/2s contraction, O and B in 9s5p/4s2p, and Si, Al and P in 11s7p/6s4p. We added a *d*-function to heavy atoms and a *p*-function (exponent = 1.0) to H to obtain the polarized DZ+*d* basis set. To rigorously expand the basis set beyond a single polarization function on each heavy atom, we employed the concept of "even-tempered" basis sets, a geometric relationship between orbital exponents introduced by Ruedenberg[59] and Van Duijeneveldt.[60] This geometric relationship converges to the complete basis set limit.[61] The *d*-orbital exponents of the DZ+2*d* basis set are related to the *d*-orbital exponent in the DZ+*d* basis set by $d_+ = \sqrt{3.5}\,d$ and $d_- = d/\sqrt{3.5}$. Thus, the original single *d*-orbital is replaced with one that is more contracted and one that is more diffuse. To reasonably compare to DFT, we must include electron correlation effects in our molecular orbital calculations. This was accomplished using 2nd order Møller-Plesset (MP2) perturbation theory.[62] The RHF calculations were done with the computer program GAMESS-UK,[63] while the MP2 calculations were done with Gaussian 92.[64]

RESULTS AND DISCUSSION

LDA Optimized Structures and Proton Affinities of Disiloxane Analogs

In a previous paper,[38] we examined the structure of disiloxane, $H_3Si\text{-}O\text{-}SiH_3$, using LDA methods. Disiloxane contains the Si-O-Si bridge, a key structural unit in zeolites. We also studied the structure and acidity of bridging hydroxyl groups in a series of disiloxane analogs, $H_3T\text{-}O(H)\text{-}TH_3$ (T = tetrahedrally coordinated atom), in which one or both of the Si atoms are replaced by Al, B, P, Ga, or Ge. These bridging hydroxyls are believed to be the source of acidic protons active in zeolite catalysis. Finally, we used silanol, $H_3Si\text{-}OH$, to model the weakly acidic terminal hydroxyl groups found on the external zeolite surface and at defect sites within the zeolite framework. Our calculations used the LDA as implemented in the DMol program. The optimizations allowed complete flexibility in the molecules except for the constraint of C_{2v} or C_s symmetry. One of our aims was to assess the sensitivity of the internal coordinates and the PA_e's of the clusters to the particular methodology employed. The LDA results displayed a pronounced sensitivity to the particular set of atomic basis functions, but a lesser dependence on the numerical grid size. To assess the accuracy of the LDA in comparison to correlated Hartree-Fock theory, we also optimized a subset of the disiloxane analogs using the MP2 method. The LDA and MP2 results closely agreed for most internal coordinates and the PA_e's. The least accurate results involved internal coordinates whose potentials are relatively weak. In these cases, we believe the inaccuracy is most probably due to the reported tendency for the LDA to overestimate weak interactions. In this section, we review our earlier results, focusing on disiloxane and the B-substituted analogs.

Tables 1-3 list the geometries and total energies for disiloxane and the B-substituted clusters, optimized with basis sets and numerical quadrature grids of increasing quality. Several key internal coordinates display a marked sensitivity to the quality of the basis set. For instance, as shown in Table 1, the Si-O-Si bond angle of disiloxane contracts from 174.1° to 130.5° as an initial set of polarization functions is added to the DN basis set. A similar, although less pronounced, contraction of the T-O-T angle occurs for all the disiloxane analogs with unprotonated bridging oxygens. Table 2 demonstrates this for the bridge angle in $H_3Si\text{-}O\text{-}BH_3^-$. With further expansion of the basis set to DNP+, these angles vary only slightly.

Several of the T-O bond lengths also change substantially as the basis set improves. Figure 3 illustrates the sensitivity of the Si-O and B-O bonds. The largest effect is the shortening of the Si-O bonds as polarization functions are added to the DN basis set. The bond lengths then negligibly decrease with further expansion of the basis set to DNP+. In contrast, the O-H bonds vary little as the basis set improves from DN to DNP, but noticably shorten

Table 1. Variation of Selected Internal Coordinates and Total Energy for $H_3Si\text{-}O\text{-}SiH_3$ with Basis Sets and Numerical Quadrature Grids of Increasing Quality[a]

	Basis Set/Grid				
	DN/ FINE	DNP/ FINE	DNP+/ FINE	DNP+/ XFINE	MP2/ DZ+2*d*
Si-O	1.681	1.666	1.657	1.657	1.653
Si-O-Si	174.1	130.5	132.9	133.0	141.1
E_t	-654.95367	-655.05070	-655.05917	-655.05923	-656.86059

[a]Bond lengths in Å. Angles in degrees. Total energy in Hartrees.

Table 2. Variation of Selected Internal Coordinates, Proton Affinity, and Total Energy for $H_3Si\text{-}O\text{-}BH_3^-$ with Basis Sets and Numerical Quadrature Grids of Increasing Quality[a]

	Basis Set/Grid				
	DN/ FINE	DNP/ FINE	DNP+/ FINE	DNP+/ XFINE	MP2/ DZ+2*d*
B-O	1.536	1.521	1.512	1.512	1.548
Si-O	1.683	1.618	1.614	1.614	1.607
Si-O-B	121.3	114.4	114.7	114.6	119.7
PA_e	322.7	308.8	312.7	312.7	326.7
E_t	-391.24249	-391.31719	-391.32643	-391.32627	-392.57286

[a]Bond lengths in Å. Angles in degrees. Proton affinity in kcal/mol. Total energy in Hartrees.

Table 3. Variation of Selected Internal Coordinates and Total Energy for $H_3Si\text{-}OH\text{-}BH_3$ with Basis Sets and Numerical Quadrature Grids of Increasing Quality[a]

	Basis Set/Grid				
	DN/ FINE	DNP/ FINE	DNP+/ FINE	DNP+/ XFINE	MP2/ DZ+2*d*
O-H	0.981	0.983	0.969	0.969	0.960
B-O	1.621	1.618	1.612	1.611	1.709
Si-O	1.760	1.707	1.701	1.701	1.698
Si-O-B	116.7	117.0	116.9	116.1	122.3
B-O-H	116.7	119.7	119.0	119.3	116.4
Si-O-H	126.6	123.3	124.1	124.6	121.3
E_t	-391.75678	-391.80925	-391.82472	-391.82457	-393.09347

[a]Bond lengths in Å. Angles in degrees. Total energy in Hartrees.

with the DNP+ basis set. For example, the O-H bond in $H_3Si\text{-}OH\text{-}BH_3$ shortens by 0.014 Å. Based on a comparison of our DMol results to basis set free calculations for H_2O,[65] we believe the DNP+ basis set is capable of giving O-H bond lengths that are very close to converged.

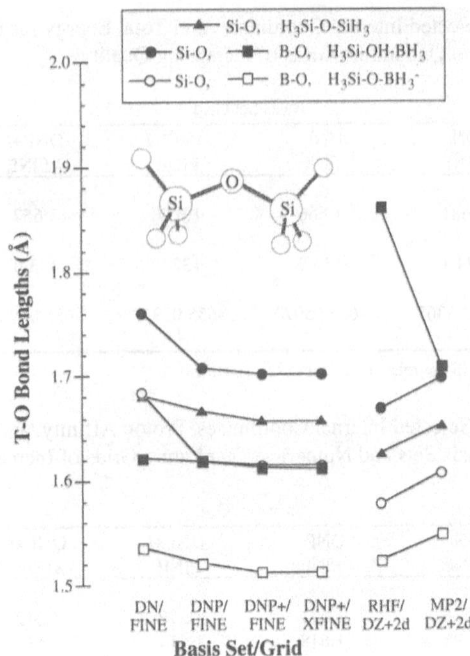

Figure 3. Variation of Si-O and B-O bond lengths with basis sets and numerical quadrature grids of increasing quality.

The PA_e's for the disiloxane analogs are also sensitive to the quality of the basis set. As shown in Table 2, the PA_e for H_3Si-O-BH_3^- has yet to converge within a few kcal/mol as the basis set size increases from DN to DNP+. This is not surprising; accurate molecular orbital calculations of PA_e's require large basis sets and extensive electron correlation to achieve convergence. In contrast to the sensitivity to basis set quality, the optimized internal coordinates and the PA_e's vary little as the grid increases from FINE to XFINE.

The behavior of the LDA results suggests that further increases in the basis set and grid size will have negligible effect. Therefore, we turn to comparing the accuracy of the LDA to molecular orbital methods. We will use only the DNP+/XFINE LDA results for this comparison. Tables 1-3 include the MP2/DZ+2*d* optimized geometries. In addition, Figure 3 compares the LDA Si-O and B-O bond lengths with both RHF/DZ+2*d* and MP2/DZ+2*d* results.

In general, the LDA and MP2 results agree reasonably well. For example, the LDA and MP2 bond angles for the disiloxane analogs typically agree within 2-3°. However, the LDA systematically predicts smaller T-O-T bond angles than the MP2 calculations. For example, as shown in Tables 1-3, the LDA Si-O-B and Si-O-Si bond angles are smaller than the MP2 values by 5-8°. As noted before, very little energy is required to expand the T-O-T bond angles from their equilibrium values. In fact, both the LDA and MP2 predict barriers to linearization of only 1.4 and 4.3 kcal/mol for the H_3Si-O-SiH_3 and H_3Si-O-BH_3^- clusters, respectively. Therefore, this floppy internal coordinate makes these clusters very sensitive test cases for

Table 4. Calculated LDA and MP2 Proton Affinities for Disiloxane Analogs and the Measured Infrared O-H Stretch Frequencies for Substituted ZSM-5 Zeolites

	LDA DNP+/XFINE PA_e (kcal/mol)	MP2/DZ+2*d* PA_e (kcal/mol)	v^a (cm^{-1})
H_3Si-OH	354.3	368.9	3740
H_3Si-O(H)-BH_3	312.7	326.7	3725
H_3Si-O(H)-GaH_3	312.3	-------	3620
H_3Si-O(H)-AlH_3	303.0	312.2	3610

[a]Observed frequencies in parentheses from IR spectra of [B]-, [Ga]-, and [Al]-ZSM-5 zeolites as reported by ref. 17. The calculated frequencies are scaled to match the observed frequency for [Al]-ZSM-5.

comparing various theoretical methods. The smaller T-O-T angles could be due to an LDA overestimate of the weak interaction between the -TH_3 groups.

The discrepancies between the LDA and MP2 bond lengths correlate with bond strength. Our previous work showed that the RHF/DZ+2*d* force constants of the O-H, Si-O, and P-O bonds are considerably larger than those of the Al-O and B-O bonds. For the stronger bonds the LDA and MP2 agree very well; the LDA bond lengths are less than 0.01 Å longer than the MP2 values. Unfortunately, the LDA and MP2 agree poorly for the weaker bonds. In particular, the LDA B-O bond lengths are as much as 0.1 Å shorter than the MP2 values. Similar discrepancies have been reported for certain metal-ligand bonds in transition metal clusters. Those studies demonstrated that gradient-corrected density functionals are essential for accurate prediction of metal-ligand bond lengths. In the next section, we will examine the effect these more accurate functionals have on the B-O bond lengths and other geometric parameters.

Since experimental measurements of the PA_e's are unavailable, we gauge the relative accuracy of the LDA and MP2 PA_e's against published GAUSSIAN-1 (G1) values.[27] The G1 procedure is reportedly capable of predicting proton affinities within 2-3 kcal/mol of experimental values. Our comparisons indicate that the LDA PA_e's are too small by ≈5-10 kcal/mol, while the MP2 PA_e's are too large by less than 5 kcal/mol. Thus, the LDA PA_e's are generally smaller than the MP2 values by 5-15 kcal/mol. For instance, the LDA PA_e for H_3Si-O-BH_3^- is 14 kcal/mol less than the MP2 value shown in Table 2.

While the absolute values for the LDA and MP2 PA_e's differ, both methods predict the same trend for the series of disiloxane analogs. The calculated PA_e's decrease in the order: H_3Si-O^- >> H_3Si-O-BH_3^- > H_3Si-O-GaH_3^- > H_3Si-O-AlH_3^-. Table 4 compares the LDA and MP2 PA_e's with IR spectroscopy measurements of the O-H stretch frequencies for terminal and bridging hydroxyls in [B]-, [Ga]-, and [Al]-ZSM-5 zeolites.[17] The vibrational frequencies for the O-H bond should decrease with increasing acidity. The relative acidity of hydroxyl groups, predicted on the basis of the calculated cluster PA_e's, compares reasonably well with the acidity trend inferred from the IR frequencies. However, the experiment suggests that the acidity of [B]-ZSM-5 should be more comparable to the terminal hydroxyls than to the other bridging hydroxyls. This disagrees with the theoretical prediction. As we will later show, this discrepancy is corrected by using larger cluster models.

In summary, this work suggests that we must at least use the DNP basis set to attain acceptable convergence in most internal coordinates. However, improvements in the O-H bond lengths can be achieved with the DNP+ basis set. In addition, the PA_e has yet to converge within a few kcal/mol even with this largest basis set. On the other hand, both the internal

Table 5. Variation of Selected Internal Coordinates, Proton Affinity, and Total Energy for $H_3Si\text{-}O\text{-}BH_3^-$ with Local and Gradient-Corrected Exchange-Correlation Potentials[a]

	Exchange-Correlation Functional			
	VWN	BP	PP	MP2/ DZ+2*d*
B-O	1.523	1.558	1.576	1.548
Si-O	1.613	1.626	1.627	1.607
Si-O-B	112.5	117.2	125.2	119.7
PA_e	318.4	325.9	326.0	326.7
E_t	-391.21329	-393.27636	-393.50807	-392.57286

[a]Bond lengths in Å. Angles in degrees. Proton affinity in kcal/mol. Total energy in Hartrees.

coordinates and the PA_e are well converged with the FINE quadrature grid. In regard to the accuracy of the LDA, the DNP+ results are generally comparable with MP2/DZ+2*d* calculations. Most bond lengths agree within 0.01 Å and most bond angles within several degrees. However, the strength of certain weak interactions appear to be significantly overestimated by the LDA. The PA_e is also typically too small by ≈10 kcal/mol. We now proceed to examine possible improvements in the description of these weak interactions through the use of gradient-corrected functionals.

The Effect of Gradient-Corrected Exchange-Correlation Potentials

Attempts to improve the LDA have focused on functionals formulated in terms of not just the electron density, but also its gradient. The increased accuracy obtained with gradient-corrected exchange-correlation potentials has been documented for a variety of systems.[39, 66, 67] In this section, we examine their effect on the structure and PA_e of zeolite cluster models. In the previous section, we showed that the most salient points of disagreement between the LDA and MP2 involve the Si-O-B angle and the B-O bond length. To assess whether gradient-corrected functionals can alleviate these discrepancies, we used the deMon program[40] to optimize the structure of the B-substituted analogs of disiloxane, using three different exchange-correlation functionals: the Vosko, Wilk, and Nusair[52] local density functional to which we add either the Becke[55] or the Perdew[54] gradient-correction for the exchange energy along with the Perdew[53] gradient-correction for the correlation energy. We denote these functionals VWN, BP, and PP, respectively.

Tables 5 and 6 list internal coordinates for the B-substituted clusters, optimized using deMon's DZVP basis set and the three different exchange-correlation potentials. Before analyzing the effect of gradient-corrections, we first compare the LDA results from the deMon and DMol programs. The optimized internal coordinates and total energies will vary somewhat due to the different numerical methodologies used by the two density functional programs. These differences largely stem from the different forms of the basis functions used to expand the atomic orbitals. In addition, the two programs use different local density functionals. Despite these differences in methodology, the deMon VWN results agree reasonably well with the DMol calculations. Specifically, the bond lengths agree within 0.01 Å and the bond angles within ≈2°. The discrepancies with MP2 that concern us are generally much larger than the variation between the DMol and deMon LDA results. Thus, our assessment of the accuracy of

Table 6. Variation of Selected Internal Coordinates and Total Energy for $H_3Si\text{-}OH\text{-}BH_3$ with Local and Gradient-Corrected Exchange-Correlation Potentials[a]

	Exchange-Correlation Potential			
	VWN	BP	PP	MP2/ DZ+2*d*
O-H	0.973	0.972	0.973	0.960
B-O	1.620	1.713	1.737	1.709
Si-O	1.702	1.720	1.719	1.698
Si-O-B	116.1	123.8	121.6	122.3
B-O-H	118.0	114.3	115.8	116.4
Si-O-H	125.9	121.8	122.6	121.3
E_t	-391.72068	-393.79563	-394.02754	-393.09347

[a]Bond lengths in Å. Angles in degrees. Total energy in Hartrees.

the LDA in comparison to MP2 is largely independent of either density functional program. In the remainder of this section, we will use the deMon VWN results for the LDA.

When either the BP or the PP gradient-corrections are added to the VWN functional, the geometric structure of the B-substituted clusters changes considerably. As expected, the greatest change involves the weak interactions identified in the previous section. The Si-O-B bond angles increase by 5° or more. In the protonated cluster, the B-O bond lengthens by ≈0.1 Å. In contrast, the stronger Si-O and O-H bonds are much less affected.

The gradient-corrected functionals largely reduce the differences between the LDA and MP2 for the optimized internal coordinates of the B-substituted disiloxane analogs. For the $H_3Si\text{-}OH\text{-}BH_3$ cluster, the LDA predicts the Si-O-B angle to be nearly 10° smaller than the Si-O-H angle. The BP gradient-corrections reverse this ordering, yielding results consistent with the MP2 predictions. The PP gradient-corrections also change the LDA results to be more in line with MP2. For these bond angles, both the BP and PP values agree with the MP2 results within ≈2°, while the LDA results differ by as much as 6°. For the deprotonated cluster, however, the agreement between the PP and MP2 Si-O-B bond angles remains rather poor. The PP functional gives a bond angle that is more than 5° larger than the MP2 value. The BP result compares more closely, although it is still 2.5° smaller than the MP2 value.

The discrepancies between the LDA and MP2 B-O bond lengths also decrease with the addition of gradient-corrections. In fact, the B-O bond is particularly sensitive to the level of electron correlation included in a calculation. The B-O bond strength is reflected in the bond length as well as in the energy required for the $H_3Si\text{-}OH\text{-}BH_3$ cluster to dissociate into $H_3Si\text{-}OH$ and BH_3 fragments. Figure 4 illustrates the differences in the description of the B-O bond by various methods. RHF calculations predict a rather weak B-O bond. At this level of theory, the bond length is 1.86 Å, while the energy difference between the molecule and its fragments is only 3.6 kcal/mol. The RHF B-O bond is so weak that for large, and more strained, values of the Si-O-B angle the cluster nearly dissociates. Inclusion of electron correlation via MP2 calculations shortens the bond length to 1.71 Å and increases the energy difference to 13.3 kcal/mol. The tendency to dissociate diminishes even further with the LDA calculations; the bond length is 1.62 Å and the energy difference is 27.8 kcal/mol. However, the LDA apparently overestimates the B-O bond strength. Much better agreement with the MP2 is attained when the gradient-corrections are added to the LDA. The BP functional gives especially good agreement with the MP2 results; the BP bond length is 1.71 Å and the energy

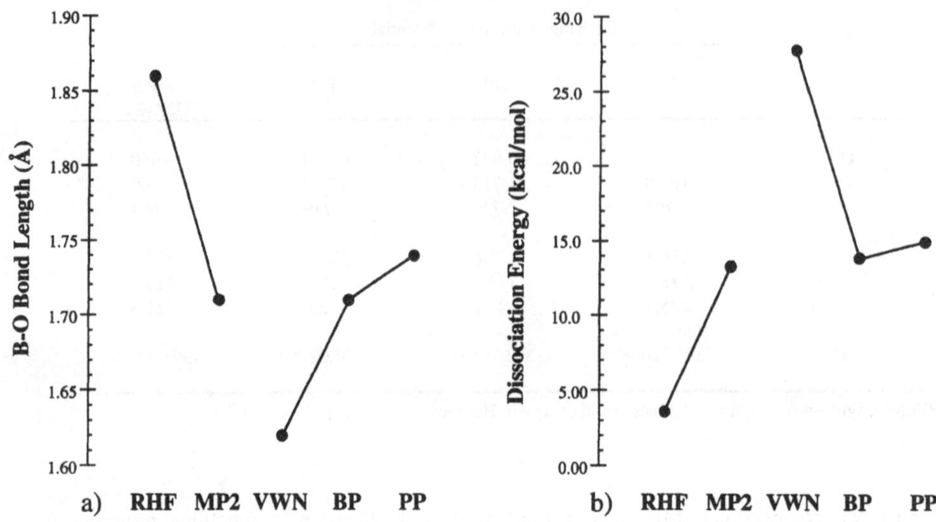

Figure 4. Variation of (a) the B-O bond length and (b) the dissociation energy for the H_3Si-OH-BH_3 cluster with different theoretical methods.

difference is 13.9 kcal/mol. Although not agreeing quite as closely, the PP values of 1.74 Å and 15.0 kcal/mol are also consistent with the MP2 description of the B-O bond.

In addition to lengthening the B-O bonds, the gradient-corrections expand the Si-O bond lengths, leading to poorer agreement with MP2. However, the change is small; the BP and PP bond lengths are still only 0.02 Å larger than the MP2 values. The O-H bond is unaffected with the addition of gradient-corrections; it remains ≈0.01 Å larger than with MP2. Therefore, when gradient-corrections are added to the LDA, no bond lengths differ with respect to MP2 results by more than 0.03 Å.

In short, the effect of gradient-corrections is most pronounced for those internal coordinates that are determined by weak interaction potentials, exemplified particularly by the large increases for the Si-O-B bond angles and the B-O bond lengths. Furthermore, the degree of change depends on the type of gradient-correction. For the H_3Si-O-BH_3^- cluster, for example, there is a difference of 8° in the Si-O-B bond angle between the BP and PP gradient-corrected results. Such discrepancies again highlight the difficulty in accurately predicting the equilibrium values for floppy internal coordinates. For the optimized internal coordinates, the BP functional gives better agreement with the MP2 results than the PP functional. Whether it is truly more accurate, however, is impossible to determine without comparison to molecular orbital calculations with larger basis sets and more accurate representations of electron correlation.

The PA_e of the B-substituted clusters also appears to improve with the use of the gradient-corrections. As discussed previously, the LDA PA_e's are generally 5-15 kcal/mol smaller than MP2 values. In the case of the B-substituted analog, both the BP and the PP exchange-correlation potentials lower the energy of the protonated cluster more than the deprotonated cluster. Therefore, the PA_e increases and now agrees within 1 kcal/mol (Table 5). The large increase in PA_e with gradient-corrections has also been reported for triazenes[68] and other molecules.[69]

In future work, we plan to extend our gradient-corrected calculations to the entire series of disiloxane analogs. Based on the results for the B-substituted clusters, the BP and the PP functionals do alleviate the deficiencies of the LDA, improving the overall agreement with MP2 results.

The Structure and Acidity of B-Substituted ZSM-5 Cluster Models

The incorporation of trivalent elements into siliceous zeolite frameworks modifies the structure and acidity of these porous materials. Using the LDA, we have studied the isomorphous substitution of B, Fe, Ga, and Al in ZSM-5 cluster models. We used the crystal structure determination of von Koningsveld[70] to specify spatial coordinates for our cluster atoms. Von Koningsveld used a highly siliceous sample (Si:Al = 299). Since the average effect of the small amount of Al should be negligible, the structure will closely correspond to silicalite. There are twelve distinct substitution sites in the unit cell of ZSM-5. In von Koningsveld's notation, we built our clusters around the T(12)-O(24)-T(12) bridge, terminating the bonds that would normally extend into the zeolite framework with hydrogens. Based on Hartree-Fock relative energies for the twelve possible $T(OH)_4$ fragments, Alvarado-Swaisgood *et al.*[71] suggest that T(12) is one of the more likely sites for Al substitution. In addition, the O(24) site lies at the intersection of the straight and zig-zag channels, where significant interaction can occur between a bridging hydroxyl and adsorbate molecules. Therefore, the T(12)-O(24)-T(12) bridge is likely catalytically active in ZSM-5. In this section, we discuss our results for B substitution in one of the T(12) sites. To better calibrate the effects of this substitution on the zeolite framework, we also present the optimized geometry for the siliceous T(12) bridge.

Many theoretical studies of isomorphous substitution in zeolites have been published using fragments of the zeolite lattice.[38, 71-78] This approach requires careful analysis of any dependence the calculated properties may have on the size of the cluster model. Two studies in particular have systematically examined this issue and are especially relevant to the work we present in this section. Using MNDO cluster calculations, Chamot[72] modeled B-substitution at the T(2) site in ZSM-5. He successively relaxed larger portions of a 70 atom crystal fragment. His work demonstrated that the structure and PA_e of the Si-OH···B bridge depends on the extent of relaxation. However, most properties did converge with relaxation of atoms within two bonds of the acidic hydroxyl group. Using Hartree-Fock cluster calculations, Brand *et al.*[73] systematically studied the dependence of the PA_e and the O-H stretch frequency on the size of ZSM-5 cluster models. Their work focused on Al substitution at the T(12) and T(2) sites of ZSM-5. Starting with a $H_3Si\text{-}OH\text{-}AlH_3$ cluster, they successively added shells of atoms up through the 4th coordination shell. The optimized structure of the central Si-OH-Al bridge appeared to converge with the addition of the 3rd shell.

To assess any sensitivity to the size of our cluster models, we optimized the structure of the central bridge in successively larger fragments. Our cluster models include atoms up through the 4th coordination shell about the central O(24) site. Specifically, we report results for clusters ranging in size from $(HO)_3Si\text{-}O\text{-}T(OH)_3$ to $((HO)_3SiO)_3Si\text{-}O\text{-}T(OSi(OH)_3)_3$, where T = Si or B. For these clusters, the geometry of the three or four atoms in the central Si-O(H)-T bridge is fully optimized. For the largest fragments, we have also fully relaxed the six O atoms of the 2nd coordination shell about the O(24) site. Tables 7-9 list our DMol LDA DNP+/FINE results. We have imposed no symmetry constraints in these calculations, although the siliceous clusters naturally have C_s symmetry. The scope of our study will allow comparisons with the work of Chamot and Brand *et al.* These cluster calculations are by far the largest published to date that explicitly account for effects due to electron correlation.

Table 7 shows the variation of selected internal coordinates for siliceous clusters of increasing size. The internal coordinate most sensitive to the size of the zeolite fragment is the Si-O-Si bond angle. In our smallest cluster (**2**), this bond angle is 10° smaller than in the crystal, while the Si-O bonds are as much as 0.03 Å longer. As the 3rd coordination shell is added (**2** $\rightarrow$ **3**), the Si-O-Si bond angle widens by 5°, while the bond lengths shorten by

Table 7. Variation of Selected Internal Coordinates and Total Energy for $X_3Si\text{-}O\text{-}SiX_3$ Clusters of Increasing Size[a]

	Cluster Size				
	2	3	4a	4b	ZSM-5 Zeolite[b]
X	OH	$OSiH_3$	$OSi(OH)_3$	$OSi(OH)_3$	
Si-O(24)	1.608	1.603	1.605	1.606	1.595
Si-O(11)	1.607	1.596	1.594	1.606	1.586
Si-O(12)	1.610	1.598	1.595	1.607	1.574
Si-O(20)	1.594	1.593	1.593	1.607	1.606
Si-O(24)-Si	135.9	141.2	141.5	141.2	146.3
O(24)-Si-O(11)	113.7	111.2	111.2	110.1	108.8
O(24)-Si-O(12)	113.3	111.6	111.3	110.1	110.2
O(24)-Si-O(20)	103.0	104.8	104.7	103.2	106.4
O(11)-Si-O(12)	107.1	108.2	108.4	109.3	109.9
O(11)-Si-O(20)	108.9	109.6	109.7	110.0	109.4
O(12)-Si-O(20)	110.8	111.5	111.6	114.0	112.0
O(20)-Si-O(24)-Si	180.2	181.8	181.2	182.7	183.1
E_t	-1103.92565	-2841.30261	-4187.74461	-4187.74876	

[a]Bond lengths in Å. Angles in degrees. Total energy in Hartrees.

[b]The internal coordinates and the atom labelling convention for the crystal structure are from ref. 70.

≈0.01 Å or less. With the addition of the 4th shell (**3** → **4a**), the geometry of the central Si-O-Si bridge converges within 0.005 Å and 0.5°. It appears likely that the geometry of the central Si-O-Si bridge will be unaffected by further increases in cluster size.

With this limited optimization, the central Si(12)-O(24) bonds are 0.01 Å longer than the bonds between Si(12) and the O atoms in the 2nd coordination shell. This asymmetry in the bond lengths results from the limited optimization. When we relax the 2nd coordination shell along with the central bridge (**4a** → **4b**), the bond lengths become much more uniform. The energy associated with the relaxation of the 2nd shell is only 2.6 kcal/mol, consistent with the minor changes in geometry.

The predicted geometry of the central Si-O-Si bridge in the siliceous zeolite fragments differs slightly from the reported crystal structure.[70] In the crystal, the average Si-O bond length about the T(12) site is 1.59 Å. Our cluster model predicts 1.61 Å. In addition, the central Si-O-Si bond angle is 5° larger in the crystal than what we predict. Given the reported accuracy of the crystal structure determination, we feel these differences are just barely significant. However, these discrepancies could reflect the weaknesses of the LDA. In particular, based on our results for the B-substituted analogs of disiloxane, we expect the T-O-T bond angle to expand with the use of gradient-corrected functionals. This could potentially lead to better agreement with the crystal data. We will report gradient-corrected results for these large clusters in later work. We also note that the relatively narrow range of Si-O bond lengths and the mean O-Si-O bond angle of 109.5° reflect the tetrahedral Si coordination in both the zeolite fragments and crystal. This will contrast sharply with the B coordination in the acidic Si-OH···B bridge.

When we substitute B for Si in one of the T(12) sites, the local geometry around the substitution site significantly changes. Moreover, the structural deformation strongly depends

Table 8. Variation of Selected Internal Coordinates and Total Energy for $X_3Si\text{-}O\text{-}BX_3^-$ Clusters of Increasing Size[a]

	Cluster Size			
	2	**3**	**4a**	**4b**
X	OH	$OSiH_3$	$OSi(OH)_3$	$OSi(OH)_3$
B-O(24)	1.517	1.491	1.497	1.516
B-O(11)	1.572	1.557	1.554	1.490
B-O(12)	1.558	1.552	1.550	1.481
B-O(20)	1.558	1.579	1.582	1.496
Si-O(24)	1.587	1.587	1.589	1.581
Si-O(11)	1.647	1.638	1.636	1.642
Si-O(12)	1.655	1.638	1.636	1.638
Si-O(20)	1.675	1.673	1.673	1.668
Si-O(24)-B	145.3	150.4	150.0	149.3
O(24)-B-O(11)	106.9	106.1	106.2	107.6
O(24)-B-O(12)	107.5	106.3	106.2	108.0
O(24)-B-O(20)	101.6	103.7	103.4	105.1
O(11)-B-O(12)	111.5	112.6	112.9	111.1
O(11)-B-O(20)	112.7	112.4	112.4	111.9
O(12)-B-O(20)	115.7	114.7	114.7	112.8
O(24)-Si-O(11)	115.8	113.5	113.6	113.9
O(24)-Si-O(12)	115.0	113.2	113.0	113.4
O(24)-Si-O(20)	113.5	115.4	115.1	113.3
O(11)-Si-O(12)	103.1	104.3	104.5	104.9
O(11)-Si-O(20)	103.3	103.8	103.9	104.6
O(12)-Si-O(20)	104.7	105.6	105.6	105.9
O(20)-Si-O(24)-B	181.3	181.5	181.0	181.7
E_t	-840.14790	-2577.58767	-3923.94111	-3923.95333

[a]Bond lengths in Å. Angles in degrees. Total energy in Hartrees.

on whether the central bridge is protonated. For the [B]-ZSM-5 clusters, we will demonstrate that the finite cluster size and, more importantly, the extent of relaxation within the cluster strongly affect the predicted structure and acidity of the zeolite framework.

Table 8 illustrates the variation of selected internal coordinates for anionic B-substituted fragments of increasing size. The structural changes display a sensitivity to cluster size similar to that which we noted for the siliceous clusters. Adding the 3rd coordination shell (**2** → **3**) significantly alters the structure of the central bridge. Specifically, the Si-O-B angle widens by 5° and all four Si-O bonds and three of the B-O bonds shorten by as much as 0.03 Å. The addition of the 4th shell (**3** → **4a**) has negligible effect on the geometry.

Relaxing the 2nd coordination shell in the largest fragment (**4a** → **4b**) affects the anionic B-substituted clusters substantially more than the siliceous clusters. The greater effect primarily results from relaxation about B. In particular, the three bonds between B and O atoms in the 2nd shell shorten by ≈0.1 Å. The additional structural relaxation lowers the energy by 7.7 kcal/mol, a considerably greater energy change than we observed for the siliceous clusters.

Figure 5 depicts the structure of the anionic B-substituted site in cluster **4b**. Both Si and B are roughly tetrahedrally coordinated. In comparison to the central Si-O-Si bridge in the

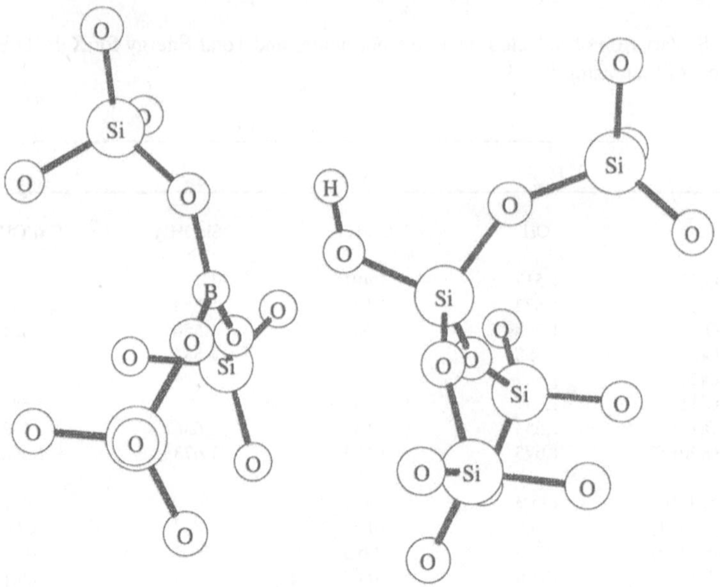

Figure 5. The anionic [B]- ZSM-5 cluster model 4b.

siliceous clusters, the Si-O-B bridge shifts towards the three 2nd shell O atoms adjacent to B (i.e., towards the left in Figure 5). As a result, the T-O bonds are not uniform in length. In cluster **4b**, the bonds between B and these three 2nd shell O atoms are slightly shorter than the B-O(24) bond. Conversely, the bonds between Si and the three other 2nd shell O atoms are longer than the Si-O(24) bond. This shift also clearly affects the bond angles. Consider, for instance, the bond angles centered on the B atom. The three involving the B-O(24) bond contract relative to the tetrahedral ideal of 109.5°, while the remaining three O-B-O angles expand. Just the opposite is true for the corresponding sets of O-Si-O bond angles. Finally, also note that the B-O bonds are substantially shorter than the Si-O bonds. As a whole, these structural changes are consistent with the reported shrinking of the unit cell that results from B-substitution in zeolites.

The geometry of the Si-OH···B bridge is also very sensitive to cluster size. Table 9 lists selected internal coordinates for the protonated fragments of increasing size. As illustrated by Figure 6, the B-O(24) bond of the central Si-OH···B bridge changes considerably more than the other geometric parameters. Once again, the geometry apparently converges with the addition of the 4th coordination shell. However, as we relax the 2nd coordination shell in the largest fragment (**4a** → **4b**), all the B-O bond distances change substantially. The B substituent recedes from the hydroxyl group by 0.2 Å, while the bonds between B and the 2nd shell O atoms shorten by ≈0.1 Å. In contrast, the Si-O bonds change much less. The large structural changes lower the energy by 18.4 kcal/mol, the largest energy change for all three T-O(H)-T bridges.

The orientation of the hydroxyl group is a unique structural feature of the protonated clusters. The Si-O-H bond angle is slightly less than 120° in each fragment. However, as shown in Table 9, the proton in the smallest cluster bends 25.9° out of the plane defined by the

Table 9. Variation of Selected Internal Coordinates and Total Energy for $X_3Si\text{-}OH\cdots BX_3$ Clusters of Increasing Size[a]

	Cluster Size			
	2	**3**	**4a**	**4b**
X	OH	$OSiH_3$	$OSi(OH)_3$	$OSi(OH)_3$
O-H	0.978	0.981	0.980	0.973
B-O(24)	1.907	1.849	1.867	2.084
B-O(11)	1.520	1.491	1.487	1.389
B-O(12)	1.479	1.485	1.480	1.381
B-O(20)	1.547	1.573	1.577	1.417
Si-O(24)	1.660	1.671	1.672	1.633
Si-O(11)	1.608	1.594	1.592	1.606
Si-O(12)	1.610	1.595	1.593	1.602
Si-O(20)	1.610	1.609	1.608	1.617
Si-O(24)-B	137.7	143.0	143.2	141.6
B-O(24)-H	98.5	98.2	97.2	95.9
Si-O(24)-H	119.1	118.8	119.5	118.8
O(24)-B-O(11)	99.9	98.7	98.3	97.3
O(24)-B-O(12)	99.3	99.1	98.3	95.2
O(24)-B-O(20)	89.0	91.1	90.6	88.1
O(11)-B-O(12)	119.2	120.8	121.4	120.0
O(11)-B-O(20)	116.2	116.4	116.4	117.4
O(12)-B-O(20)	121.2	119.1	119.2	121.5
O(24)-Si-O(11)	114.6	110.7	110.7	111.9
O(24)-Si-O(12)	111.9	110.8	110.3	107.8
O(24)-Si-O(20)	105.1	107.4	107.3	104.2
O(11)-Si-O(12)	107.0	108.4	108.6	108.9
O(11)-Si-O(20)	108.1	108.8	109.0	109.6
O(12)-Si-O(20)	110.0	110.8	110.9	114.3
O(20)-Si-O(24)-H	28.8	2.3	4.2	20.5
O(20)-Si-O(24)-B	178.8	181.5	180.8	172.5
OH out-of-plane[b]	25.9	0.7	3.0	24.3
E_t	-840.66051	-2578.05447	-3924.46141	-3924.49076

[a]Bond lengths in Å. Angles in degrees. Total energy in Hartrees.

[b]This angle is defined relative to the plane defined by the Si-O-B bridge.

Si-O-B bridge. In structures **3** and **4a**, the orientation changes so that the proton nearly lies in this plane. Finally, the proton orients out-of-plane once again in structure **4b**. Apparently, the orientation of the hydroxyl group is very sensitive to the long range electrostatic interaction between the proton and atoms in distant coordination shells.

Clearly, the geometry of the B-substituted acid site strongly depends on both the size of the cluster model and the extent of relaxation. Due to computational limitations, we cannot extend these calculations to confirm whether the structure of the protonated central bridge converges with structure **4b**. However, Chamot's MNDO calculations[72] suggest little additional change would occur with further relaxation beyond the 2nd coordination shell.

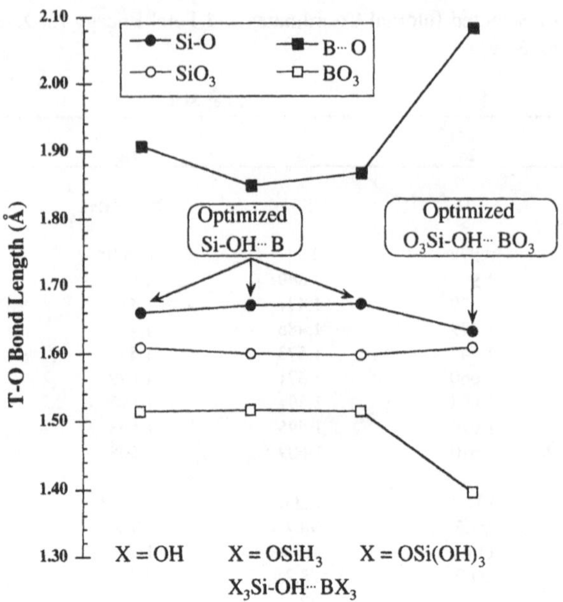

Figure 6. Variation of the B-O and Si-O bond lengths in protonated ZSM-5 fragments of increasing size.

In each of the protonated [B]-ZSM-5 clusters, B is unequivocally trigonally coordinated. The internal coordinates of the largest fragment especially delineate this feature. In structure **4b**, B is 2.1 Å from the hydroxyl O atom. This large distance contrasts sharply with the 1.4 Å bonds between B and the 2nd shell O atoms. These short bonds are essentially the same length as those we obtain for the isolated $B(OH)_3$ molecule, optimized using the LDA DNP+/FINE method. The six bond angles centered on the B atom also reflect the B trigonal coordination. Each of the three bond angles formed by B and the 2nd shell O atoms are ≈120°, while the other three O-B-O bond angles approach 90°. In contrast, the O-Si-O bond angles indicate that the tetrahedral arrangement of O atoms around the other T(12) site is only slightly perturbed by the B substitution. This predicted geometry is consistent with the presence of the doublet in IR spectra of [B]-ZSM-5 that is generally associated with the B-O asymmetric vibrational mode in trigonal BO_3 units.[15]

Figure 7 clearly depicts the B substituent sitting near the center of three O atoms. The large distance between the substituent and the acid site suggests that B interacts very weakly with the hydroxyl group. As we will now discuss, this markedly affects the acid strength of the hydroxyl group.

The Brønsted acidity of the hydroxyl group depends on the cluster model. In particular, the PA_e's do not converge with increasing fragment size, but rather oscillate. Consider, for instance, the PA_e of the B-substituted clusters listed in Table 10. As a shell of Si atoms is added (**2** → **3**), the PA_e decreases nearly 30 kcal/mol. Then, as a shell of O atoms is added (**3** → **4a**), the PA_e increases by more than 30 kcal/mol. Brand *et al.*[73] observe similar

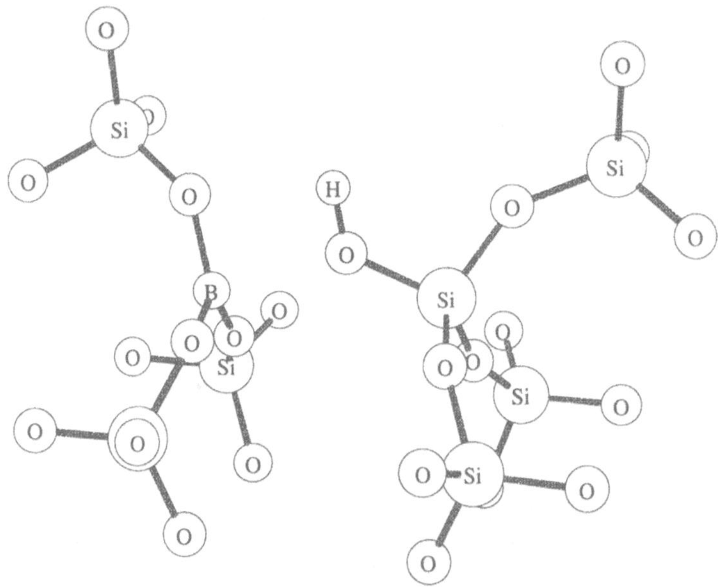

Figure 7. The protonated [B]- ZSM-5 cluster model 4b.

variation in the PA_e for Al-substituted clusters. They ascribe these changes to electrostatic effects arising from the different partial charges associated with Si and O shells.

In addition to oscillating with increasing cluster size, the PA_e increases by 11 kcal/mol when the 2nd coordination shell about the acid site is relaxed (**4a** $\rightarrow$ **4b**). We observe similar PA_e increases for Fe-, Ga-, and Al-substituted clusters. These large energy changes conflict with the results of Brand *et al.*[73] They only observe a PA_e increase of 0.3 kcal/mol when the 2nd shell O atoms in an Al-substituted cluster are allowed to partially relax.

Predictions for the relative acidity of substituted zeolites also strongly depend on the cluster model. In a future report, we will thoroughly compare the effects of B, Fe, Ga, and Al substitution on the structure and acidity of ZSM-5 cluster models. In this paper, we briefly report the PA_e's of the B, Ga, and Al substituents to emphasize their size dependence and to highlight the low acidity of the B-substituted clusters. Table 10 lists the calculated PA_e's. For all sizes of the zeolite clusters, our LDA calculations predict the B-substituted PA_e to be 13-19 kcal/mol greater than the values for the Ga- or Al-substituted clusters. In contrast, the Ga- and Al-substituted PA_e's differ by only 2-4 kcal/mol. This gap in the calculated PA_e's correlates with the sharp break in IR vibrational frequencies between bridging hydroxyls in [B]-ZSM-5 and those in the other substituted ZSM-5 zeolites (Table 4).

We should also note that relaxation of the 2nd shell O atoms is necessary to correctly predict the slightly greater acidity of [Al]-ZSM-5 relative to [Ga]-ZSM-5.[17] Chamot reports a similar dependence on the extent of relaxation near the substitution site.[72] In his MNDO calculations, the B-substituted cluster is actually more acidic than the Al-substituted cluster with relaxation of just the hydroxyl group. Only with relaxation of both the 1st and 2nd coordination

Table 10. Variation of LDA Proton Affinities for $X_3Si\text{-}O\text{-}TX_3^-$ (T = B, Ga, Al) Clusters of Increasing Size[a]

	Cluster Size			
	2	**3**	**4a**	**4b**
X	OH	$OSiH_3$	$OSi(OH)_3$	$OSi(OH)_3$
Si-O-B	321.7	292.9	326.5	337.3
Si-O-Ga	302.6	276.3	308.5	322.1
Si-O-Al	304.6	279.6	312.1	319.1

[a]Proton affinity in kcal/mol.

shells does he predict the borosilicate to be less acidic. Evidently, all the O atoms around a substituent must be relaxed to correctly distinguish small differences in acidity.

The substituted ZSM-5 clusters accurately model the relatively low acidity of B-substituted zeolites. The predictions based on the disiloxane analogs failed in this regard. The larger cluster models allow the trivalent B substituent to fully relax into a planar geometry when the central bridge is protonated. Thus, B incorporation in zeolite frameworks essentially introduces a defect in the network of TO_4 tetrahedra. The hydroxyl group near the B substitution site consequently resembles those found at true terminal points in the zeolite framework. In fact, the O-H vibrational frequency of 3725 cm^{-1} observed for bridging hydroxyls in [B]-ZSM-5 differs only slightly from the 3740 cm^{-1} frequency usually assigned to terminal hydroxyl groups.[17] Other trivalent substituents like Al are too large to move far from the center of the T(12) site and, therefore, more strongly affect the nearby hydroxyl group. These structural differences associated with the size of the substituent account for the markedly lower [B]-ZSM-5 acidity.

To expand our study of isomorphous substitution in zeolites, we could consider substitution sites other than the T(12) site in ZSM-5. In addition to T(12), Alvarado-Swaisgood *et al.*[71] predicted that T(6) and T(9) are likely sites for Al substitution in ZSM-5. Their predictions, based on relative Hartree-Fock energies for substituting Si by Al in the 12 possible $T(OH)_4$ fragments, do not account for any effects due to relaxation. For all 12 sites, the reported energies differ by less than 4 kcal/mol. This energy range is rather small. In fact, we typically observe relaxation energies in our [B]-, [Fe]-, [Ga]-, and [Al]-ZSM-5 cluster models that are almost an order of magnitude greater. Therefore, it is debatable whether energetically favored substitution sites can be predicted without relaxing the zeolite framework. It may even be the case that the flexible zeolite framework can accomodate certain substituents equally well in any of the possible sites.

The Adsorption of Trimethylamine

The geometry around B likely changes when a Brønsted base interacts with the hydroxyl group. In fact, the vibrational bands associated with the trigonal BO_3 unit disappear with NH_3 adsorption.[15, 16] In addition, NMR, IR, and neutron scattering results indicate that bases like NH_3 are not weakly physisorbed to the zeolite framework. Typically, NH_3 adsorption energies are 20-40 kcal/mol.[75] The acidic protons within the zeolite are probably transfered to form ionic adsorption complexes ($ZO^- \cdots BH^+$). Using DMol LDA DNP+/FINE calculations, we have examined the adsorption of trimethylamine (TMA) on the smallest

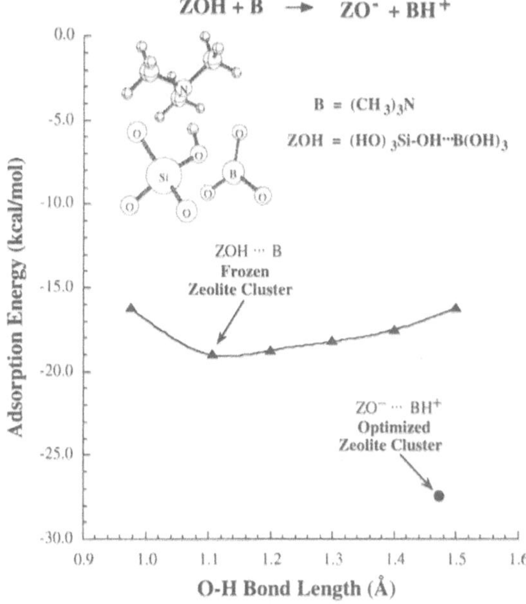

Figure 8. TMA adsorption energy on [B]-ZSM-5 cluster model 2.

[B]-ZSM-5 cluster model, $(HO)_3Si\text{-}OH\cdots B(OH)_3$. Although the geometry of the Si-OH···B bridge is not yet converged with this cluster size, the fragment is large enough to demonstrate a change in coordination for the B substituent. We have also examined the adsorption of NH_3 on the [B]-ZSM-5 clusters. NH_3 adsorption, however, is more complicated because multiple hydrogen bonds can form between the adsorbate and framework O atoms. Due to the methyl groups, this is not the case in TMA. TMA is also a stronger Brønsted base than NH_3. Using the DNP+/FINE method, we calculate the LDA PA_e for NH_3 and TMA to be 206.4 and 225.9 kcal/mol, respectively. Therefore, TMA will more likely form an ionic adsorption complex than NH_3.

TMA abstracts the proton from the zeolite acid site with no apparent energetic barrier to chemisorption. Figure 8 illustrates the adsorption energy at various stages of geometry optimization. Initially, we adsorbed TMA on the rigid cluster in which the O-H bond length is 0.98 Å. In this physisorbed state, TMA sits 2.7 Å above the hydroxyl O atom and its adsorption energy is 16.3 kcal/mol. Allowing the proton to relax lengthens the O-H bond by more than 0.1 Å and stabilizes the complex by 2.8 kcal/mol. When the entire Si-OH···B bridge is allowed to relax, the O-H bond lengthens even further to 1.47 Å and the adsorption energy increases to 27.4 kcal/mol. As the Si-OH···B bridge relaxes, TMA draws 0.1 Å closer to the hydroxyl O atom, while the distance between the N and the proton decreases from 1.7 to 1.1 Å. In the optimized adsorption complex, the N-H bond length is only 0.1 Å longer than in the fully optimized, isolated $(CH_3)_3NH^+$ ion. Therefore, the proton readily transfers from the zeolite to TMA.

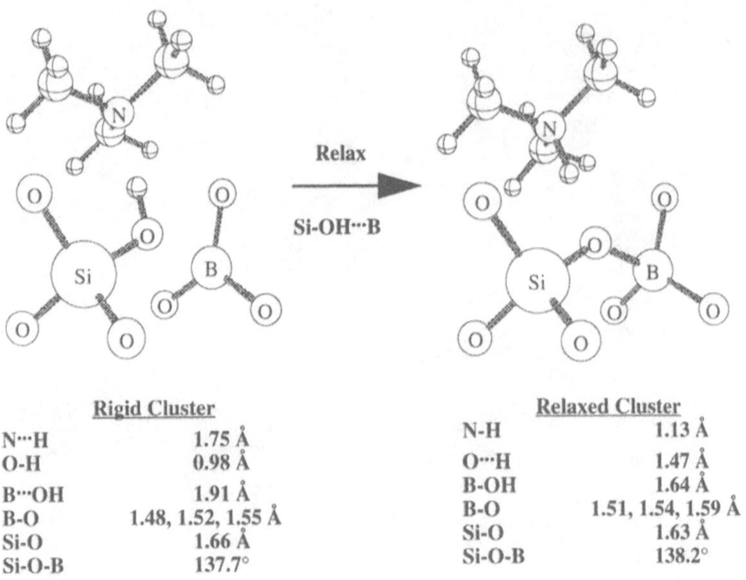

Figure 9. Schematic representation of the structural changes in [B]-ZSM-5 cluster model 2 that occur with TMA adsorption.

In addition to the proton transfer, TMA adsorption considerably alters the zeolite structure around the B substituent. Figure 9 shows changes in several geometric parameters when the zeolite interacts with the adsorbate. As the Si-OH···B bridge relaxes, the distance between the O and B atoms shortens significantly and their interaction increases. Thus, TMA adsorption causes the B coordination to change from threefold to fourfold. The geometry of the central bridge now more closely resembles the anionic cluster. This change, of course, accounts for the decrease in the IR peak intensity for the BO_3 vibrations with NH_3 adsorption.

Further investigations might examine the size dependence of the interaction of adsorbates with the zeolite clusters. However, given the oscillation of the PA_e with increasing cluster size, we expect the adsorption energy to be very sensitive to the size of the cluster model. Other theoretical studies indicate that long range electrostatic interactions stabilize the ionic (ZO^-···BH^+) adsorption complex.[73, 76, 77]

CONCLUSION

This review demonstrates the valuable role DFT can play in studies of zeolite chemistry. The LDA provides accurate predictions of zeolite structure and acidity. Gradient-corrected exchange-correlation potentials offer improvements that are important for particular cases. In future studies, we will extend our use of these apparently more accurate functionals. The computational efficiency of DFT methods allows us to consider large cluster models. This is very important since many of the calculated properties display a substantial size dependence. For the ZSM-5 clusters considered in this paper, we observed significant variation in the predicted structure and relative acidity with both increasing fragment size and the extent of relaxation within the cluster. Converged results appear to require at least 3-4 shells of atoms around the central bridge, of which at least the first two must be fully relaxed.

With this approach, we were able to elucidate the reasons for the unique properties of [B]-ZSM-5. The small size of the B substituent allows it to relax into a planar geometry. This leads to very weak interaction between B and the nearby hydroxyl group. The B geometry accounts for (1) the presence of the BO_3 vibrational bands in IR spectra of [B]-ZSM-5 and (2) the markedly low acidity of B-substituted zeolites. In addition, we also were able to demonstrate the structural change in the Si-OH···B bridge predicted by the disappearance of the BO_3 vibrational bands with the adsorption of Brønsted bases.

We are presently developing a density functional computer program. Our objective is to develop an efficient, parallel Gaussian-based code to study molecular and periodic systems. This will allow us to extend our studies of zeolites to both larger cluster models and to calculations of the complete crystalline unit cell.

ACKNOWLEDGEMENTS

This work is supported in part by the Advanced Industrial Concepts Division of the DOE Office of Conservation and Renewable Energies under contract No. 16697. One of the authors (MSS) is supported by a fellowship administered by the Associated Western Universities-Northwest Division under grant DE-FG06-89ER-75522 or DE-FG06-92RL-12451 with the Department of Energy (CAS). This work was done in part to support the Environmental and Molecular Sciences Laboratory (EMSL) project at Pacific Northwest Laboratory. Funding for this aspect of the work was provided by the EMSL Construction Project, Office of Health and Environmental Research, Office of Energy Research. Computer resources were provided by the Scientific Computing Staff, Office of Energy Research, at the National Energy Research Supercomputer Center (NERSC), Livermore, CA. Pacific Northwest Laboratory is operated by Battelle Memorial Institute for the U.S. Department of Energy under Contract DE-AC06-76RLO 1830.

REFERENCES

1. M. E. Davis, *Acc. Chem. Res.* **26**, 111-115 (1993).
2. W. M. H. Sachtler, *Acc. Chem. Res.* **26**, 383-387 (1993).
3. S. L. Suib, *Chem. Rev.* **93**, 803-826 (1993).
4. B. C. Gates, *Catalytic Chemistry* (John Wiley & Sons, Inc., New York, 1992).
5. J. C. Vedrine, *Stud. Surf. Sci. Catal.* **69**, 25 (1991).
6. G. T. Kerr, *Sci. Am.* **261**, 100-105 (1989).
7. D. E. W. Vaughan, *Chemical Engineering Progress* **84**, 25-41 (1988).
8. S. N. Chaudhuri, C. Halik and J. A. Lercher, *J. Mol. Catal.* **62**, 289-295 (1990).
9. K. Yanagisawa, S. Kanahara, M. Nishioka and N. Yamasaki, *J. Nucl. Sci. Technol.* **21**, 558-560 (1984).
10. P. B. Weisz, *CHEMTECH* 498-505 (1973).
11. J. Datka and M. Kawalek, *J. Chem. Soc. Faraday Trans.* **89**, 1829-1831 (1993).
12. M. W. Simon, S. S. Nam, W. Xu, S. L. Suib, J. C. Edwards and C. O'Young, *J. Phys. Chem.* **96**, 6381-6388 (1992).
13. B. Sulikowski and J. Klinowski, *J. Phys. Chem.* **96**, 5030-5035 (1992).
14. M. Derewinski, S. Dzwigaj, J. Haber, R. Mostowicz and B. Sulikowski, *Zeit. Physik. Chem.* **171**, 53-73 (1991).
15. J. Datka and Z. Piwowarska, *J. Chem. Soc. Faraday Trans.* **85**, 47-53 (1989).
16. G. Coudurier and J. C. Vedrine, *Pure & Appl. Chem.* **58**, 1389-1396 (1986).
17. C. T.-W. Chu and C. D. Chang, *J. Phys. Chem.* **89**, 1569-1571 (1985).

18. K. F. M. G. J. Scholle, A. P. M. Kentgens, W. S. Veeman, P. Frenken and G. P. M. van der Velden, *J. Phys. Chem.* **88**, 5-8 (1984).
19. H. Kosslick, V. A. Tuan, B. Parlitz, R. Fricke, C. Peuker and W. Storek, *J. Chem. Soc. Faraday Trans.* **89**, 1131-1138 (1993).
20. C. R. Bayense, A. P. M. Kentgens, J. W. de Haan, L. J. M. van de Ven and J. H. C. van Hooff, *J. Phys. Chem* **96**, 775-782 (1992).
21. X. Liu and J. Klinowski, *J. Phys. Chem.* **96**, 3403-3408 (1992).
22. E. Lalik, X. Liu and J. Klinowski, *J. Phys. Chem.* **96**, 805-809 (1992).
23. H. Kosslick, V. A. Tuan, R. Fricke, C. Peuker, W. Pilz and W. Storek, *J. Phys. Chem.* **97**, 5678-5684 (1993).
24. P. N. Joshi, S. V. Awate and V. P. Shiralkar, *J. Phys. Chem.* **97**, 9749-9753 (1993).
25. J. Sauer, *Chem. Rev.* **89**, 199-255 (1989).
26. J. B. Nicholas and A. C. Hess, to be submitted to *J. Phys. Chem.* (1994).
27. L. A. Curtiss, H. Brand, J. B. Nicholas and L. E. Iton, *Chem. Phys. Lett.* **184**, 215-220 (1991).
28. J. Sauer and R. Ahlrichs, *J. Chem. Phys.* **93**, 2575-2583 (1990).
29. D. A. Dixon and S. G. Lias, in *Molecular Structure and Energetics* (eds. J. F. Liebman and A. Greenberg) 269-313 (VCH, New York, 1987).
30. W. J. Hehre, L. Radom, P. Schleyer and J. A. Pople, *Ab Initio Molecular Orbital Theory* (Wiley, New York, 1986).
31. P. Hohenberg and W. Kohn, *Phys. Rev. B* **136**, 864 (1964).
32. W. Kohn and L. J. Sham, *Phys. Rev. A* **140**, 1133 (1965).
33. R. G. Parr and W. Yang, *Density-Functional Theory of Atoms and Molecules* (Oxford University, New York, 1989).
34. E. S. Kryachko and E. V. Ludena, *Energy Density Functional Theory of Many-Electron Systems* (Kluwer Academic Publishers, Boston, 1990).
35. *Density Functional Methods in Chemistry*; J. Labanowski and J. Andzelm, Ed.; Springer-Verlag: New York, 1991.
36. R. O. Jones and O. Gunnarsson, *Rev. Mod. Phys.* **61**, 689 (1990).
37. T. Ziegler, *Chem. Rev.* **91**, 651 (1991).
38. M. S. Stave and J. B. Nicholas, *J. Phys. Chem.* **97**, 9630-9641 (1993).
39. J. Andzelm and E. Wimmer, *J. Chem. Phys.* **96**, 1280 (1992).
40. A. St-Amant and D. R. Salahub, *Chem. Phys. Lett.* **169**, 387 (1990).
41. E. J. Baerends, D. E. Ellis and P. Ros, *Chem. Phys.* **2**, 41 (1973).
42. A. Rosen and D. E. Ellis, *J. Chem. Phys.* **65**, 3629 (1976).
43. F. W. Averill and D. E. Ellis, *J. Chem. Phys.* **59**, 6412 (1973).
44. B. Delley and D. E. Ellis, *J. Chem. Phys.* **76**, 1949 (1982).
45. B. Delley, *J. Chem. Phys.* **92**, 508 (1990).
46. R. Car and M. Parrinello, *Phys. Rev. Lett.* **55**, 2471 (1985).
47. M. P. Teter, M. C. Payne and D. C. Allan, *Phys. Rev. B* **40**, 12255 (1989).
48. E. Wimmer, H. Krakauer, M. Weinert and A. J. Freeman, *Phys. Rev. B* **24**, 864 (1981).
49. H. J. F. Jansen and A. J. Freeman, *Phys. Rev. B* **30**, 561 (1984).
50. A. D. Becke, *Int. J. Quant. Chem.* **S23**, 599 (1989).
51. U. von Barth and L. Hedin, *J. Phys. C* **5**, 1629 (1972).
52. S. H. Vosko, L. Wilk and M. Nusair, *Can. J. Phys.* **58**, 1200 (1980).
53. J. P. Perdew, *Phys. Rev. B* **33**, 8822 (1986).
54. J. P. Perdew and Y. Wang, *Phys. Rev. B* **33**, 8800 (1986).
55. A. D. Becke, *Phys. Rev. A* **38**, 3098 (1988).
56. J. B. Nicholas, R. E. Winans, R. J. Harrison, L. E. Iton, L. A. Curtiss and A. J. Hopfinger, *J. Phys. Chem.* **96**, 7958-7965 (1992).
57. T. H. Dunning Jr. and P. J. Hay, in *Modern Theoretical Chemistry* (eds. H. F. Schaefer) 1- (Plenum, New York, 1977).

58. T. H. Dunning Jr., *J. Chem. Phys.* **53**, 2823-2833 (1970).
59. K. Ruedenberg, R. C. Raffenetti and R. D. Bardo, Energy, Structure, and Reactivity, D. W. Smith and W. B. McRaes, Eds., Proc. 1972 Boulder Res. Conf. Theor. Chem. (Wiley, Boulder, 1973), pp. 164-169.
60. F. Van Duijeneveldt, (IBM, 1971).
61. S. Wilson, *Adv. Chem. Phys.* **67**, 439-500 (1987).
62. C. Møller and M. S. Plesset, *Phys. Rev.* **46**, 618-622 (1934).
63. M. F. Guest, R. J. Harrison, J. H. van Lenthe and L. C. H. van Corler, *Theor. Chim. Acta* **71**, 117-148 (1987).
64. M. J. Frisch, G. W. Trucks, M. Head-Gordon, P. M. W. Gill, M. W. Wong, J. B. Foresman, B. G. Johnson, H. B. Schlegel, M. Robb, E. S. Replogle, R. Gomperts, J. L. Andres, K. Raghavachari, J. S. Binkley, C. Gonzalez, R. L. Martin, D. J. Fox, D. J. Defrees, J. Baker, J. J. P. Stewart and J. A. Pople, *GAUSSIAN 92* (Gaussian, Inc., Pittsburgh, PA, 1992).
65. R. M. Dickson, Personal Communication (1992).
66. A. D. Becke, *J. Chem. Phys.* **98**, 5648 (1993).
67. B. G. Johnson, P. M. W. Gill and J. A. Pople, *J. Chem. Phys.* **98**, 5612 (1993).
68. A. M. Schmiedekamp, I. A. Topol, K. B. Burt, H. Razafinjanahary, H. Chermette, T. Pfaltzgraff and C. J. Michejda, *J. Comp. Chem.* (1993).
69. G. Fitzgerald and J. Andzelm, *J. Phys. Chem.* **95**, 10531-10534 (1991).
70. H. van Koningsveld, H. van Bekkum and J. C. Jansen, *Acta Crystallogr.* **B43**, 127-132 (1987).
71. A. E. Alvarado-Swaisgood, M. K. Barr, P. J. Hay and A. Redondo, *J. Phys. Chem.* **95**, 10031-10036 (1991).
72. E. Chamot, Modeling Acid Sites in MFI Zeolites with Realistic Geometric Constraints, Molecular Modeling of Petroleum Processes and Catalysis (Division of Petroleum Chemistry, Inc., American Chemical Society, San Francisco, 1992), vol. 37.
73. H. V. Brand, L. A. Curtiss and L. E. Iton, *J. Phys. Chem.* **96**, 7725-7732 (1992).
74. E. Kassab, J. Fouquet, M. Allavena and E. M. Evleth, *J. Phys. Chem.* **97**, 9034-9039 (1993).
75. E. H. Teunissen, R. A. van Santen, A. P. J. Jansen and F. B. van Duijneveldt, *J. Phys. Chem.* **97**, 203-210 (1993).
76. M. Allavena, K. Seiti, E. Kassab, G. Ferenczy and J. G. Angyan, *Chem. Phys. Lett.* **168**, 461-467 (1990).
77. E. Kassab, K. Seiti and M. Allavena, *J. Phys. Chem.* **95**, 9425-9431 (1991).
78. S. J. Cook, A. K. Chakraborty, A. T. Bell and D. N. Theodorou, *J. Phys. Chem.* **97**, 6679-6685 (1993).

INDEX

Activation barrier
for reactions on metal surfaces 41
H_2 dissociation on Ni(111) 48
Activation energy
41, 42
for surface diffusion of H on Ni(100) 29
Alkane diffusion 120
Aluminum cluster calculations 205
Amoco process 220
Ammonia adsorption on ZSM-5 clusters 239
Assoreni process 220
Atomic heats of chemisorption
for various metals 38
Binding energies 194, 195
Biological interface 138
Bond order conservation 36
Bond strength 194
Born-Oppenheimer-Huang approximation 2
Bulk solids 210
Cassie Equation 165
Cation adsorption 95
Chemisorption
of alkane thiols 50-53
of CO on Rh and Pd 202
of CO_2 on Pd 198
of formate 200
of polyatomics 36-50
of water on Pt 114
Copper cluster monoligand complexes 197
Data/information integration 143
Databases 142
deMon 188, 223
Density functional studies 187, 219
Differential capacitance 78
Diffusion 1
of H on Cu and Ni 11, 22
of D on Cu 11
Diffusion coefficient
of H on Cu 14
of H on Ni 25
of D on Cu 15
Diffusive motion 120
Direchlet boundary conditions 59
Dissociation 47-49, 51, 52
DMol 222
Domain structure 151
Dropping mercury electronde 76
Dynamic Morse potential 42-44
Effective reduced mass 8
Electric double layer 75
Electronic adiabatic approximation 2
Embedded cluster method 10
Equations of motion 85
Euler-Lagrange equation 172
Exchange correlation potential 189
Fast multipole method 81
Ferroelectric packing arrangement 156
Flip-up molecule 108
Flop-down molecule 108
Fluoride solutions 89
Frequencies for Ni-acetylene/ethylene 197
GAMESS-UK 224
Gaussian 92, 224
Genome modeling 138
Giant squid axon 57
Heterogeneous chemical catalysis 119
Hexadecane wettability calculations 180
Hopping motions 120
Ice-like structures in water 105, 106
Imaginary action integral 6, 7
Immersed electrode 80

Infinite surfaces 210
Interfacial free energy 162, 170
Interfacial ordering 162
Iron cluster adiabatic IP 208
Iron cluster calculations 207
Isoinertial coordinates 3
Kirkwood fluctuation integral 131
Kohn-Sham theory 188
Landau-Ginzburg free energy
density 171
Langmuir-Blodgett monolayers 149
Linear distance plot of trypsin 144
Mean field approximations 171
Mean field free energy 161
Metal clusters 49, 204
Metal cluster models
rules for 49
Au 51
Metal surface catalysis 35
Methyl terminated chains
on Au(111) 150
on Au(100) 151
Molecular dynamics simulations 101
with bond order conservation 44-46
Molecular heats of chemisorption 39
Monoligand complexes of
transition metal complexes 194
Morse potential methods 35
Muffin-tin potential 188
Neumann boundary conditions 59
Neural networks 147
Niobium cluster calculations 207
Nonlocal spin-density corrections 192
Pauling type bond order 35
(n-) Pentane surface diffusion
simulation 124
Periodic Hartree-Fock methods 220
Periodic systems 210
Phase separation 169
POLYRATE 11
Potential energy
functions 13
surfaces 35
of a bulk atom 13
Pre-exponential factor
of H diffusion on Ni 29
Protein characteristics 141
Proton affinity 220
Rate constant 48, 49
tunneling 9
overbarrier 9
Reactive flux 3
Reaction path 2
Self-assembled monolayers 150
SH kinetic model 57
Sodium fluoride solutions 93
Solubility models 131
Solubility in supercritical CO_2 133, 134
Stimulus pulse 70
Stockholm rule 50
Stroboscopic map 58
Subthreshold behavior 70
Supercritical fluids 131
Superthreshold behavior 70
Surface tension 173
Taylor series expansion 42, 62
Transition metal ligand bonding 194
Transmission coefficient 6
Trimethylamine adsorption on
zeolite 238
Tunneling 2
Unified atom model 121
Unimolecular reaction 7
van der Waals complexes 50
Variational determination of wettability
181
Variational transition state theory 3
Water hydrocarbon interface 102
Water metal interface 102
Water models 81
Water Pt surface potential 103
Wettability 165
Wetting models 168
$X\alpha$ - scattered wave methods 188
Young equation 164
Zeolite acidity 231
Zeolite acid sites 220
Zeolite ZSM5 219

SPRINGER NATURE

GPSR Compliance

The European Union's (EU) General Product Safety Regulation (GPSR) is a set of rules that requires consumer products to be safe and our obligations to ensure this.

If you have any concerns about our products, you can contact us on ProductSafety@springernature.com

In case Publisher is established outside the EU, the EU authorized representative is:

Springer Nature Customer Service Center GmbH
Europaplatz 3
69115 Heidelberg, Germany

Zeitfracht Medien GmbH
Ferdinand-Jühlke-Straße 7
99095 Erfurt, Deutschland
produktsicherheit@kolibri360.de